LONDON MATHEMATICAL SOCIETY LECTURE NOTE SERIES

Managing Editor: Professor J.W.S. Cassels, Department of Pure Mathematics and Mathematical Statistics, University of Cambridge, 16 Mill Lane, Cambridge CB2 1SB, England

The books in the series listed below are available from booksellers, or, in case of difficulty, from Cambridge University Press.

4 Algebraic topology, J.F. ADAMS
17 Differential germs and catastrophes, Th. BROCKER & L. LANDER
20 Sheaf theory, B.R. TENNISON
27 Skew field constructions, P.M. COHN
34 Representation theory of Lie groups, M.F. ATIYAH *et al*
36 Homological group theory, C.T.C. WALL (ed)
39 Affine sets and affine groups, D.G. NORTHCOTT
40 Introduction to H_p spaces, P.J. KOOSIS
43 Graphs, codes and designs, P.J. CAMERON & J.H. VAN LINT
45 Recursion theory: its generalisations and applications, F.R. DRAKE & S.S. WAINER (eds)
46 p-adic analysis: a short course on recent work, N. KOBLITZ
49 Finite geometries and designs, P. CAMERON, J.W.P. HIRSCHFELD & D.R. HUGHES (eds)
50 Commutator calculus and groups of homotopy classes, H.J. BAUES
51 Synthetic differential geometry, A. KOCK
57 Techniques of geometric topology, R.A. FENN
58 Singularities of smooth functions and maps, J.A. MARTINET
59 Applicable differential geometry, M. CRAMPIN & F.A.E. PIRANI
60 Integrable systems, S.P. NOVIKOV *et al*
62 Economics for mathematicians, J.W.S. CASSELS
65 Several complex variables and complex manifolds I, M.J. FIELD
66 Several complex variables and complex manifolds II, M.J. FIELD
68 Complex algebraic surfaces, A. BEAUVILLE
69 Representation theory, I.M. GELFAND *et al*
72 Commutat: e algebra: Durham 1981, R.Y. SHARP (ed)
76 Spectral theory of linear differential operators and comparison algebras, H.O. CORDES
77 Isolated singular points on complete intersections, E.J.N. LOOIJENGA
78 A primer on Riemann surfaces, A.F. BEARDON
79 Probability, statistics and analysis, J.F.C. KINGMAN & G.E.H. REUTER (eds)
80 Introduction to the representation theory of compact & locally compact groups, A. ROBE
81 Skew fields, P.K. DRAXL
83 Homogeneous structures on Riemannian manifolds, F. TRICERRI & L. VANHECKE
84 Finite group algebras and their modules, P. LANDROCK
85 Solitons, P.G. DRAZIN
86 Topological topics, I.M. JAMES (ed)
87 Surveys in set theory, A.R.D. MATHIAS (ed)
88 FPF ring theory, C. FAITH & S. PAGE
89 An F-space sampler, N.J. KALTON, N.T. PECK & J.W. ROBERTS
90 Polytopes and symmetry, S.A. ROBERTSON
91 Classgroups of group rings, M.J. TAYLOR
92 Representation of rings over skew fields, A.H. SCHOFIELD
93 Aspects of topology, I.M. JAMES & E.H. KRONHEIMER (eds)
94 Representations of general linear groups, G.D. JAMES
95 Low-dimensional topology 1982, R.A. FENN (ed)
96 Diophantine equations over function fields, R.C. MASON
97 Varieties of constructive mathematics, D.S. BRIDGES & F. RICHMAN
98 Localization in Noetherian rings, A.V. JATEGAONKAR
99 Methods of differential geometry in algebraic topology, M. KAROUBI & C. LERUSTE
100 Stopping time techniques for analysts and probabilists, L. EGGHE
101 Groups and geometry, ROGER C. LYNDON
103 Surveys in combinatorics 1985, I. ANDERSON (ed)

104 Elliptic structures on 3-manifolds, C.B. THOMAS
105 A local spectral theory for closed operators, I. ERDELYI & WANG SHENGWANG
106 Syzygies, E.G. EVANS & P. GRIFFITH
107 Compactification of Siegel moduli schemes, C-L. CHAI
108 Some topics in graph theory, H.P. YAP
109 Diophantine Analysis, J. LOXTON & A. VAN DER POORTEN (eds)
110 An introduction to surreal numbers, H. GONSHOR
111 Analytical and geometric aspects of hyperbolic space, D.B.A.EPSTEIN (ed)
112 Low-dimensional topology and Kleinian groups, D.B.A. EPSTEIN (ed)
114 Lectures on Bochner-Riesz means, K.M. DAVIS & Y-C. CHANG
115 An introduction to independence for analysts, H.G. DALES & W.H. WOODIN
116 Representations of algebras, P.J. WEBB (ed)
117 Homotopy theory, E. REES & J.D.S. JONES (eds)
118 Skew linear groups, M. SHIRVANI & B. WEHRFRITZ
119 Triangulated categories in the representation theory of finite-dimensional algebras, D. HAPPEL
121 Proceedings of *Groups - St Andrews 1985*, E. ROBERTSON & C. CAMPBELL (eds)
122 Non-classical continuum mechanics, R.J. KNOPS & A.A. LACEY (eds)
123 Surveys in combinatorics 1987, C. WHITEHEAD (ed)
124 Lie groupoids and Lie algebroids in differential geometry, K. MACKENZIE
125 Commutator theory for congruence modular varieties, R. FREESE & R. MCKENZIE
127 New Directions in Dynamical Systems, T.BEDFORD & J. SWIFT (eds)
128 Descriptive set theory and the structure of sets of uniqueness, A.S. KECHRIS & A. LOUVEAU
129 The subgroup structure of the finite classical groups, P.B. KLEIDMAN & M.W. LIEBECK
130 Model theory and modules, M. PREST
131 Algebraic, extremal & metrical combinatorics, M-M. DEZA, P. FRANKL & I.G. ROSENBERG (eds)
132 Whitehead groups of finite groups, ROBERT OLIVER

London Mathematical Society Lecture Note Series. 132

WHITEHEAD GROUPS OF FINITE GROUPS

Robert Oliver
Matematisk Institut, Aarhus University

CAMBRIDGE UNIVERSITY PRESS
Cambridge
New York New Rochelle Melbourne Sydney

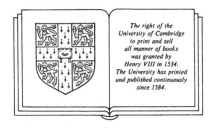

CAMBRIDGE UNIVERSITY PRESS
Cambridge, New York, Melbourne, Madrid, Cape Town, Singapore, São Paulo

Cambridge University Press
The Edinburgh Building, Cambridge CB2 8RU, UK

Published in the United States of America by Cambridge University Press, New York

www.cambridge.org
Information on this title: www.cambridge.org/9780521336468

First published 1988
Re-issued in this digitally printed version 2008

A catalogue record for this publication is available from the British Library

Library of Congress Cataloguing in Publication data
Oliver, R. (Robert)
Whitehead groups of finite groups / R. Oliver.
p. cm. – (London Mathematical Society lecture
note series: 132)
Bibliography: p. Includes Index
1. Whitehead groups. 2. Finite groups.
3. Induction (Mathematics)
I. Title II. Series
QA171.044 1988
512'.22 – dc 19 87 – 27725

ISBN 978-0-521-33646-8 paperback

PREFACE

This book is written with the intention of making more easily accessible techniques for studying Whitehead groups of finite groups, as well as a variety of related topics such as induction theory and p-adic logarithms. It developed out of a realization that most of the recent work in the field is scattered over a large number of papers, making it very difficult even for experts already working with K- and L-theory of finite groups to find and use them. The book is aimed, not only at such experts, but also at nonspecialists who either need some specific application involving Whitehead groups, or who just want to get an overview of the current state of knowledge in the subject. It is especially with the latter group in mind that the lengthy introduction — as well as the separate introductions to Parts I, II, and III — have been written. They are designed to give a quick orientation to the contents of the book, and in particular to the techniques available for describing Whitehead groups.

I would like to thank several people, in particular Jim Davis, Erkki Laitinen, Jim Schafer, Terry Wall, and Chuck Weibel, for all of their helpful suggestions during the preparation of the book. Also, my many thanks to Ioan James for encouraging me to write the book, and for arranging its publication.

LIST OF NOTATION

The following is a list of some of the notation used throughout the book. In many cases, these are defined again where used.

Groups:

$N_G(H)$, $C_G(H)$ denote the normalizer and centralizer of H in G

$G^{ab} = G/[G,G]$ (the abelianization) for any group G

$S_p(G)$ denotes a p-Sylow subgroup of G

C_n denotes a (multiplicative) cyclic group of order n

$D(2n)$, $Q(2n)$, $SD(2n)$ denote the dihedral, quaternion, and semidihedral groups of order $2n$

S_n, A_n denote the symmetric and alternating groups on n letters

$H \rtimes G$ denotes a semidirect product where H is normal

$G \wr C_n$ and $G \wr S_n$ denote the wreath products $G^n \rtimes C_n$ and $G^n \rtimes S_n$

$$\left.\begin{array}{ll} M^G = \{x \in M : Gx = x\} & \cong H^0(G;M) \\ M_G = M/\langle gx-x : g \in G,\ x \in M\rangle & \cong H_0(G;M) \end{array}\right\} \text{ if } G \text{ acts linearly on } M \text{ (groups of invariants and coinvariants)}$$

Fields and rings:

$\hat{K}_p = \hat{\mathbb{Q}}_p \otimes_{\mathbb{Q}} K$ if K is any number field and p a rational prime (so $\hat{K}_p$ is possibly a product of fields)

$\hat{R}_p = \hat{\mathbb{Z}}_p \otimes_{\mathbb{Z}} R$ if R is the ring of integers in a number field

μ_K, $(\mu_K)_p$ (K any field) denote the groups of roots of unity, and p-th power roots of unity, in K

ζ_n $(n \geq 1)$ denotes a primitive n-th root of unity

ξ_n $(n \geq 0)$, when some prime p is fixed, denotes the root of unity $\exp(2\pi i/p^n) \in \mathbb{C}$.

$K\zeta_n$ (for any field K and any $n \geq 1$) denotes the smallest field extension of K containing ζ_n

$J(R)$ denotes the Jacobson radical of the ring R

$\langle - \rangle$ means "subgroup (or $\hat{\mathbb{Z}}_p$-module) generated by"

$\langle - \rangle_R$ means "R-ideal or R-module generated by"

$e_{ij}^r = e_{ij}(r)$ (where $i,j \geq 1$, $i \neq j$, and $r \in R$) denote the elementary matrix with single off-diagonal entry r in the (i,j)-position

<u>K-theory:</u>

$SK_1(\mathfrak{A}) = \mathrm{Ker}[K_1(\mathfrak{A}) \longrightarrow K_1(A)]$
$K_1'(\mathfrak{A}) = K_1(\mathfrak{A})/SK_1(\mathfrak{A})$
for any $\mathbb{Z}$- or $\hat{\mathbb{Z}}_p$-order $\mathfrak{A}$ in a semisimple $\mathbb{Q}$- or $\hat{\mathbb{Q}}_p$-algebra A

$Cl_1(\mathfrak{A}) = \mathrm{Ker}[SK_1(\mathfrak{A}) \twoheadrightarrow \oplus_p SK_1(\hat{\mathfrak{A}}_p)]$ for any $\mathbb{Z}$-order $\mathfrak{A}$

$C(A) = \varprojlim_I SK_1(\mathfrak{A},I)$ for any semisimple $\mathbb{Q}$-algebra A and any $\mathbb{Z}$-order $\mathfrak{A} \subseteq A$, where the limit is taken over all ideals of finite index (see Definition 3.7)

$C_p(A)$ denotes the p-power torsion in the finite group $C(A)$

$Wh(R[G]) = K_1(R[G])/\langle rg\colon r \in \mu_K,\ g \in G\rangle$ and $Wh'(R[G]) = Wh(R[G])/SK_1(R[G])$ whenever R is the ring of integers in any finite extension K of $\mathbb{Q}$ or $\hat{\mathbb{Q}}_p$ (and G is any finite group)

$Wh'(G) = Wh(G)/SK_1(\mathbb{Z}[G]) = K_1'(\mathbb{Z}[G])/\langle \pm g\rangle$ for any finite group G

$K_2(R,I) = \mathrm{Ker}[K_2(R) \longrightarrow K_2(R/I)]$ for any ring R and any ideal $I \subseteq R$ (see remarks in Section 3a)

CONTENTS

INTRODUCTION

For any associative ring R with unit, an abelian group $K_1(R)$ is defined as follows. For each $n > 0$, let $GL_n(R)$ denote the group of invertible $n\times n$-matrices with entries in R. Regard $GL_n(R)$ as a subgroup of $GL_{n+1}(R)$ by identifying $A \in GL_n(R)$ with $\begin{pmatrix} A & 0 \\ 0 & 1 \end{pmatrix} \in GL_{n+1}(R)$; and set $GL(R) = \bigcup_{n=1}^{\infty} GL_n(R)$. For each n, let $E_n(R) \subseteq GL_n(R)$ be the subgroup generated by all *elementary* $n\times n$-matrices — i. e., all those which are the identity except for one nonzero off-diagonal entry — and set $E(R) = \bigcup_{n=1}^{\infty} E_n(R)$. Then by Whitehead's lemma (Theorem 1.13 below), $E(R) = [GL(R), GL(R)]$, the commutator subgroup of $GL(R)$. In particular, $E(R) \lhd GL(R)$; and the quotient group

$$K_1(R) = GL(R)/E(R)$$

is an abelian group.

One family of rings to which this applies is that of group rings. If G is any group, and if R is any commutative ring, then the group ring R[G] is the free R-module with basis G, where ring multiplication is induced by the group product. In particular, group elements $g \in G$, and units $u \in R^*$, can be regarded as invertible 1×1-matrices over R[G], and hence represent elements in $K_1(R[G])$. The *Whitehead group* of G is defined by setting

$$Wh(G) = K_1(\mathbb{Z}[G])/\langle \pm g \colon g \in G \rangle.$$

By construction, $K_1(R)$ (or $Wh(G)$) measures the obstruction to taking an arbitrary invertible matrix over R (or $\mathbb{Z}[G]$), and reducing it to the identity (or to some $\pm g$) via a series of *elementary operations*. Here, an elementary operation is one of the familiar matrix operations of adding a multiple of one row or column to another; and these elementary operations are very closely related to Whitehead's "elementary deformations" of finite CW complexes. This relationship leads to the definition of the *Whitehead torsion*

$$\tau(f) \in Wh(\pi_1(X))$$

of any homotopy equivalence $f: X \longrightarrow Y$ between finite CW complexes; where $\tau(f) = 1$ (i. e., the identity element) if and only if f is induced by a series of elementary deformations which transform X into Y. A homotopy equivalence f such that $\tau(f) = 1$ is called a *simple* homotopy equivalence.

Whitehead torsion plays a role, not only in studying homotopy equivalences of finite CW complexes, but also when classifying manifolds. The *s-cobordism theorem* (see Mazur [1]) says that if M and N are smooth closed n-dimensional manifolds, where $n \geq 5$, and if W is a compact (n+1)-dimensional manifold such that $\partial W = M \amalg N$, and such that the inclusions $M \hookrightarrow W$ and $N \hookrightarrow W$ are simple homotopy equivalences, then W is diffeomorphic to $M \times [0,1]$. In particular, M and N are diffeomorphic in this situation; and this theorem is one of the important tools for proving that two manifolds are diffeomorphic. This theorem is also one of the reasons for the importance of Whitehead groups when computing surgery obstructions.

When G is a finite group, then $K_1(\mathbb{Z}[G])$ and $Wh(G)$ are finitely generated abelian groups, whose rank was described by Bass (see the section on algorithms below, or Theorem 2.6). The main goal of this book is to develop techniques which allow a more complete description of $Wh(G)$ for finite G; and in particular which describe the subgroup

$$SK_1(\mathbb{Z}[G]) = \mathrm{Ker}\Big[K_1(\mathbb{Z}[G]) \longrightarrow K_1(\mathbb{Q}[G])\Big].$$

This is a finite subgroup (Theorem 2.5), and is in fact by a theorem of Wall (Theorem 7.4 below) isomorphic to the full torsion subgroup of $Wh(G)$. When G is abelian, then $SK_1(\mathbb{Z}[G]) = SL(\mathbb{Z}[G])/E(\mathbb{Z}[G])$, where $SL(\mathbb{Z}[G])$ denotes the group of matrices of determinant 1.

Most of the general background results have been presented here without proofs — especially when they can be referenced in standard textbooks such as Bass [2], Curtis & Reiner [1], Janusz [1], Milnor

[2], and Reiner [1]. Also, some of the longer and more technical proofs have been omitted when they are well documented in the literature, or are not needed for the central results. Proofs are included, or at least sketched, for most results which deal directly with the problem of describing Whitehead groups.

Historical survey

Whitehead groups were first defined by Whitehead [1], in order to find an algebraic analog to his "elementary deformations" of finite CW complexes, and to simple homotopy equivalences between finite CW complexes. Whitehead also showed in [1] that $Wh(G) = 1$ if $|G| \leq 4$ or if $G \cong \mathbb{Z}$; and that $Wh(C_5) \neq 1$. (Note that C_n always denotes a multilicative cyclic group of order n.)

A more systematic understanding of the structure of the groups $Wh(G)$ came only with the development of algebraic K-theory. Bass' theorem [1, Corollary 20.3], showing that the $Wh(G)$ are finitely generated and computing their rank, has already been mentioned. This made it natural to focus attention on the torsion subgroup of $Wh(G)$: shown by Higman [1] and Wall [1] to be isomorphic to $SK_1(\mathbb{Z}[G])$.

Milnor, in [1, Appendix A], noted that if the "congruence subgroup problem" could be proven, then it would follow that $SK_1(\mathbb{Z}[G]) = 1$ for all finite abelian groups G. This conjecture was shown by Bass, Milnor, and Serre [1] to be false (see Section 4c below); but their results were still sufficient to show that $SK_1(\mathbb{Z}[G])$ vanishes for many abelian groups. In particular, it was shown that $SK_1(\mathbb{Z}[G]) = 1$ if G is cyclic (Bass et al [1, Proposition 4.14]), if $G \cong C_{p^n} \times C_p$ for any prime p and any n (Lam, [1, Theorem 5.1.1]), or if $G \cong (C_2)^n$ for some n (Bass et al [1, Corollary 4.13]).

The first examples of finite groups for which $SK_1(\mathbb{Z}[G]) \neq 1$ were constructed by Alperin, Dennis, and Stein [1]. They combined earlier results from the solution to the congruence subgroup problem with theorems about generators for K_2 of finite rings, to explicitly describe $SK_1(\mathbb{Z}[G])$ when $G \cong (C_p)^n$, $n \geq 3$, and p is an odd prime. In particular, $SK_1(\mathbb{Z}[G])$ is nonvanishing for all such G. Their methods were then carried further, and used to show that for finite abelian G, $SK_1(\mathbb{Z}[G]) = 1$ if and only if either $G \cong (C_2)^n$, or each Sylow subgroup of G has the form C_{p^n} or $C_{p^n} \times C_p$.

Later results of Obayashi [1,2], Keating [1,2], and Magurn [1,2], showed that $SK_1(\mathbb{Z}[G])$ vanishes for many nonabelian metacyclic groups G, and in particular when G is any dihedral group. These were proven using various ad hoc methods, which did not give much hope for having generalizations to arbitrary finite groups. To get general results, a more systematic approach using localization sequences is needed — extending the methods of Alperin, Dennis, and Stein — and it is that approach which is the main focus of this book.

Algorithms for describing Wh(G)

If R is any commutative ring, then the usual matrix determinant induces a homomorphism

$$\det : K_1(R) = GL(R)/E(R) \longrightarrow R^*.$$

This is split surjective — split by the homomorphism $R^* \longrightarrow K_1(R)$ induced by identifying $R^* = GL_1(R)$. Hence, in this case, $K_1(R)$ factors as a product

$$K_1(R) = R^* \times SK_1(R),$$

where

$$SK_1(R) = SL(R)/E(R) \quad \text{and} \quad SL(R) = \{A \in GL(R) : \det(A) = 1\}.$$

If $R = \mathbb{Z}[G]$, then this coincides with the definition of $SK_1(\mathbb{Z}[G])$ given earlier: $\mathbb{Q}[G]$ is a product of fields, so $K_1(\mathbb{Q}[G]) \cong (\mathbb{Q}[G])^*$.

Determinants are not, in general, defined for noncommutative rings. However, in the case of the group rings $\mathbb{Z}[G]$ and $\mathbb{Q}[G]$ for finite groups G, they can be replaced by certain analogous homomorphisms: the reduced norm homomorphisms. One way to do this is to consider, for fixed G, the Wedderburn decomposition

$$\mathbb{C}[G] \cong \prod_{i=1}^{k} M_{r_i}(\mathbb{C})$$

of the complex group ring as a product of matrix rings (see Theorem 1.1). For each n, the reduced norm on $GL_n(\mathbb{Q}[G])$ is then defined to be the composite

$$nr : GL_n(\mathbb{Q}[G]) \xrightarrow{\text{incl}} GL_n(\mathbb{C}[G]) \cong \prod_{i=1}^{k} GL_{n\cdot r_i}(\mathbb{C}) \xrightarrow{\Pi \det} \prod_{i=1}^{k} \mathbb{C}^*.$$

These then factor through homomorphisms

$$nr_{\mathbb{Z}[G]} : K_1(\mathbb{Z}[G]) \longrightarrow \prod_{i=1}^{k} \mathbb{C}^* \quad \text{and} \quad nr_{\mathbb{Q}[G]} : K_1(\mathbb{Q}[G]) \longrightarrow \prod_{i=1}^{k} \mathbb{C}^*.$$

Also, $nr_{\mathbb{Q}[G]}$ is injective (Theorem 2.3), and so

$$SK_1(\mathbb{Z}[G]) = \mathrm{Ker}\Big[K_1(\mathbb{Z}[G]) \longrightarrow K_1(\mathbb{Q}[G])\Big] \qquad \text{(by definition)}$$

$$= \mathrm{Ker}(nr_{\mathbb{Z}[G]}). \tag{1}$$

Note that when G is commutative, then

$$\mathrm{Ker}(nr_{\mathbb{Z}[G]}) = \mathrm{Ker}\Big[\det\colon K_1(\mathbb{Z}[G]) \longrightarrow (\mathbb{Z}[G])^*\Big];$$

so that the two definitions of $SK_1(\mathbb{Z}[G])$ coincide in this case. For more details about reduced norms (and in more generality), see Section 2a.

Reduced norm homomorphisms are also the key to computing the ranks of the finitely generated groups $K_1(\mathbb{Z}[G])$ and $Wh(G)$. Not only is $SK_1(\mathbb{Z}[G]) = \mathrm{Ker}(nr_{\mathbb{Z}[G]})$ finite, but — once the target group has been restricted appropriately — $\mathrm{Coker}(nr_{\mathbb{Z}[G]})$ is also finite (Theorem 2.5).

A straightforward computation using Dirichlet's unit theorem then yields the formula

$$\mathrm{rk}(K_1(\mathbb{Z}[G])) = \mathrm{rk}(Wh(G))$$

$$= \#\Big(\text{irreducible } \mathbb{R}[G]\text{-modules}\Big) - \#\Big(\text{irreducible } \mathbb{Q}[G]\text{-modules}\Big).$$

Furthermore, by the theorem of Higman [1] (for commutative G) and Wall [1] (in the general case),

$$\mathrm{tors}(K_1(\mathbb{Z}[G])) = \{\pm 1\} \times G^{ab} \times SK_1(\mathbb{Z}[G])$$

(see Theorem 7.4 below). Thus, as abstract groups, at least, the structure of $K_1(\mathbb{Z}[G])$ and $Wh(G)$ will be completely understood once the structure of the finite group $SK_1(\mathbb{Z}[G])$ is known.

A much more difficult problem arises if one needs to construct explicit generators for the torsion free group $Wh'(G) = Wh(G)/SK_1(\mathbb{Z}[G])$. One case where it is possible to get relatively good control of this is when G is a p-group, for some regular prime p (including the case $p=2$). In this case, logarithmic methods can be used to identify the p-adic completion $\hat{\mathbb{Z}}_p \otimes Wh'(G)$ with a certain subgroup of $H_0(G;\hat{\mathbb{Z}}_p[G])$ (i. e., the free $\hat{\mathbb{Z}}_p$-module with basis the set of conjugacy classes in G). This is explained, and some applications are given, in Chapter 10; based on Oliver & Taylor [1, Section 4].

$SK_1(\mathbb{Z}[G])$: When studying $SK_1(\mathbb{Z}[G])$, it is convenient to first define a certain subgroup $Cl_1(\mathbb{Z}[G]) \subseteq SK_1(\mathbb{Z}[G])$. For each prime p, let $\hat{\mathbb{Z}}_p[G]$ and $\hat{\mathbb{Q}}_p[G]$ denote the p-adic completions of $\mathbb{Z}[G]$ and $\mathbb{Q}[G]$ (see Section 1b); and set $SK_1(\hat{\mathbb{Z}}_p[G]) = \mathrm{Ker}[K_1(\hat{\mathbb{Z}}_p[G]) \longrightarrow K_1(\hat{\mathbb{Q}}_p[G])]$. Then set

$$Cl_1(\mathbb{Z}[G]) = \mathrm{Ker}\Big[SK_1(\mathbb{Z}[G]) \xrightarrow{\ \ell\ } \bigoplus_p SK_1(\hat{\mathbb{Z}}_p[G])\Big].$$

The sum $\oplus_p SK_1(\hat{\mathbb{Z}}_p[G])$ is, in fact, a finite sum — $SK_1(\hat{\mathbb{Z}}_p[G]) = 1$

whenever $p \nmid |G|$ — and the localization homomorphism ℓ is onto (Theorem 3.9). Note that $Cl_1(\mathbb{Z}[G]) = SK_1(\mathbb{Z}[G])$ if G is abelian: $K_1(\hat{\mathbb{Z}}_p[G]) \cong SK_1(\hat{\mathbb{Z}}_p[G]) \times (\hat{\mathbb{Z}}_p[G])^*$ in this case, and matrices over a $\hat{\mathbb{Z}}_p$-algebra can always be diagonalized using elementary row and column operations (see Theorem 1.14(i)).

In particular, $SK_1(\mathbb{Z}[G])$ sits in an extension

$$1 \longrightarrow Cl_1(\mathbb{Z}[G]) \longrightarrow SK_1(\mathbb{Z}[G]) \xrightarrow{\ \ell\ } \bigoplus_p SK_1(\hat{\mathbb{Z}}_p[G]) \longrightarrow 1. \qquad (2)$$

The groups $SK_1(\hat{\mathbb{Z}}_p[G])$ and $Cl_1(\mathbb{Z}[G])$ are described independently, using very different methods, and it is difficult to find a way of handling them both simultaneously. In fact, one of the remaining problems is to understand the extension (2) in 2-torsion (it does have a natural splitting in odd torsion).

$SK_1(\hat{\mathbb{Z}}_p[G])$: By a theorem of Wall [1, Theorem 2.5], $SK_1(\hat{\mathbb{Z}}_p[G])$ is a p-group for any prime p and any finite group G, and $SK_1(\hat{\mathbb{Z}}_p[G]) = 1$ if the p-Sylow subgroup of G is abelian. In fact, for most "familiar" groups G, $SK_1(\hat{\mathbb{Z}}_p[G]) = 1$ for all p.

If G is a p-group, then

$$SK_1(\hat{\mathbb{Z}}_p[G]) \cong H_2(G)/H_2^{ab}(G); \qquad (3)$$

where

$$H_2^{ab}(G) = \operatorname{Im}\left[\sum \{H_2(K) : K \subseteq G,\ K \text{ abelian}\} \xrightarrow{\text{ind}} H_2(G)\right]$$

$$= \langle g\wedge h \in H_2(G) : g,h \in G,\ gh = hg\rangle$$

(see Section 8a). Formula (3) is shown in Theorem 8.6, and the isomorphism itself is described in Proposition 8.4.

If G is an arbitrary finite group, and if p is a fixed prime, then set $G_r = \{g \in G\colon p \nmid |g|\}$ (the "p-regular" elements). Consider the homology group $H_2(G;\hat{\mathbb{Z}}_p(G_r))$, where G acts on $\hat{\mathbb{Z}}_p(G_r)$ by conjugation. Let

$$\Phi : H_2(G;\hat{\mathbb{Z}}_p(G_r)) \longrightarrow H_2(G;\hat{\mathbb{Z}}_p(G_r))$$

be induced by the endomorphism $\Phi(\sum r_i g_i) = \sum r_i g_i^p$ on $\hat{\mathbb{Z}}_p(G_r)$; and let

$$H_2(G;\hat{\mathbb{Z}}_p(G_r))_\Phi = H_2(G;\hat{\mathbb{Z}}_p(G_r))/\mathrm{Im}(1-\Phi)$$

be the group of Φ-coinvariants. Then

$$SK_1(\hat{\mathbb{Z}}_p[G]) \cong H_2(G;\hat{\mathbb{Z}}_p(G_r))_\Phi / H_2^{ab}(G;\hat{\mathbb{Z}}_p(G_r))_\Phi \tag{4}$$

(see Theorem 12.10). Here, in analogy with the p-group case:

$$H_2^{ab}(G;\hat{\mathbb{Z}}_p(G_r))_\Phi = \mathrm{Im}\left[\sum_{\substack{K \subseteq G \\ \text{abelian}}} H_2(K;\hat{\mathbb{Z}}_p(K_r)) \xrightarrow{\text{ind}} H_2(G;\hat{\mathbb{Z}}_p(G_r))_\Phi\right]$$

$$= \langle (g\wedge h)\otimes k : g,h \in G,\ k \in G_r,\ \ g,h,k \text{ commute pairwise}\rangle.$$

The following alternative description of $SK_1(\hat{\mathbb{Z}}_p[G])$, for a non-p-group G, is often easier to use. Let $g_1,\ldots,g_k \in G$ be "$\hat{\mathbb{Q}}_p$-conjugacy" class representatives for elements of G of order prime to p — where two elements $g,h \in G$ are $\hat{\mathbb{Q}}_p$-conjugate if g is conjugate to h^{p^n} for some n. Set $Z_i = C_G(g_i)$ (the centralizer), and

$$N_i = \{x \in G : xg_ix^{-1} = g_i^{p^n},\ \text{ some } n\}.$$

Then by Theorem 12.5 below,

$$SK_1(\hat{\mathbb{Z}}_p[G]) \cong \bigoplus_{i=1}^{k} H_0\Big(N_i/Z_i;\ H_2(Z_i)/H_2^{ab}(Z_i)\Big)_{(p)}. \tag{5}$$

$Cl_1(\mathbb{Z}[G])$: The subgroup $Cl_1(\mathbb{Z}[G]) \subseteq SK_1(\mathbb{Z}[G])$ can be thought of as the part of $K_1(\mathbb{Z}[G])$ which is hit from behind in localization sequences. One way to study this is to consider, for any ideal $I \subseteq \mathbb{Z}[G]$ of finite index, the relative exact sequence

$$K_2(\mathbb{Z}[G]/I) \longrightarrow SK_1(\mathbb{Z}[G],I) \longrightarrow SK_1(\mathbb{Z}[G]) \longrightarrow K_1(\mathbb{Z}[G]/I)$$

of Milnor [2, Lemma 4.1 and Theorem 6.2]. After taking inverse limits over all such I, this takes the form of a new exact sequence

$$\bigoplus_p K_2^c(\hat{\mathbb{Z}}_p[G]) \longrightarrow \varprojlim_I SK_1(\mathbb{Z}[G],I) \xrightarrow{\partial} SK_1(\mathbb{Z}[G]) \xrightarrow{\ell} \bigoplus_p SK_1(\hat{\mathbb{Z}}_p[G]). \qquad (6)$$

We now have another characterization of $Cl_1(\mathbb{Z}[G])$: it is the set of elements in $SK_1(\mathbb{Z}[G])$ which can be represented by matrices congruent to 1 mod I, for arbitrarily small ideals $I \subseteq \mathbb{Z}[G]$ of finite index.

The second term in (6) remains unchanged when $\mathbb{Z}[G]$ is replaced by any other $\mathbb{Z}$-order in $\mathbb{Q}[G]$. Hence, it is convenient to define

$$C(\mathbb{Q}[G]) = \varprojlim_I SK_1(\mathbb{Z}[G],I) \qquad \text{(all } I \subseteq \mathbb{Z}[G] \text{ of finite index)}$$

$$\cong \operatorname{Coker}\Big[K_2(\mathbb{Q}[G]) \longrightarrow \bigoplus_p K_2^c(\hat{\mathbb{Q}}_p[G])\Big]. \qquad \text{(Theorem 3.12)}$$

This is a finite group; and $C(-)$ is a functor on the category of finite dimensional semisimple $\mathbb{Q}$-algebras. See Section 3c for more details.

The computation of $C(\mathbb{Q}[G])$ is based on the solution to the congruence subgroup problem. In Theorem 4.13, it will be seen that for each simple summand A of $\mathbb{Q}[G]$ with center K,

$$C(A) \cong \begin{cases} 1 & \text{if for some } v\colon K \hookrightarrow \mathbb{R},\ \mathbb{R}\otimes_{vK} A \cong M_r(\mathbb{R}) \text{ (some } r) \\ \mu_K & \text{otherwise.} \end{cases} \qquad (7)$$

Here, μ_K denotes the group of roots of unity in K. One convenient way to use this involves the complex representation ring $R_{\mathbb{C}}(G)$.

Fix a group G, and fix any *even* n such that $\exp(G)|n$. Then $K = \mathbb{Q}(\zeta_n)$ is a splitting field for G, where ζ_n is a primitive n-th root of unity, and we can identify the representation rings $R_{\mathbb{C}}(G) = R_K(G)$. The group $Gal(K/\mathbb{Q}) \cong (\mathbb{Z}/n)^*$ thus acts both on $R_{\mathbb{C}}(G)$ (via Galois automorphisms) and on $\mathbb{Z}/n$ (by multiplication). Regard $R_{\mathbb{R}}(G)$ (the real representation ring) as a subgroup of $R_{\mathbb{C}}(G)$ in the usual way, and set $R_{\mathbb{C}/\mathbb{R}}(G) = R_{\mathbb{C}}(G)/R_{\mathbb{R}}(G)$ for short. Then we will see in Lemma 5.9 that

$$C(\mathbb{Q}[G]) \cong \left[R_{\mathbb{C}/\mathbb{R}}(G) \otimes \mathbb{Z}/n\right]_{(\mathbb{Z}/n)^*} \qquad \text{(i. e., } (\mathbb{Z}/n)^*\text{-coinvariants)}$$

$$= R_{\mathbb{C}/\mathbb{R}}(G)/\langle [V] - a\cdot[\gamma_a(V)] : V \in R_{\mathbb{C}}(G),\ (a,n)=1,\ \gamma_a \in \mathrm{Gal}(K/\mathbb{Q})\rangle. \tag{8}$$

This description, while somewhat complicated, has the advantage of being natural in that the induced epimorphisms

$$\begin{array}{ccc} R_{\mathbb{C}}(G) & \xrightarrow{\ \tilde{\mathcal{J}}_G\ } & C(\mathbb{Q}[G]) \\ & \searrow^{\mathcal{J}_G} & \downarrow \partial \\ & & Cl_1(\mathbb{Z}[G]) \end{array}$$

commute both with maps induced by group homomorphisms and with maps induced by restriction to subgroups (Proposition 5.2). For example, one immediate consequence of this is that $Cl_1(\mathbb{Z}[G])$ is generated by induction from elementary subgroups of G (i. e., products of cyclic groups with p-groups) — since $R_{\mathbb{C}}(G)$ is generated by elementary induction by Brauer's induction theorem.

Odd torsion in $Cl_1(\mathbb{Z}[G])$ and $SK_1(\mathbb{Z}[G])$: For any finite group G, the short exact sequence (2) has a natural splitting in odd torsion, to give a direct sum decomposition

$$SK_1(\mathbb{Z}[G])[\tfrac{1}{2}] \cong Cl_1(\mathbb{Z}[G])[\tfrac{1}{2}] \oplus \bigoplus_{p\neq 2} SK_1(\hat{\mathbb{Z}}_p[G]). \tag{9}$$

Furthermore, for odd p, there is a close relationship between the groups $K_2^c(\hat{\mathbb{Z}}_p[G])$ and $H_1(G;\hat{\mathbb{Z}}_p[G]) \cong H_1(G;\mathbb{Z}[G])_{(p)}$ (where G again acts by conjugation) — close enough so that (6) can be replaced by an isomorphism

$$Cl_1(\mathbb{Z}[G])[\tfrac{1}{2}] \cong \mathrm{Coker}\left[H_1(G;\mathbb{Z}[G]) \xrightarrow{\ \psi_G\ } C(\mathbb{Q}[G])\right][\tfrac{1}{2}]. \tag{10}$$

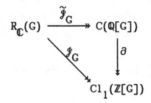

When G is a p-group (and p is odd), this formula is shown in Theorem 9.5, and an explicit definition of ψ_G is given in Definition 9.2. If G is arbitrary, then the formula, as well as the definition of ψ_G, are derived in the discussion following Theorem 13.9.

When G is not a p-group, then an alternative description of $Cl_1(\mathbb{Z}[G])_{(p)}$, for any finite G and any odd prime p, is given in Theorem 13.9. This takes the form

$$Cl_1(\mathbb{Z}[G])_{(p)} \cong \bigoplus_{i=1}^{k} H_0\Big(N_i/Z_i;\ \varinjlim_{\pi\in\mathcal{P}(Z_i)} Cl_1(\mathbb{Z}[\pi])\Big); \tag{11}$$

where $\sigma_1,\ldots,\sigma_k \subseteq G$ is a set of conjugacy class representatives for cyclic subgroups of order prime to p; and $N_i = N_G(\sigma_i)$, $Z_i = C_G(\sigma_i)$, and $\mathcal{P}(Z_i)$ is the set of p-subgroups.

2-torsion in $SK_1(\mathbb{Z}[G])$: The description of $Cl_1(\mathbb{Z}[G])_{(2)}$ — even when G is a 2-group — is still rather mysterious. If G is abelian, then $SK_1(\mathbb{Z}[G]) = Cl_1(\mathbb{Z}[G])$ can be described via formulas analogous to (10) above (see Theorem 9.6 for the case of an abelian 2-group, and Theorem 13.13 for the general abelian case). When G is an arbitrary 2-group, we conjecture that $SK_1(\mathbb{Z}[G])$ can be (mostly) described via a pushout square

$$\begin{array}{ccc}
H_1(G;\mathbb{Z}[G]) & \xrightarrow{\ \ \upsilon\ \ } & H_2(G) \\
\Big\downarrow{\scriptstyle \psi_G} & & \Big\downarrow{\scriptstyle \widetilde{\Theta}_G} \\
C(\mathbb{Q}[G])/C^{Q}(\mathbb{Q}[G]) & \xrightarrow{\ \ \partial\ \ } & SK_1(\mathbb{Z}[G])/Q(G).
\end{array}$$

Here, $C^{Q}(\mathbb{Q}[G]) \subseteq C(\mathbb{Q}[G])$ denotes the subgroups of elements coming from quaternionic summands: i. e., simple summands A of $\mathbb{Q}[G]$ which are matrix algebras over division algebras of the form $\mathbb{Q}(\xi_n, j)$ $(\subseteq \mathbb{H})$, where $\xi_n = \exp(2\pi i/2^n) \in \mathbb{C}$ (see Theorem 9.1). Also, $Q(G) \subseteq Cl_1(\mathbb{Z}[G])$ is the image of $C^{Q}(\mathbb{Q}[G])$ under $\partial\colon C(\mathbb{Q}[G]) \twoheadrightarrow Cl_1(\mathbb{Z}[G])$; and υ is defined by setting $\nu(g \otimes h) = g \wedge h \in H_2(G)$ for commuting $g,h \in G$. Note that

$$\mathrm{Coker}(\nu) = H_2(G)/H_2^{ab}(G) \cong SK_1(\hat{\mathbb{Z}}_2[G])$$

by (2). The most interesting point here is the conjectured existence of a lifting $\tilde{\Theta}_G$ of the isomorphism in (2). This is currently the only hope for constructing examples where extension (1) is not split. For more details, see Conjecture 9.7, as well as Theorems 9.6, 13.4, 13.12, and 13.14.

Induction theory: Each of the functors $SK_1(\hat{\mathbb{Z}}_p[G])$, and $Cl_1(\mathbb{Z}[G])_{(p)}$ for odd p, has been given two descriptions above. The direct sum formulas (5) and (11) are based on a general decomposition formula in Theorem 11.8, and are usually the easiest to apply when computing $SK_1(\hat{\mathbb{Z}}_p[G])$ or $Cl_1(\mathbb{Z}[G])_{(p)}$ as an abstract group. The other formulas ((4) and (10)) seem more natural, and are easier to use to determine whether or not a given element vanishes.

In both cases — $SK_1(\hat{\mathbb{Z}}_p[G])$ and $Cl_1(\mathbb{Z}[G])_{(p)}$ — these formulas are derived from those in the p-group case with the help of induction theory as formulated by Dress [2]. In the terminology of Chapter 11, these two functors are "computable" with respect to induction from p-elementary subgroups (i. e., subgroups of the form $C_n \times \pi$ when π is a p-group). See Chapter 11, and Theorems 12.4 and 13.5, for more details.

Detecting and constructing explicit elements: For simplicity, the above algorithms have been stated so as to describe Wh(G) and $SK_1(\mathbb{Z}[G])$ as abstract groups. But in fact, they can in many cases be used to determine whether or not a given invertible matrix over $\mathbb{Z}[G]$ vanishes in Wh(G); or to construct matrices representing given nonvanishing elements.

The procedures for constructing explicit nontrivial elements in $SK_1(\mathbb{Z}[G])$ are fairly straightforward. One example of this, for the group $G = C_4 \times C_2 \times C_2$, is worked out in detail in Example 5.1; and essentially the same procedure can be used to construct elements in $Cl_1(\mathbb{Z}[G])$ for any finite G (once the group $Cl_1(\mathbb{Z}[G])$ itself is known, that is).

Explicit elements in $SK_1(\hat{\mathbb{Z}}_p[G])$ can be constructed using Proposition 8.4, or Theorem 12.5 or 12.10; although Theorem 8.13 provides a much simpler way of doing this in many cases. The procedure for lifting an element $[A] \in SK_1(\hat{\mathbb{Z}}_p[G])$ to $SK_1(\mathbb{Z}[G])$ can be found in the proof of Theorem 3.9; note however that this depends on finding an explicit decomposition of A as a product of elementary matrices over $\hat{\mathbb{Q}}_p[G]$.

If $A \in GL(\mathbb{Z}[G])$ is given, then the first step when determining whether it vanishes in $Wh(G)$ is to compute its reduced norm, and determine (using (1)) whether or not $[A] \in SK_1(\mathbb{Z}[G])$. Once this is done, if G is abelian, then $SK_1(\mathbb{Z}[G]) = Cl_1(\mathbb{Z}[G])$, and the procedure for determining whether $[A] = 1$ is fairly straightforward. The details are described in the proof of Example 5.1, and in the discussion afterwards.

If G is nonabelian, and if $[A]$ is known to lie in $SK_1(\mathbb{Z}[G])$, then one must next check whether or not it vanishes in $SK_1(\hat{\mathbb{Z}}_p[G])$ for primes $p \mid |G|$. The procedure for doing this is described in Proposition 8.4 when G is a p-group, and in Theorem 12.10 for general finite G. In both cases, this involves first choosing some group extension $\alpha: \tilde{G} \twoheadrightarrow G$ such that $SK_1(\hat{\mathbb{Z}}_p[\tilde{G}])$ maps trivially to $SK_1(\hat{\mathbb{Z}}_p[G])$; then lifting A to $[\tilde{A}] \in K_1(\hat{\mathbb{Z}}_p[\tilde{G}])$, taking its logarithm (more precisely, its integral logarithm $\Gamma(\tilde{A}) \in H_0(\tilde{G};\hat{\mathbb{Z}}_p[\tilde{G}])$); and then composing that by a certain explicit homomorphism to $SK_1(\hat{\mathbb{Z}}_p[G])$ using formula (3) or (4) above.

The general procedure for detecting elements in $Cl_1(\mathbb{Z}[G])$ for nonabelian G is much less clear, although there are some remarks about that at the end of Section 5a. The main problem (once the group $Cl_1(\mathbb{Z}[G])$ itself is understood) is to lift $[M] \in Cl_1(\mathbb{Z}[G])$ to $C(\mathbb{Q}[G])$ along the boundary map in sequence (6).

In some specialized cases, there are other ways of doing this. The proofs Propositions 16-18 in Oliver [1] give one example, and can be used to detect (certain) nonvanishing elements in $Cl_1(\mathbb{Z}[G])$ for many nonabelian groups G. Another such example is given by the procedure in Oliver [5] for detecting the Whitehead torsion of homotopy equivalences of S^1-bundles.

Survey of computations

The examples listed here give both a survey of the type of computations which can be made using the techniques sketched in the last section, as well as an idea of some of the patterns which arise from the computations. The first few examples give some conditions which are necessary or sufficient for $SK_1(\mathbb{Z}[G])$ to vanish.

<u>Example 1</u> (Theorem 5.6, or Alperin et al [3, Theorem 3.3]) $SK_1(R[C_n]) = 1$ *for any finite cyclic group* C_n, *when* R *is the ring of integers in any finite extension of* $\mathbb{Q}$.

<u>Example 2</u> (Theorem 14.2 and Example 14.4) $SK_1(\mathbb{Z}[G]) = 1$ *if* $G \cong C_{p^n}$ *or* $C_{p^n} \times C_p$ *(for any prime* p *and any* n*), if* $G \cong (C_2)^n$ *(any* n*), or if* G *is any dihedral, quaternion, or semidihedral 2-group. Conversely, if* G *is a p-group and* $Cl_1(\mathbb{Z}[G]) = 1$, *then either* G *is one of the above groups, or* $p = 2$ *and* $G^{ab} \cong (C_2)^n$ *for some* n.

The next example (as well as Example 12) helps to illustrate the role played by the p-Sylow subgroup $S_p(G)$ in determining the p-torsion in $SK_1(\mathbb{Z}[G])$.

<u>Example 3</u> (Theorem 14.2(i), or Oliver [1, Theorem 2]) $SK_1(\mathbb{Z}[G])_{(p)} = 1$ *if* $S_p(G) \cong C_{p^n}$ *or* $C_{p^n} \times C_p$ *(any* n*).*

The next example gives some completely different criteria for $SK_1(\mathbb{Z}[G])$ (or Wh(G)) to vanish. This, together with the first three examples, helps to show the hopelessness of finding general necessary and sufficient conditions for $SK_1(\mathbb{Z}[G]) = 1$ (or $Wh(G) = 1$). Note in particular that $Wh(G) = 1$ if G is any symmetric group.

<u>Example 4</u> (Theorem 14.1) *Let* $\mathscr{C} \subseteq \mathscr{C}'$ *be the smallest classes of finite groups which are closed under direct product and under wreath product with any symmetric group* S_n; *and such that* $\mathscr{C}$ *contains the*

trivial group and $\mathscr{C}'$ *also contains all dihedral groups. (Note that* $\mathscr{C}$ *contains* $D(8)$, *as well as all symmetric groups.) Then* $Wh(G) = 1$ *for all* $G \in \mathscr{C}$, *and* $SK_1(\mathbb{Z}[G]) = 1$ *for all* $G \in \mathscr{C}'$.

Note that the classes of finite groups G for which $Wh(G) = 1$, or $SK_1(\mathbb{Z}[G]) = 1$, are *not* closed under products (see Example 6). A slightly stronger version of Example 4 is given in Theorem 14.1.

We now consider examples where $SK_1(\mathbb{Z}[G]) \neq 1$. The easiest case is that of abelian groups. In fact, the exponent of $SK_1(\mathbb{Z}[G]) = Cl_1(\mathbb{Z}[G])$ can be explicitly determined in this case.

Example 5 (Alperin et al [3, Theorem 4.8]) *Let* G *be a finite abelian group, and let* $k(G)$ *be the product of the distinct primes* p *dividing* $|G|$ *for which* $S_p(G)$ *is not cyclic. Then*

$$\exp(SK_1(\mathbb{Z}[G])) = \epsilon\cdot\gcd\left(\exp(G)\ ,\ \frac{|G|}{k(G)\cdot\exp(G)}\right);$$

where $\epsilon = \frac{1}{2}$ *if*

(*i*) $G \cong (C_2)^n$ *for some* $n \geq 3$, *or*

(*ii*) $S_2(G) \cong C_{2^n} \times C_{2^n}$ *for some* $n \geq 3$, *or*

(*iii*) $S_2(G) \cong C_{2^n} \times C_{2^n} \times C_2$ *for some* $n \geq 2$;

and $\epsilon = 1$ *otherwise.*

We now consider some more precise computations of $SK_1(\mathbb{Z}[G])$ in cases where it is nonvanishing.

Example 6 (Example 9.8, and Alperin et al [3, Theorems 2.4, 5.1, 5.5, 5.6, and Corollary 5.9]) *The following are examples of computations of* $SK_1(\mathbb{Z}[G]) = Cl_1(\mathbb{Z}[G])$ *for some abelian p-groups* G:

(i) If p *is odd, then*

$$SK_1(\mathbb{Z}[(C_p)^k]) \cong (\mathbb{Z}/p)^N, \quad \textit{where} \quad N = \frac{p^k-1}{p-1} - \binom{p+k-1}{p}$$

(ii) $SK_1(\mathbb{Z}[C_{p^2} \times C_{p^n}]) \cong (\mathbb{Z}/p)^{(p-1)(n-1)}$ *(any prime* p*)*

$$(iii)\ SK_1(\mathbb{Z}[(C_p)^2 \times C_{p^n}]) \cong \begin{cases} (\mathbb{Z}/p)^{np(p-1)/2} & \textit{if } p \textit{ is odd} \\ (\mathbb{Z}/2)^{n-1} & \textit{if } p = 2 \end{cases}$$

$$(iv)\ SK_1(\mathbb{Z}[C_{p^3} \times C_{p^3}]) \cong \begin{cases} (\mathbb{Z}/p)^{p^2-1} \times (\mathbb{Z}/p^2)^{p-1} & \textit{if } p \textit{ is odd} \\ (\mathbb{Z}/2)^4 & \textit{if } p = 2 \end{cases}$$

$$(v)\ SK_1(\mathbb{Z}[(C_2)^k \times C_{2^n}]) \cong \Big[\bigoplus_{r=2}^{k} \binom{k}{r}\cdot(\mathbb{Z}/2^{r-1})\Big] \oplus \Big[\bigoplus_{s=2}^{n} (\mathbb{Z}/2^s)\Big]$$

We now look at some nonabelian p-groups: first for odd p and then for p = 2.

Example 7 (Example 9.9, and Oliver [7, Section 4]) *Let* p *be an odd prime, and let* G *be a nonabelian p-group. Then* $SK_1(\mathbb{Z}[G]) = Cl_1(\mathbb{Z}[G]) \cong (\mathbb{Z}/p)^{p-1}$ *if* $|G| = p^3$. *If* $|G| = p^4$, *then* $SK_1(\mathbb{Z}[G]) = Cl_1(\mathbb{Z}[G])$ *and:*

$$SK_1(\mathbb{Z}[G]) \cong \begin{cases} (\mathbb{Z}/p)^{2(p-1)} & \textit{if } G^{ab} \cong C_p \times C_{p^2} \\ (\mathbb{Z}/p)^{(p^2+3p-6)/2} & \textit{if } G^{ab} \cong (C_p)^3, \quad \exp(G) = p \\ (\mathbb{Z}/p)^{(p^2+p-2)/2} & \textit{if } G^{ab} \cong (C_p)^3, \quad \exp(G) = p^2 \\ (\mathbb{Z}/p)^{3(p-1)/2} & \textit{if } G^{ab} \cong C_p \times C_p, \quad \exists\ (C_p)^3 \subseteq G \\ (\mathbb{Z}/p)^{p-1} & \textit{if } G^{ab} \cong C_p \times C_p, \quad \exists\ C_p \times C_{p^2} \subseteq G. \end{cases}$$

Note that the p- and p^2-rank of $Cl_1(\mathbb{Z}[G])$ is a polynomial in p for each of the families listed in Examples 6 and 7 above. Presumably, this holds in general, and is a formal consequence of Theorem 9.5 below; but we know of no proof.

Example 8 (Examples 9.9 and 9.10) *If* $|G| = 16$, *then*

$$SK_1(\mathbb{Z}[G]) = Cl_1(\mathbb{Z}[G]) \cong \begin{cases} 1 & \text{if } G^{ab} \cong (C_2)^2 \text{ or } (C_2)^3 \\ \mathbb{Z}/2 & \text{if } G^{ab} \cong C_4 \times C_2. \end{cases}$$

If G *is any (nonabelian) quaternion or semidihedral 2-group, then for all* $k \geq 0$:

$$SK_1(\mathbb{Z}[G \times (C_2)^k]) \cong Cl_1(\mathbb{Z}[G \times (C_2)^k]) \cong (\mathbb{Z}/2)^{2^k-k-1}.$$

We next give some examples of computations for three specific classes of non-p-groups.

Example 9 (Example 14.4) Assume G *is a finite group whose 2-Sylow subgroups are dihedral, quaternionic, or semidihedral. Then*

$$SK_1(\mathbb{Z}[G])_{(2)} = Cl_1(\mathbb{Z}[G])_{(2)} \cong (\mathbb{Z}/2)^k,$$

where k *is the number of conjugacy classes of cyclic subgroups* $\sigma \subseteq G$ *such that* (a) $|\sigma|$ *is odd,* (b) $C_G(\sigma)$ *has nonabelian 2-Sylow subgroup, and* (c) *there is no* $g \in N_G(\sigma)$ *with* $gxg^{-1} = x^{-1}$ *for all* $x \in \sigma$.

Note, in the next two examples, the peculiar way in which 3-torsion (and only 3-torsion) appears.

Example 10 (Theorem 14.5) *For any prime* p *and any* $k \geq 1$,

$$SK_1(\mathbb{Z}[PSL(2,p^k)]) \cong \begin{cases} \mathbb{Z}/3 & \text{if } p = 3,\ 2\nmid k,\ k \geq 5 \\ 1 & \text{otherwise.} \end{cases}$$

and

$$SK_1(\mathbb{Z}[SL(2,p^k)]) \cong \begin{cases} \mathbb{Z}/3 \times \mathbb{Z}/3 & \text{if } p = 3,\ 2\nmid k,\ k \geq 5 \\ 1 & \text{otherwise.} \end{cases}$$

Example 11 (Theorem 14.6) *For any* $n > 1$, *let* A_n *be the alternating group on* n *letters. Then* $SK_1(\mathbb{Z}[G]) = Cl_1(\mathbb{Z}[G])$, *and*

$$SK_1(\mathbb{Z}[A_n]) \cong \begin{cases} \mathbb{Z}/3 & \text{if } n = \sum_{i=1}^{r} 3^{m_i} \geq 27, \quad m_1 > m_2 > \ldots > m_r \geq 0, \quad \sum m_i \text{ odd} \\ 1 & \text{otherwise.} \end{cases}$$

The next example involves the groups $SK_1(\hat{\mathbb{Z}}_p[G])$. Constructing a group G for which $SK_1(\hat{\mathbb{Z}}_p[G]) \neq 1$ is rather complicated (note that $SK_1(\mathbb{Z}[G]) = Cl_1(\mathbb{Z}[G])$ in all of the examples above); so instead of doing that here we refer to Example 8.11 and the discussion after Theorem 14.1. For now, we just note the following condition for $SK_1(\hat{\mathbb{Z}}_p[G])$ to vanish.

Example 12 (Proposition 12.7) $SK_1(\hat{\mathbb{Z}}_p[G]) = 1$ *if the p-Sylow subgroup of* G *has an abelian normal subgroup with cyclic quotient.*

To end the section, we note two specific examples of concrete matrices or units representing nontrivial elements in $Cl_1(\mathbb{Z}[G])$.

Example 13 (i) *Set* $G = C_4 \times C_2 \times C_2 = \langle g \rangle \times \langle h_1 \rangle \times \langle h_2 \rangle$. *Then* $SK_1(\mathbb{Z}[G]) \cong \mathbb{Z}/2$, *and the nontrivial element is represented by the matrix*

$$\begin{pmatrix} 1 + 8(1-g^2)(1+h_1)(1+h^2)(1-g) & -(1-g^2)(1+h_1)(1+h_2)(3+g) \\ -13(1-g^2)(1+h_1)(1+h_2)(3-g) & 1 + 8(1-g^2)(1+h_1)(1+h_2)(1+g) \end{pmatrix} \in GL_2(\mathbb{Z}[G])$$

(ii) *Set* $G = C_3 \times Q(8) = \langle g \rangle \times \langle a,b \rangle$, *where* $Q(8)$ *is a quaternion group of order* 8. *Then* $SK_1(\mathbb{Z}[G]) \cong \mathbb{Z}/2$, *and the nontrivial element is represented by the unit*

$$1 + (2-g-g^2)(1-a^2)\left(3g + a + 4g^2a + 4(g^2-g)b + 8ab\right) \in (\mathbb{Z}[G])^*.$$

The matrix in (i) is constructed in Example 5.1. In (ii), $SK_1(\mathbb{Z}[G])$ is computed as a special case of Example 9, and the explicit unit representing its nontrivial element can be constructed using the proof of

Oliver [1, Proposition 17]. The general problem of determining whether or not a given element of Wh(G) can be represented by a unit in $\mathbb{Z}[G]$ is studied in Magurn et al [1], and is discussed briefly in Chapter 10 (Theorems 10.6 to 10.8) below.

PART I: GENERAL THEORY

These first six chapters give a general introduction to the tools used when studying K_1 of $\mathbb{Z}$- and $\hat{\mathbb{Z}}_p$-orders, and in particular of integral group rings. While some concrete examples of computations of $SK_1(\mathbb{Z}[G])$ are given in Sections 5a and 5b, the systematic algorithms for making such computations are not developed until Parts II and III.

The central chapters in Part I are Chapters 2, 3, and 4. The torsion free part of $K_1(\mathfrak{A})$, for any $\mathbb{Z}$- or $\hat{\mathbb{Z}}_p$-order $\mathfrak{A}$, is studied in Chapter 2 using reduced norm homomorphisms and p-adic logarithms. In Chapter 3, the continuous K_2 for p-adic algebras and orders is defined, and then used to construct the localization sequences which will be used later to study $SK_1(\mathbb{Z}[G])$ for finite G. Chapter 4 is centered around the congruence subgroup problem: the computation of one term

$$C(\mathbb{Q}[G]) = \varprojlim_n SK_1(\mathbb{Z}[G], n\mathbb{Z}[G]) \cong \text{Coker}\Big[K_2(\mathbb{Q}[G]) \longrightarrow \bigoplus_p K_2^c(\hat{\mathbb{Q}}_p[G])\Big]$$

in the localization sequence of Chapter 3.

In addition, Chapter 1 provides a survey of some general background material on such subjects as semisimple algebras and orders, number theory, and K-theory of finite and semilocal rings. Chapter 5 collects some miscellaneous quick applications of the results in Chapter 4: for example, the results that $Cl_1(\mathbb{Z}[G]) = 1$ whenever G is cyclic, dihedral, or quaternionic. Also, the "standard involution" on $K_1(\mathbb{Z}[G])$, $K_2^c(\hat{\mathbb{Z}}_p[G])$, etc., studied in Section 5c, is the key to many of the later results involving odd torsion in $Cl_1(\mathbb{Z}[G]) \subseteq SK_1(\mathbb{Z}[G])$. The integral p-adic logarithm (Chapter 6), which at first glance seems useful only for getting an additive description of $K_1(\hat{\mathbb{Z}}_p[G])/\text{torsion}$, will be seen later to play a central role in the computations of both $SK_1(\hat{\mathbb{Z}}_p[G])$ and $Cl_1(\mathbb{Z}[G])$.

Chapter 1 BASIC ALGEBRAIC BACKGROUND

By a $\mathbb{Z}$-*order* $\mathfrak{A}$ in a semisimple $\mathbb{Q}$-algebra A is meant a $\mathbb{Z}$-lattice (i. e., $\mathfrak{A}$ is a finitely generated $\mathbb{Z}$-module and $A = \mathbb{Q}\cdot\mathfrak{A}$) which is a subring. One of the reasons why Whitehead groups are more easily studied for finite groups than for infinite groups is that strong structure theorems for semisimple $\mathbb{Q}$-algebras and their orders are available as tools. In fact, it is almost impossible to study the K-theory of group rings $\mathbb{Z}[G]$ without considering some orders which are not themselves group rings. Furthermore, the use of localization sequences as a tool for studying $K_1(\mathfrak{A})$ for $\mathbb{Z}$-orders $\mathfrak{A}$ makes it also important to study the K-theory of orders over the p-adic integers $\hat{\mathbb{Z}}_p$.

This chapter summarizes some of the basic background material about semisimple algebras, orders, p-adic localization, semilocal rings, and similar topics, which will be needed later on. The results are presented mostly without proof. The first two sections are independent of K-theory. Section 1c includes some results about K_1 of semilocal or finite rings, as well as Quillen's localization sequence for a maximal order. Section 1d contains a short discussion about bimodule-induced homomorphisms for $K_i(-)$, and in particular about Morita equivalences.

Recall that a *number field* is any finite field extension of $\mathbb{Q}$. The *ring of integers* in a number field K is the integral closure of $\mathbb{Z}$ in K: i. e., the set of elements in K which are roots of monic polynomials over $\mathbb{Z}$.

1a. Semisimple algebras and maximal orders

The definition of a semisimple algebra (or ring) varies somewhat; the most standard is to define it to be a ring which is semisimple (i. e., a direct sum of modules with no proper submodules) as a (left or right) module over itself. Then a simple algebra is a semisimple algebra which has no proper 2-sided ideals. *Throughout this book, whenever "semisimple algebra" is used, it is always assumed to mean finite dimensional over the base field.* Our main references for this topic are Curtis & Reiner [1, Section 3] and Reiner [1, Section 7].

For any field K of characteristic zero, and any finite group G, standard representation theory shows that $K[G]$ is a semisimple K-algebra. In particular, the structure of $\mathbb{Q}[G]$ as a semisimple $\mathbb{Q}$-algebra plays an important role when studying $K_1(\mathbb{Z}[G])$.

The center of any algebra A will be denoted $Z(A)$.

Theorem 1.1 *Let K be a field, and let A be any semisimple K-algebra. Then the following hold:*

(i) (Wedderburn theorem) There are division algebras $D_1,\ldots,D_k$ over K, and numbers $r_1,\ldots,r_k > 0$, such that $A \cong \prod_{i=1}^{k} M_{r_i}(D_i)$. Here, each $M_{r_i}(D_i)$ is a simple algebra, and has a unique irreducible module isomorphic to $(D_i)^{r_i}$. Furthermore,

$$Z(A) \cong \prod_{i=1}^{k} Z(M_{r_i}(D_i)) \cong \prod_{i=1}^{k} Z(D_i);$$

and A is simple if and only if $Z(A)$ is a field.

(ii) If A is simple and $Z(A)/K$ is separable, then for any field extension $L \supseteq K$, $L \otimes_K A$ is semisimple with center $L \otimes_K Z(A)$. In particular, $L \otimes_K A$ is simple if $K = Z(A)$.

(iii) If A is a central simple K-algebra (i. e., $K = Z(A)$), then $[A:K] = n^2$ for some $n \in \mathbb{Z}$.

(iv) (Skolem-Noether theorem) If A is a central simple K-algebra, and if $B \subseteq A$ is a simple subalgebra which contains K, then any ring homomorphism $f\colon B \longrightarrow A$ which fixes K is the restriction of an inner automorphism of A.

Proof The Wedderburn theorem is shown, for example, in Curtis & Reiner [1, Theorems 3.22 and 3.28]; and the other statements in (i) are easy consequences of that. The other three points are shown in Reiner [1, Theorem 7.18 and Corollary 7.8, Theorem 7.15, and Theorem 7.21]. □

Note in particular that a commutative semisimple K-algebra always factors as a product of fields. In the case of group rings of abelian groups, one can be more specific. Recall that for any n, ζ_n denotes a primitive n-th root of unity; and that for any field K, $K\zeta_n$ denotes the smallest field extension of K which contains the n-th roots of unity.

__Example 1.2__ *For any* $n \geq 1$, $\mathbb{Q}[C_n] \cong \prod_{d|n}\mathbb{Q}\zeta_d$. *More generally, for any field* K *of characteristic zero and any finite abelian group* G, $K[G] \cong \prod_{i=1}^{k} K_i$, *where for each* i, $K_i \cong K(\zeta_{n_i})$ *for some* $n_i|\exp(G)$.

__Proof__ For any n, a homomorphism $\alpha = \prod\alpha_d\colon \mathbb{Q}[C_n] \longrightarrow \prod_{d|n}\mathbb{Q}\zeta_d$ is induced by setting $\alpha_d(g) = \zeta_d$ for some fixed generator $g \in C_n$. Each α_d induces an irreducible $\mathbb{Q}[C_n]$-representation $\mathbb{Q}\zeta_d$, and they are distinct since C_n acts on $\mathbb{Q}\zeta_d$ with order d. So α is surjective. Since $[\mathbb{Q}\zeta_d:\mathbb{Q}] = \varphi(d)$ for each d (see Janusz [1, Theorem I.9.2]), a dimension count shows that α is an isomorphism.

The last point is clear: $K[G]$ is a product of fields by Theorem 1.1, and each field component is generated by K and the images of the elements of G, which must be roots of unity. □

As another example, consider group rings $\mathbb{C}[G]$ and $\mathbb{R}[G]$ for a finite group G. The only (finite dimensional) division algebra over $\mathbb{C}$ is $\mathbb{C}$ itself (this is the case for any algebraically closed field); and the only division algebras over $\mathbb{R}$ are $\mathbb{R}$, $\mathbb{C}$, and $\mathbb{H}$ (the quaternion algebra). Note in particular that $\mathbb{H}$ is *not* a $\mathbb{C}$-algebra, since $\mathbb{C}$ is not central in $\mathbb{H}$. The Wedderburn theorem thus implies that for any finite G, $\mathbb{C}[G]$ is a product of matrix algebras over $\mathbb{C}$, and $\mathbb{R}[G]$ is a product of matrix algebras over $\mathbb{R}$, $\mathbb{C}$, and $\mathbb{H}$. Also, by (ii), if A is a simple $\mathbb{Q}$-algebra with center K, and if $K \hookrightarrow \mathbb{R}$ is any embedding, then $\mathbb{R}\otimes_K A$ is a matrix algebra over either $\mathbb{R}$ or $\mathbb{H}$. This last point will play an important role later, for example when describing the image of the reduced norm in Theorem 2.3.

If A is a simple algebra with center K, and $A = M_r(D)$ for some division algebra D, then the *index* of A is defined by setting

$$\operatorname{ind}(A) = \operatorname{ind}(D) = [D:K]^{1/2}$$

(an integer by Theorem 1.1(iii)). A field $L \supseteq K$ is called a *splitting field* for A if $L \otimes_K A$ is a matrix algebra over L.

<u>Proposition 1.3</u> *Let A be a simple algebra with center K. Then for any splitting field $L \supseteq K$ for A, $\operatorname{ind}(A) | [L:K]$. If A is a division algebra, then any maximal subfield $L \subseteq A$ is a splitting field for A and satisfies $[L:K] = [A:K]^{1/2}$.*

<u>Proof</u> See Reiner [1, Theorem 28.5, and Theorem 7.15]. □

We now consider orders in semisimple algebras. If R is a Dedekind domain with field of fractions K, an *R-order* $\mathfrak{A}$ in a semisimple K-algebra A is defined to be an R-lattice (i. e., $\mathfrak{A}$ is a finitely generated R-module and $K \cdot \mathfrak{A} = A$) which is a subring. A *maximal R-order* in A is just an order which is not contained in any larger order. Our main reference for orders and maximal orders is Reiner [1]. The most important properties of maximal orders needed when studying Whitehead groups are listed in the next theorem (and Theorems 1.9 and 1.19 below).

<u>Theorem 1.4</u> *Fix a Dedekind domain R with field of fractions K of characteristic zero, and let A be a semisimple K-algebra. Then the following hold.*

(i) A contains at least one maximal R-order, and any R-order in A is contained in a maximal order.

(ii) If $A = \prod_{i=1}^{k} A_i$, where the A_i are simple and $\mathfrak{M} \subseteq A$ is a maximal R-order, then $\mathfrak{M}$ splits as a product $\mathfrak{M} = \prod_{i=1}^{k} \mathfrak{M}_i$, where for all i, $\mathfrak{M}_i$ is a maximal order in A_i.

(iii) If A *is commutative, then there is a unique maximal* R*-order* $\mathfrak{M} \subseteq A$. *If* $A = \prod_{i=1}^{k} K_i$ *where the* K_i *are finite field extensions of* K, *then* $\mathfrak{M} = \prod_{i=1}^{k} R_i$, *where* R_i *is the ring of* R*-integers in* K_i*: i. e., the integral closure of* R *in* K_i.

(iv) Any maximal R*-order* $\mathfrak{M} \subseteq A$ *is hereditary: all left (or right) ideals in* $\mathfrak{M}$ *are projective as* $\mathfrak{M}$*-modules, and all finitely generated* R*-torsion free* $\mathfrak{M}$*-modules are projective.*

(v) If G *is any finite group, and if* $\mathfrak{M} \subseteq R[G]$ *is a maximal order containing* $R[G]$, *then* $|G|\cdot\mathfrak{M} \subseteq R[G]$.

Proof These are shown in Reiner [1]: (i) in Corollary 10.4, (ii) in Theorem 10.5(i), (iii) in Theorem 10.5(iii), (iv) in Theorem 21.4 and Corollary 21.5, and (v) in Theorem 41.1. □

Note that point (i) above is false if A is not semisimple. For example, for $n \geq 2$, the ring of upper triangular $n{\times}n$ matrices over $\mathbb{Q}$ has no maximal $\mathbb{Z}$-orders.

Example 1.2 has already hinted at the important role played by cyclotomic extensions when working with group rings. The following properties will be useful later.

Theorem 1.5 *Fix a field* K, *and* $n > 1$ *such that* $\mathrm{char}(K) \nmid n$. *Let* $K\zeta_n$ *denote a field extension of* K *by a primitive n-th root of unity. Then* $K\zeta_n/K$ *is an abelian Galois extension, and* $\mathrm{Gal}(K\zeta_n/K)$ *can be identified as a subgroup of* $(\mathbb{Z}/n)^*$*: each* $\gamma \in \mathrm{Gal}(K\zeta_n/K)$ *has the form* $\gamma(\zeta_n) = (\zeta_n)^a$ *for some unique* $a \in (\mathbb{Z}/n)^*$. *Furthermore:*

(i) $(K = \mathbb{Q})$ $\mathrm{Gal}(\mathbb{Q}\zeta_n/\mathbb{Q}) = (\mathbb{Z}/n)^*$, *and* $\mathbb{Z}\zeta_n \subseteq \mathbb{Q}\zeta_n$ *is the ring of integers. In particular, under the identification* $\mathbb{Q}[C_n] \cong \prod_{d|n}\mathbb{Q}\zeta_d$, *the maximal* $\mathbb{Z}$*-order in* $\mathbb{Q}[C_n]$ *is* $\prod_{d|n}\mathbb{Z}\zeta_d$.

(ii) (Brauer) If G *is a finite group, and* $\mathrm{char}(K) \nmid \exp(G) | n$, *then* K *is a splitting field for* G*: i. e.,* $K\zeta_n[G]$ *is a product of*

matrix algebras over $K\zeta_n$.

<u>Proof</u> The embedding $\mathrm{Gal}(K\zeta_n/K) \subseteq (\mathbb{Z}/n)^*$ is clear. When $K = \mathbb{Q}$, $\mathrm{Gal}(\mathbb{Q}\zeta_n/\mathbb{Q}) = (\mathbb{Z}/n)^*$ since $[\mathbb{Q}\zeta_n:\mathbb{Q}] = \varphi(n)$ (see Janusz [1, Theorem I.9.2]); and $\mathbb{Z}\zeta_n$ is the ring of integers in $\mathbb{Q}\zeta_n$ by Janusz [1, §I.9, Exercise 2]. The last statement in (i) then follows from Theorem 1.4(iii). Brauer's splitting theorem is shown in Curtis & Reiner [1, Corollary 15.18 and Theorem 17.1]. □

Note that when R is the ring of integers in an arbitrary number field K, then $R\zeta_n$ need *not* be the integral closure of R in $K\zeta_n$. For example, $\mathbb{Z}[\sqrt{3}]$ is the ring of integers in $\mathbb{Q}(\sqrt{3})$, but $\mathbb{Z}[\sqrt{3},i]$ is not the ring of integers in $\mathbb{Q}(\sqrt{3},i) = \mathbb{Q}(\zeta_{12})$.

For any field K and any finite group G, two elements $g,h \in G$ of order n prime to $\mathrm{char}(K)$ are called *K-conjugate* if $g^a = xhx^{-1}$ for some $x \in G$ and some $a \in \mathrm{Gal}(K\zeta_n/K) \subseteq (\mathbb{Z}/n)^*$. For example, g and h are $\mathbb{C}$-conjugate if and only if they are conjugate; and they are $\mathbb{Q}$-conjugate if and only if $\langle g\rangle$ and $\langle h\rangle$ are conjugate subgroups of G. The importance of K-conjugacy lies in the following theorem.

<u>Theorem 1.6</u> (Witt-Berman theorem) *For any field* K *of characteristic zero, and for any finite group* G, *the number of irreducible* $K[G]$*-modules — i. e., the number of simple summands of* $K[G]$ *— is equal to the number of K-conjugacy classes of elements in* G.

<u>Proof</u> The characters of the irreducible $K[G]$-modules form a basis for the vector space of all functions $(G \longrightarrow \mathbb{C})$ which are constant on K-conjugacy classes. This is shown, for example, in Curtis & Reiner [1, Theorem 21.5] and Serre [2, §12.4, Corollary 2]. □

Note that there also is a version of the Witt-Berman theorem when $\mathrm{char}(K) > 0$: the number of distinct irreducible $K[G]$-modules is equal to the number of K-conjugacy classes in G of elements of order prime to $\mathrm{char}(K)$. See Curtis & Reiner [1,Theorem 21.25] for details.

1b. P-adic completion

Let R be the ring of integers in any number field K. For any maximal ideal $p \subseteq R$, the *p-adic completions* of R and K are defined by setting

$$\hat{R}_p = \varprojlim_n R/p^n; \qquad \hat{K}_p = \hat{R}_p[\tfrac{1}{p}] \qquad (p = \mathrm{char}(R/p)).$$

Then $\hat{R}_p$ is a local ring with unique maximal ideal $p\hat{R}_p$, and $\hat{K}_p$ is its field of fractions. Furthermore, $\hat{K}_p$ is a finite extension of $\hat{\mathbb{Q}}_p$, and $\hat{R}_p$ is the integral closure of $\hat{\mathbb{Z}}_p$ in $\hat{K}_p$.

Alternatively, $\hat{R}_p$ and $\hat{K}_p$ can be constructed using the *p-adic valuation* $v_p : K \longrightarrow \mathbb{Z} \cup \infty$. This is defined by setting

$$v_p(r) = \max\{n \geq 0 : r \in p^n\}$$

for $r \in R$, and $v_p(r/s) = v_p(r) - v_p(s)$ in general. This induces a topology on K — based on the norm $|x|_p = p^{-v_p(x)}$ — and $\hat{K}_p$ and $\hat{R}_p$ are the corresponding completions of K and R. Note that $\hat{R}_p$ is compact under this p-adic topology, since it is an inverse limit of finite groups.

If $p \subseteq R \subseteq K$ are as above, then for any semisimple K-algebra A any R-order $\mathfrak{A} \subseteq A$, the p-adic completions of A and $\mathfrak{A}$ are defined by setting

$$\hat{\mathfrak{A}}_p = \varprojlim_n \mathfrak{A}/p^n\mathfrak{A} \cong \hat{R}_p \otimes_R \mathfrak{A}, \qquad \hat{A}_p = \hat{K}_p \otimes_K A \cong K \otimes_R \hat{\mathfrak{A}}_p.$$

Then $\hat{A}_p$ is a semisimple $\hat{\mathbb{Q}}_p$-algebra (where $p = \mathrm{char}(R/p)$), and $\hat{\mathfrak{A}}_p$ is a $\hat{\mathbb{Z}}_p$-order in $\hat{A}_p$. Note that if we regard A as a $\mathbb{Q}$-algebra, then for any rational prime p (i. e., $p \in \mathbb{Z}$), and any $\mathfrak{A} \subseteq A$, $\hat{\mathfrak{A}}_p = \hat{\mathbb{Z}}_p \otimes_{\mathbb{Z}} \mathfrak{A}$ and $\hat{A}_p = \hat{\mathbb{Q}}_p \otimes_{\mathbb{Q}} A$.

The importance of using p-adic completions when studying $K_1(\mathfrak{A})$ for a $\mathbb{Z}$-order $\mathfrak{A}$ is due partly because it is far easier to identify the units (and invertible matrices) in the $\hat{\mathfrak{A}}_p$ than in $\mathfrak{A}$, and partly because analytic tools such as logarithms and exponents can be used when working in $K_1(\hat{\mathfrak{A}}_p)$. For example, results in Chapters 6 and 8 illustrate how much more simply the groups $K_1'(\hat{\mathbb{Z}}_p[G])$ and $SK_1(\hat{\mathbb{Z}}_p[G])$ are described than the groups $K_1'(\mathbb{Z}[G])$ and $SK_1(\mathbb{Z}[G])$.

Now let E/F be any pair of finite extensions of $\hat{\mathbb{Q}}_p$, let $R \subseteq F$ and $S \subseteq E$ be the rings of integers, and let $\mathfrak{p} \subseteq R$ and $\mathfrak{q} \subseteq S$ be the maximal ideals. Then E/F is *unramified* if $\mathfrak{q} = \mathfrak{p}S$, and is *totally ramified* if $S/\mathfrak{q} \cong R/\mathfrak{p}$.

<u>Theorem 1.7</u> *Fix an algebraic number field* K, *and let* R *be its ring of integers. Let* A *be any semisimple* K-*algebra, and let* $\mathfrak{A}$ *be any* R-*order in* A.

(i) For any rational prime p,

$$\hat{K}_p \cong \prod_{\mathfrak{p}|p} \hat{K}_{\mathfrak{p}}, \quad \hat{R}_p \cong \prod_{\mathfrak{p}|p} \hat{R}_{\mathfrak{p}}, \quad \hat{A}_p \cong \prod_{\mathfrak{p}|p} \hat{A}_{\mathfrak{p}}, \quad \textit{and} \quad \hat{\mathfrak{A}}_p \cong \prod_{\mathfrak{p}|p} \hat{\mathfrak{A}}_{\mathfrak{p}}.$$

Here, the products are taken over all maximal ideals $\mathfrak{p} \subseteq R$ *which divide* p *(i.e.,* $\mathfrak{p} \supseteq pR$).

(ii) $\hat{K}_{\mathfrak{p}}/\hat{\mathbb{Q}}_p$ $(p = \mathrm{char}(R/\mathfrak{p}))$ *is unramified for all but finitely many maximal ideals* $\mathfrak{p} \subseteq R$.

(iii) $\hat{A}_{\mathfrak{p}}$ *is a product of matrix rings over fields for all but finitely many* $\mathfrak{p} \subseteq R$

(iv) $\hat{\mathfrak{A}}_{\mathfrak{p}}$ *is a maximal* $\hat{R}_{\mathfrak{p}}$-*order in* $\hat{A}_{\mathfrak{p}}$ *for almost all* $\mathfrak{p}$ *in* R; *and* $\mathfrak{A}$ *is a maximal* R-*order in* A *if and only if* $\hat{\mathfrak{A}}_{\mathfrak{p}}$ *is a maximal* $\hat{R}_{\mathfrak{p}}$-*order in* $\hat{A}_{\mathfrak{p}}$ *for all* $\mathfrak{p} \subseteq R$.

<u>Proof</u> The first two points are shown in Janusz [1]: (i) in Theorem II.5.1, and (ii) in Theorem I.7.3. Points (iii) and (iv) are shown in

Reiner [1, Theorem 25.7 and Corollary 11.6]. □

The next proposition gives information about the groups F^* and R^*, when R is the ring of integers in a finite extension F of $\hat{\mathbb{Q}}_p$. In particular, the reciprocity map $F^*/N(E^*) \cong \mathrm{Gal}(E/F)^{ab}$ in (ii) below is the key to defining norm residue symbols in $K_2(F)$ (Section 3a).

Proposition 1.8 *Let F be any finite extension of $\hat{\mathbb{Q}}_p$, and let $\mathfrak{p} \subseteq R \subseteq F$ be the maximal ideal and ring of integers.*

(i) Let $\mu \subseteq R^*$ be the group of all roots of unity of order prime to p. Then projection mod $\mathfrak{p}$ induces an isomorphism $\mu \cong (R/\mathfrak{p})^*$; and for any generator π of $\mathfrak{p}$:

$$R^* = \mu \times (1+\mathfrak{p}) \qquad \text{and} \qquad F^* = \mu \times (1+\mathfrak{p}) \times \langle\pi\rangle.$$

(ii) For any finite Galois extension E/F, there is a canonical isomorphism

$$s : F^*/N_{E/F}(E^*) \xrightarrow{\ \cong\ } \mathrm{Gal}(E/F)^{ab}$$

(the reciprocity map). If E/F is not Galois, then $F^*/N_{E/F}(E^*) \cong \mathrm{Gal}(E'/F)$, where E' denotes the maximal abelian Galois extension of F contained in E.

(iii) If E/F is unramified, and if $\mathfrak{q} \subseteq S \subseteq E$ are the maximal ideal and ring of integers, then the norm and trace homomorphisms

$$N = N_{S/R} : S^* \twoheadrightarrow R^* \qquad \text{and} \qquad Tr = Tr_{S/R} : S \twoheadrightarrow R$$

are surjective. Also, for all $n \geq 1$, $N(1+\mathfrak{q}^n) = 1+\mathfrak{p}^n$ and $Tr(\mathfrak{q}^n) = \mathfrak{p}^n$.

<u>Proof</u> To see (i), just note that

$$R^* = \varprojlim_n (R/\mathfrak{p}^n)^* \cong \varprojlim_n \Big((1+\mathfrak{p})/(1+\mathfrak{p}^n) \times (R/\mathfrak{p})^*\Big) \cong (1+\mathfrak{p}) \times (R/\mathfrak{p})^*;$$

since $(1+p)/(1+p^n)$ is a p-group for each n, and $p \nmid |(R/p)^*|$. Point (ii) is shown in Cassels & Fröhlich [1, §VI.2.2, and §VI.2.6, Proposition 4]. The surjectivity of $N_{S/R}$ and $Tr_{S/R}$ in (iii) is shown in Serre [1, Section V.2]: by filtering R^* and R by the p^n, and then using analogous results about norms and traces for finite fields. □

The following very powerful structure theorem for p-adic division algebras and their maximal orders is due to Hasse [1].

<u>**Theorem 1.9**</u> *Fix a finite extension F of $\hat{\mathbb{Q}}_p$, let $R \subseteq F$ be the ring of integers, and let D be a division algebra with center F. Set $n = [D:F]^{1/2}$. Then there exists a maximal subfield $E \subseteq D$, with ring of integers $S \subseteq E$, and an element $\pi \in D$ such that $\pi E \pi^{-1} = E$, for which the following hold:*

(i) E/F is unramified, and $D = \bigoplus_{i=0}^{n-1} E\cdot\pi^i$

(ii) $\Lambda = \bigoplus_{i=0}^{n-1} S\cdot\pi^i$ *is the unique maximal $\hat{\mathbb{Z}}_p$-order in D*

(iii) $\pi\Lambda$ is the unique maximal ideal in Λ

(iv) $\pi^n R$ is the maximal ideal in R.

Furthermore, for any $r \geq 1$ and any maximal $\hat{\mathbb{Z}}_p$-order $\mathfrak{M}$ in $M_r(D)$, $\mathfrak{M}$ is conjugate (in $M_r(D)$) to $M_r(\Lambda)$.

<u>**Proof**</u> See Hasse [1, Sätze 10 & 47], or Reiner [1, Section 14 and Theorem 17.3]. □

We end the section by noting the following more specialized properties of p-adic group rings.

<u>**Theorem 1.10**</u> *Fix a prime p, let F be any finite extension of $\hat{\mathbb{Q}}_p$, and let $R \subseteq F$ be the ring of integers.*

(i) For any n *such that* $p \nmid n$, $F[C_n] \cong \prod_{i=1}^{k} F_i$, *where the* F_i *are finite unramified extensions of* F. *Under this identification,* $R[C_n] = \prod_{i=1}^{k} R_i$, *where* R_i *is the ring of integers in* F_i. In particular, $F\zeta_n/F$ is unramified, and $R[\zeta_n]$ is the ring of integers in $F\zeta_n$.

(ii) For any finite group G, $F[G]$ *is a product of matrix algebras over division algebras of index dividing* 2 *(if* $p = 2$*) or* $p-1$ *(if* p *is odd). More precisely, if* p *is odd and* $\zeta_p \in F$, *or if* $p = 2$ *and* $i \in F$, *then* $F[G]$ *is a product of matrix algebras over fields.*

<u>Proof</u> (i) Since $\frac{1}{n} \in R$, $R[C_n]$ is a maximal $\hat{\mathbb{Z}}_p$-order in $F[C_n]$ by Theorem 1.4(v). Hence, $R[C_n] = \prod_{i=1}^{k} R_i$ where the R_i are the rings of integers in F_i. If $\mathfrak{p} \subseteq R$ is the maximal ideal, then $R/\mathfrak{p}[C_n]$ is a product of finite fields (since $p = \mathrm{char}(R/\mathfrak{p}) \nmid n$), so $\mathfrak{p}R_i$ is the maximal ideal in R_i for each i, and F_i/F is unramified. The last statement follows since $F_i \cong F\zeta_n$ for some i.

(ii) For any field K of characteristic zero, a *cyclotomic algebra* over K is a twisted group ring of the form $A = L^{\beta}[G]^t$, where L is a finite cyclotomic extension of K, $G = \mathrm{Gal}(L/K)$, and $\beta \in H^2(G;\mu_L)$ (so A is a central simple K-algebra). By the Brauer-Witt theorem (see Witt [1], or Yamada [1, Theorem 3.9]), any simple summand A of $F[G]$ is similar to a cyclotomic algebra over its center. Then by another theorem of Witt [1, Satz 12] (see also Yamada [1, Proposition 4.8 and Corollary 5.4]), for any finite extension $E \supseteq \hat{\mathbb{Q}}_p(\zeta_p)$ (p odd) or $E \supseteq \hat{\mathbb{Q}}_2(i)$ ($p = 2$), any cyclotomic algebra over E is a matrix algebra. This proves the last statement in (ii). The first statement then follows from Proposition 1.3: for any simple summand A of $F[G]$ with center $E \supseteq F$, $\mathrm{ind}(A) \mid [E\zeta_p:E] \mid p-1$ if p is odd, and $\mathrm{ind}(A) \mid [E(i):E] \mid 2$ if $p = 2$. □

Many of the elementary properties of $K_1(\mathfrak{A})$, when $\mathfrak{A}$ is a $\hat{\mathbb{Z}}_p$-order in a semisimple $\hat{\mathbb{Q}}_p$-algebra, are special cases of results about semilocal rings. These will be discussed in the next section.

1c. Semilocal rings and the Jacobson radical

For any ring R, the *Jacobson radical* $J(R)$ is defined to be the intersection of all maximal left ideals in R; or, equivalently, the intersection of all maximal right ideals in R (see Bass [2, Section III.2]). For example, the Jacobson radical of a local ring is its unique maximal ideal, and the Jacobson radical of a semisimple ring is trivial.

An ideal $I \subseteq R$ is called a *radical ideal* if it is contained in $J(R)$. If R is finite, then $I \subseteq R$ is a radical ideal if and only if it is nilpotent (see Reiner [1, Theorem 6.9]). If $\mathfrak{A}$ is a $\hat{\mathbb{Z}}_p$-order in a semisimple $\hat{\mathbb{Q}}_p$-algebra, then $J(\mathfrak{A}) \supseteq p\mathfrak{A}$ and $J(\mathfrak{A})/p\mathfrak{A} = J(\mathfrak{A}/p\mathfrak{A})$; so $I \subseteq \mathfrak{A}$ is radical if and only if $\lim_{n\to\infty} I^n = 0$. The next theorem helps to explain the importance of radical ideals when working in K-theory.

Theorem 1.11 *For any ring R with Jacobson radical $J = J(R)$, and any $n \geq 1$, a matrix $M \in M_n(R)$ is invertible if and only if it becomes invertible in $M_n(R/J)$. In particular, $1+J \subseteq R^*$.*

Proof See Bass [2, Proposition III.2.2 and Corollary III.2.7]. □

The next example shows that p-adic group rings of p-groups are, in fact, local rings.

Example 1.12 *If R is the ring of integers in any finite extension of $\hat{\mathbb{Q}}_p$, if $\mathfrak{p} \subseteq R$ is the maximal ideal, and if G is any p-group, then $R[G]$ is a local ring with unique maximal ideal*

$$J(R[G]) = \{\textstyle\sum r_i g_i \; : \; r_i \in R, \;\; g_i \in G, \;\; \sum r_i \in \mathfrak{p}\}.$$

In particular, $R[G]/J(R[G]) \cong R/\mathfrak{p}$.

Proof See Curtis & Reiner [1, Corollary 5.25]. □

We now recall some of the basic definitions and properties of $K_1(-)$. For any ring R, let $GL_n(R)$ be the group of invertible $n{\times}n$ matrices over R (any $n \geq 1$); and set $GL(R) = \bigcup_{n=1}^{\infty} GL_n(R)$. For any $i \neq j$ and any $r \in R$, let $e_{ij}^r \in GL(R)$ denote the elementary matrix which is the identity except for the entry r in the (i,j)-position; and let $E(R) \subseteq GL(R)$ be the subgroup generated by the e_{ij}^r. If $I \subseteq R$ is any (2-sided) ideal, then $GL(R,I)$ denotes the group of invertible matrices which are congruent to the identity modulo I; and $E(R,I)$ denotes the smallest *normal* subgroup of $GL(R)$ containing all e_{ij}^r for $r \in I$. Finally, set $K_1(R) = GL(R)/E(R)$ and $K_1(R,I) = GL(R,I)/E(R,I)$. That these are, in fact, abelian groups will follow from the next theorem.

For the purposes of this chapter, we define $K_2(R)$, for any ring R, by setting $K_2(R) = H_2(E(R))$. The usual definition (involving the Steinberg group), as well as some of the basic properties of, e. g., Steinberg symbols in $K_2(R)$, will be given in Section 3a.

__Theorem 1.13__ (Whitehead's lemma) *For any ring* R, *and any ideal* $I \subseteq R$,

$$E(R) = [GL(R),GL(R)] = [E(R),E(R)], \quad \textit{and}$$

$$E(R,I) = [GL(R),GL(R,I)] = [E(R),E(R,I)].$$

For any $A,B \in GL_n(R,I)$, $\begin{pmatrix} A & 0 \\ 0 & A^{-1} \end{pmatrix} \in E_{2n}(R,I)$, *and* $[A]\cdot[B] = \begin{bmatrix} A & 0 \\ 0 & B \end{bmatrix}$ *in* $K_1(R,I)$. *Furthermore, there is an exact sequence*

$$K_2(R) \longrightarrow K_2(R/I) \longrightarrow K_1(R,I) \longrightarrow K_1(R) \longrightarrow K_1(R/I).$$

__Proof__ The commutator relations are due to Whitehead and Bass, and are shown in Milnor [2, Lemmas 3.1 and 4.3]. The relation

$$\begin{pmatrix} A & 0 \\ 0 & A^{-1} \end{pmatrix} = \begin{pmatrix} I & A \\ 0 & I \end{pmatrix}\begin{pmatrix} I & 0 \\ -A^{-1} & I \end{pmatrix}\begin{pmatrix} I & A \\ 0 & I \end{pmatrix}\begin{pmatrix} I & -I \\ 0 & I \end{pmatrix}\begin{pmatrix} I & 0 \\ I & I \end{pmatrix}\begin{pmatrix} I & -I \\ 0 & I \end{pmatrix} \in E_{2n}(R,I);$$

is clear from the definition of $E_{2n}(R,I)$ (and is part of the proof that $[GL(R),GL(R,I)] \subseteq E(R,I)$); and $[A]\cdot[B] = [A]\cdot[\mathrm{diag}(I,B)] = [\mathrm{diag}(A,B)]$ as an immediate consequence. The exact sequence is constructed in Milnor [2,

Lemma 4.1 and Theorem 6.2]; and can also be derived from the five term exact homology sequence (see Theorem 8.2 below). □

A ring R is called *semilocal* if $R/J(R)$ is semisimple; or equivalently, if $R/J(R)$ is artinian (see Bass [2, §III.2]). Thus, any finite ring, and any $\hat{\mathbb{Z}}_p$-order, are semilocal. As one might guess, given Theorem 1.11 above, the functor K_1 behaves particularly nicely for semilocal rings.

<u>**Theorem 1.14**</u> *The following hold for any semilocal ring* R.

(i) Any element of $K_1(R)$ *is represented by a unit (i. e., by a one-by-one matrix).*

(ii) If R *is commutative, then* $K_1(R) \cong R^*$. *In particular,* $SK_1(\mathfrak{A}) = 1$ *if* $\mathfrak{A}$ *is any commutative* $\hat{\mathbb{Z}}_p$*-order.*

(iii) If S *is another semilocal ring, and* $\alpha: R \twoheadrightarrow S$ *is an epimorphism, then the maps*

$$GL_n(\alpha): GL_n(R) \longrightarrow GL_n(S) \quad \text{and} \quad K_1(\alpha): K_1(R) \longrightarrow K_1(S)$$

(any $n \geq 1$*) are all surjective.*

<u>**Proof**</u> These are all shown in Bass [2]: (i) in Theorem V.9.1, (ii) in Corollary V.9.2, and (iii) in Corollary III.2.9. □

The following relation in $K_1(R,I)$, due to Vaserstein [1], is often useful, and helps to simplify some of the proofs in later chapters. Swan's presentation of $K_2(R,I)$ below (when I is a radical ideal) will be used in this book only in the case when $I^2 = 0$.

<u>**Theorem 1.15**</u> *For any ring* R *and any ideal* $I \subseteq R$, *if* $r \in R$ *and* $x \in I$ *are such that* $(1+rx) \in R^*$, *then* $(1+xr) \in R^*$ *and*

$$(1+rx)(1+xr)^{-1} \in E(R,I).$$

If I *is a radical ideal (i. e.,* $I \subseteq J(R)$*), then*

$$K_1(R,I) \cong (1+I)/\langle (1+rx)(1+xr)^{-1} : r \in R,\ x \in I\rangle.$$

In particular, $K_1(R,I) \cong I/\langle rx - xr : r \in R,\ x \in I\rangle$ *if* $I^2 = 0$.

<u>Proof</u> Recall that $E(R,I) = [GL(R), GL(R,I)]$ (Theorem 1.13). Using this, the relation

$$\begin{pmatrix}1 & r\\ 0 & 1\end{pmatrix}\begin{pmatrix}1 & 0\\ x & 1\end{pmatrix}\begin{pmatrix}1 & 0\\ 0 & 1+xr\end{pmatrix} = \begin{pmatrix}1+rx & 0\\ 0 & 1\end{pmatrix}\begin{pmatrix}1 & 0\\ x & 1\end{pmatrix}\begin{pmatrix}1 & r\\ 0 & 1\end{pmatrix} \qquad (1)$$

shows that $(1+xr) \in R^*$, and that $\begin{pmatrix}1+rx & 0\\ 0 & (1+xr)^{-1}\end{pmatrix} \in E(R,I)$. In particular, since $\begin{pmatrix}u & 0\\ 0 & u^{-1}\end{pmatrix} \in E(R,I)$ for any $u \in (1+I)^*$ (Theorem 1.13 again), this shows that $(1+rx)(1+xr)^{-1} \in E(R,I)$.

The presentation for $K_1(R,I)$, when $I \subseteq J(R)$, is due to Swan [2,Theorem 2.1]. The last presentation (when $I^2 = 0$) is a special case of Swan's presentation, but is also an easy consequence of Vaserstein's identity. □

When studying $K_1(\mathfrak{A})$, for a $\hat{\mathbb{Z}}_p$-order $\mathfrak{A}$, it is often necessary to get information about $K_1(\mathfrak{A}/I)$ and $K_2(\mathfrak{A}/I)$ for ideals $I \subseteq \mathfrak{A}$ of finite index. The next result is a first step towards doing this.

<u>Theorem 1.16</u> *Let* R *be a finite ring. Then* $K_1(R)$ *and* $K_2(R)$ *are finite. Furthermore, (i)* $K_2(R) = 1$ *if* R *is semisimple; (ii)* $K_2(R)$ *is a p-group if* R *has p-power order (for any prime* p*); and (iii)* $p \nmid |K_1(R)|$ *if* R *is semisimple and has p-power order for some prime* p.

<u>Proof</u> By Theorem 1.14(i), R^* surjects onto $K_1(R)$. By Dennis [1, Theorem 1], there is a surjection of $H_2(E_5(R))$ onto $K_2(R)$. So $K_1(R)$ and $K_2(R)$ are both finite. If R is semisimple, then by the Wedderburn theorem, $R \cong \prod_{i=1}^k M_{r_i}(D_i)$, where the D_i are finite division algebras and hence fields. Then $GL(R) \cong \prod_{i=1}^k GL(D_i)$, $E(R) \cong \prod_{i=1}^k E(D_i)$; and hence

$K_n(R) \cong \prod_{i=1}^k K_n(D_i)$ for $n = 1,2$. So $K_2(R) = 1$ by Milnor [2, Corollary 9.13]; and $p \nmid |K_1(R)|$ if R (and hence the D_i) have p-power order.

By Theorem 1.11, if $J \subseteq R$ is the Jacobson radical, then the group

$$X = \operatorname{Ker}\Big[E(R) \twoheadrightarrow E(R/J)\Big] \subseteq \bigcup_{n\geq 1} (1+M_n(J))$$

is a union of finite p-groups. Hence $H_i(X)[\frac{1}{p}] = 0$ for all $i > 0$; and the Hochschild-Serre spectral sequence (see Brown [1, Theorem VII.6.3]) applies to show that

$$K_2(R)[\tfrac{1}{p}] \cong H_2(E(R))[\tfrac{1}{p}] \cong H_2(E(R/J))[\tfrac{1}{p}] \cong K_2(R/J)[\tfrac{1}{p}].$$

Since R/J is semisimple, $K_2(R/J) = 1$; and hence $K_2(R)[\frac{1}{p}] = 1$. □

We end the section with some localization exact sequences which help to describe $K_i(\mathfrak{M})$ when $\mathfrak{M}$ is a maximal $\mathbb{Z}$- or $\hat{\mathbb{Z}}_p$-order. They are special cases of Quillen's localization sequences for regular rings (or abelian categories).

Theorem 1.17 *(i) For any prime* p, *if* $\mathfrak{M}$ *is a maximal* $\hat{\mathbb{Z}}_p$*-order in a semisimple* $\hat{\mathbb{Q}}_p$*-algebra* A, *and if* $J \subseteq \mathfrak{M}$ *is the Jacobson radical, then there is for all* $n \geq 0$ *an exact sequence*

$$\cdots \longrightarrow K_{i+1}(\mathfrak{M}) \longrightarrow K_{i+1}(A) \longrightarrow K_i(\mathfrak{M}/J) \longrightarrow K_i(\mathfrak{M}) \longrightarrow K_i(A) \longrightarrow \cdots.$$

In particular, $p \nmid |SK_1(\mathfrak{M})|$.

(ii) Fix a subring $\Lambda \subseteq \mathbb{Q}$ *and a maximal* Λ*-order* $\mathfrak{M}$ *in a semisimple* $\mathbb{Q}$*-algebra* A, *and let* $\mathcal{P}$ *be any set of primes not invertible in* Λ. *Set* $\mathfrak{M}[\frac{1}{\mathcal{P}}] = \mathfrak{M}[\frac{1}{p}: p\in\mathcal{P}]$, *and let* $J_p \subseteq \hat{\mathfrak{M}}_p$ *(for* $p \in \mathcal{P}$*) be the Jacobson radical. Then there is an exact sequence*

$$\cdots \to K_{i+1}(\mathfrak{M}) \to K_{i+1}(\mathfrak{M}[\tfrac{1}{\mathcal{P}}]) \to \bigoplus_{p\in\mathcal{P}} K_i(\hat{\mathfrak{M}}_p/J_p) \to K_i(\mathfrak{M}) \to K_i(\mathfrak{M}[\tfrac{1}{\mathcal{P}}]) \to \cdots.$$

Proof These follow from Quillen [1, Theorems 4 and 5]. For example, in case (ii), if $\underline{\underline{M}}(\mathfrak{M})$, $\underline{\underline{M}}(\mathfrak{M}[\frac{1}{\mathcal{P}}])$, and $\underline{\underline{M}}^t(\mathfrak{M})$ denote the categories of finitely generated $\mathfrak{M}$-modules, $\mathfrak{M}[\frac{1}{\mathcal{P}}]$-modules, and $\mathcal{P}$-torsion $\mathfrak{M}$-modules, respectively; then by Quillen [1, Theorem 5] there is an exact localization sequence

$$\cdots \longrightarrow K_{i+1}(\underline{\underline{M}}(\mathfrak{M}[\tfrac{1}{\mathcal{P}}])) \longrightarrow K_i(\underline{\underline{M}}^t(\mathfrak{M})) \longrightarrow K_i(\underline{\underline{M}}(\mathfrak{M})) \longrightarrow K_i(\underline{\underline{M}}(\mathfrak{M}[\tfrac{1}{\mathcal{P}}])) \longrightarrow \cdots$$

Since $\mathfrak{M}$ and $\mathfrak{M}[\frac{1}{\mathcal{P}}]$ are hereditary (Theorem 1.4(iv)), all finitely generated $\mathfrak{M}$- or $\mathfrak{M}[\frac{1}{\mathcal{P}}]$-modules have finite projective resolutions. It follows that $K_i(\underline{\underline{M}}(\mathfrak{M})) \cong K_i(\mathfrak{M})$ and $K_i(\underline{\underline{M}}(\mathfrak{M}[\frac{1}{\mathcal{P}}])) \cong K_i(\mathfrak{M}[\frac{1}{\mathcal{P}}])$. For any $\mathcal{P}$-torsion $\mathfrak{M}$-module M, $M = \oplus_{p\in\mathcal{P}} M_{(p)}$, and each $M_{(p)}$ has a filtration by $\mathfrak{M}_p/J_p$-modules. So by devissage (Quillen [1, Theorem 4]),

$$K_i(\underline{\underline{M}}^t(\mathfrak{M})) \cong \bigoplus_{p\in\mathcal{P}} K_i(\hat{\mathfrak{M}}_p/J_p).$$

In case (i), $SK_1(\mathfrak{M}) = \operatorname{Im}\Big[K_1(\mathfrak{M}/J) \longrightarrow K_1(\mathfrak{M})\Big]$. Since $\mathfrak{M}/J$ is semisimple, $K_1(\mathfrak{M}/J)$ has order prime to p by Theorem 1.16(iii), and so $p \nmid |SK_1(\mathfrak{M})|$. □

1d. Bimodule-induced homomorphisms and Morita equivalence

Define the category of "rings with bimodule morphisms" to be the category whose objects are rings; and where Mor(R,S), for any rings R and S, is the Grothendieck group modulo short exact sequences of all isomorphism classes of (S,R)-bimodules ${}_SM_R$ such that M is finitely generated and projective as a left S-module. Composition of morphisms is given by tensor product. The usual category of rings with homomorphisms is mapped to this category by sending any $f\colon R \longrightarrow S$ to the bimodule ${}_SS_R$, where $s_1\cdot(s_2)\cdot r = s_1s_2f(r)$. The importance of this category for our purposes here follows from the following proposition.

Proposition 1.18 *For each* i, K_i *is an additive functor on the category of rings with bimodule morphisms.*

Proof Any (S,R)-bimodule M which is finitely generated and projective as a left S-module induces a functor

$$M\otimes_R : \underline{P}(R) \longrightarrow \underline{P}(S);$$

where $\underline{P}(-)$ denotes the category of finitely generated projective modules. So the proposition follows immediately from Quillen's definition in [1] of $K_i(R)$ using the Q-construction on $\underline{P}(R)$.

In the case of $K_1(-)$ and $K_2(-)$, this can be seen much more directly. Let ${}_SM_R$ be any bimodule as above, and fix some isomorphism $M\oplus P \cong S^k$ of left S-modules. For each $n \geq 1$, define homomorphisms

$$[M\otimes_R]_n : GL_n(R) \cong Aut_R(R^n) \longrightarrow Aut_S(S^{nk}) \cong GL_{nk}(S)$$

by setting $[M\otimes_R]_n(\alpha) = (M\otimes_R \alpha)\oplus Id(P^n)$ for each $\alpha \in Aut(R^n)$. The $[M\otimes_R]_n$ are easily seen to be (up to inner automorphism and stabilizing) independent of the choice of isomorphism $M\oplus P\cong S^k$; and hence induce unique homomorphisms on $K_1(R) \cong H_1(GL(R))$ and $K_2(R) \cong H_2(E(R))$. □

As one example, transfer homomorphisms in K-theory can be defined in terms of bimodules. If $R \subseteq S$ is any pair of rings such that S is projective and finitely generated as an R-module, then

$$trf_R^S = [S\otimes_S]_* : K_i(S) \longrightarrow K_i(R);$$

when S is regarded as an (R,S)-bimodule in the obvious way. The above proposition is often useful when verifying the commutativity of K-theoretic diagrams which mix transfer homomorphisms, maps induced by ring homomorphisms, and others: commutativity is checked by constructing isomorphisms of bimodules. Examples of this can be seen in the proofs of Proposition 5.2 and Theorem 12.3, as well as throughout Chapter 11.

Another setting in which it is useful to regard $K_i(-)$ as a functor defined on rings with bimodule morphisms is that of *Morita equivalence*. A Morita equivalence between two rings R and S is an "invertible" bimodule ${}_SM_R$: i. e., for some bimodule ${}_RN_S$, $M \otimes_R N \cong S$ and $N \otimes_S M \cong R$ as bimodules. In particular, $[M\otimes_R]_*$ and $[N\otimes_S]_*$ are inverse isomorphisms between $K_i(R)$ and $K_i(S)$.

The simplest example of this is a matrix algebra. For any ring R and any $n > 1$, R^n is invertible when regarded as an $(M_n(R),R)$-bimodule. In this case, the induced isomorphisms $K_i(M_n(R)) \cong K_i(R)$ are precisely those induced by identifying $GL_m(S)$ with $GL_{mn}(R)$.

In Theorem 1.9, we saw that any maximal $\hat{\mathbb{Z}}_p$-order in a simple $\hat{\mathbb{Q}}_p$-algebra $M_n(D)$ (D a division algebra) is conjugate to a matrix algebra over the maximal order in D. This is not the case for maximal $\mathbb{Z}$-orders in simple $\mathbb{Q}$-algebras; but a result which is almost as good can be stated in terms of Morita equivalence.

<u>Theorem 1.19</u> *Fix a Dedekind domain* R *with field of fractions* K. *Let* A *be any simple* K-*algebra. Write* $A = M_n(D)$, *where* D *is a division algebra, and identify* $A = \mathrm{End}_D(V)$ *for some* n-*dimensional* D-*module* V. *Let* $\Delta \subseteq D$ *be any maximal* R-*order. Then* $M_n(\Delta)$ *is a maximal* R-*order in* A; *and any maximal* R-*order in* A *has the form* $\mathfrak{M} = \mathrm{End}_\Delta(\Lambda)$ *for some* Δ-*lattice* Λ *in* V. *Furthermore,* Λ *is invertible as an* $(\mathfrak{M},\Delta)$-*bimodule, and so* Λ *and* V *induce for all* i *a commutative square*

$$\begin{array}{ccc} K_i(\Delta) & \xrightarrow{\text{incl}} & K_i(D) \\ \cong\big\downarrow [\Lambda]_* & & \cong\big\downarrow [V]_* \\ K_i(\mathfrak{M}) & \xrightarrow{\text{incl}} & K_i(A) = K_i(M_n(D)). \end{array}$$

<u>Proof</u> See Reiner [1, Theorem 21.6 & Corollary 21.7]. □

Chapter 2 STRUCTURE THEOREMS FOR K_1 OF ORDERS

This chapter presents some of the basic applications of the reduced norm and logarithm homomorphisms to describe $K_1(\mathfrak{A})$ and $K_1(A)$, when $\mathfrak{A}$ is a $\mathbb{Z}$-order or $\hat{\mathbb{Z}}_p$-order in a semisimple $\mathbb{Q}$- or $\hat{\mathbb{Q}}_p$-algebra A. For example, $K_1(\mathfrak{A})$ is shown to be finitely generated whenever $\mathfrak{A}$ is a $\mathbb{Z}$-order; and is shown to be a product of a finite group with a finitely generated $\hat{\mathbb{Z}}_p$-module in the $\hat{\mathbb{Z}}_p$-order case. In both cases, the rank of $K_1(\mathfrak{A})$ is determined. Also, $SK_1(\mathfrak{A})$ is shown (for both $\mathbb{Z}$- and $\hat{\mathbb{Z}}_p$-orders) to be the kernel of the "reduced norm" homomorphism from $K_1(\mathfrak{A})$ to units in the center of A.

The results about reduced norms are dealt with in Section 2a. These include all of the results about $\mathbb{Z}$-orders mentioned above, as well as some properties of $\hat{\mathbb{Z}}_p$-orders. Then, in Section 2b, p-adic logarithms are applied to show, for example, that for any $\hat{\mathbb{Z}}_p$-order $\mathfrak{A}$, $E(\mathfrak{A})$ is p-adically closed in $GL(\mathfrak{A})$ (i. e., that $K_1(\mathfrak{A})$ is Hausdorff in the p-adic topology).

2a. Applications of the reduced norm

For any field F, and any central simple F-algebra A, the *reduced norm homomorphism* $nr_{A/K}: A^* \longrightarrow F^*$ for A is defined as follows. Let $E \supseteq F$ be any extension which splits A, and fix an isomorphism $E\otimes_F A \xrightarrow[\cong]{\varphi} M_n(E)$. Then for any $a \in A^*$, set $nr_{A/F}(a) = \det_E(\varphi(1\otimes a)) \in E^*$. This is independent of the choice of φ: any two such isomorphisms differ by an inner automorphism of $M_n(E)$ by the Skolem-Noether theorem (Theorem 1.1(iv) above). Furthermore, $nr_{A/F}(a) \in F^*$ (it is fixed by the action of $Gal(E/F)$ if E/F is Galois); and is independent of the choice of splitting field E. For more details, see Reiner [1, Section 9a].

As one easy example, consider the quaternion algebra $\mathbb{H}$ with center $\mathbb{R}$. A $\mathbb{C}$-linear ring isomorphism $\varphi\colon \mathbb{C}\otimes_{\mathbb{R}}\mathbb{H} \xrightarrow{\cong} M_2(\mathbb{C})$ is defined by setting

$$\varphi(1\otimes 1) = \begin{pmatrix}1&0\\0&1\end{pmatrix}, \quad \varphi(1\otimes i) = \begin{pmatrix}i&0\\0&-i\end{pmatrix}, \quad \varphi(1\otimes j) = \begin{pmatrix}0&1\\-1&0\end{pmatrix}, \quad \varphi(1\otimes k) = \begin{pmatrix}0&i\\i&0\end{pmatrix}.$$

Then, for any $\xi = a+bi+cj+dk \in \mathbb{H}$,

$$nr_{\mathbb{H}/\mathbb{R}}(\xi) = \det\begin{pmatrix}a+bi & c+di\\ -c+di & a-bi\end{pmatrix} = a^2+b^2+c^2+d^2.$$

It is immediate from the definition that $nr_{A/F}\colon A^* \longrightarrow F^*$ is a homomorphism. For any $n > 1$, $M_n(A)$ is again a central simple F-algebra, and $nr_{M_n(A)/F}$ is the extension to $GL_n(A)$ of $nr_{M_{n-1}(A)/F}$. So the reduced norm extends to a homomorphism defined on $GL(A)$, and hence factors through its abelianization $K_1(A)$. For example, $nr_{\mathbb{H}/\mathbb{R}}$ induces an isomorphism between $K_1(\mathbb{H})$ and the multiplicative group of positive real numbers.

The first lemma lists some of the immediate properties of reduced norms.

<u>Lemma 2.1</u> *Fix a field* F, *and let* A *be a central simple* F-*algebra. Set* $n = [A:K]^{1/2}$. *Then the following hold.*

(i) $\det_F(A \xrightarrow{\cdot u} A) = nr_{A/F}(u)^n$ *for any* $u \in A^*$.

(ii) $nr_{A/F}(u) = u^n$ *for any* $u \in F^*$.

(iii) If $A \cong M_n(F)$, *then* $nr_{A/F}\colon A^* \longrightarrow F^*$ *is the determinant homomorphism.*

(iv) If $E \subseteq A$ *is a subfield containing* F, and *if* B *is the centralizer of* E *in* A, *then* $nr_{A/F}(u) = N_{E/F}(nr_{B/E}(u))$ *for any* $u \in B^*$.

<u>Proof</u> The first three points are shown in Reiner [1, Section 9a]. Point (iv) is shown in Draxl [1, Corollary 22.5]; and also follows easily

from the relations among (reduced) characteristic polynomials in Reiner [1, Theorems 9.5, 9.6a, and 9.10(iii)]. □

Now, for any semisimple $\mathbb{Q}$- or $\hat{\mathbb{Q}}_p$-algebra $A = \prod_{i=1}^k A_i$, where each A_i is simple with center F_i, and any $\mathbb{Z}$- or $\hat{\mathbb{Z}}_p$-order $\mathfrak{A}$, we let

$$nr_A: K_1(A) \longrightarrow \prod_i (F_i)^* = Z(A)^* \quad \text{and} \quad nr_{\mathfrak{A}}: K_1(\mathfrak{A}) \longrightarrow Z(A)^*$$

denote the homomorphisms induced by the product of the reduced norm maps for the A_i. Note that nr_A and $nr_{\mathfrak{A}}$ are used here to denote the homomorphisms defined on $K_1(-)$, while $nr_{A/F}$ denotes the reduced norm as a map on A^*.

The following lemma will be useful when computing $\mathrm{Ker}(nr_A)$.

Lemma 2.2 *Let* $E \supseteq F$ *be any finite field extension of degree* n, *and let* A *be any* F-*algebra. Then*

$$\mathrm{Ker}\Big[K_1(i): K_1(A) \longrightarrow K_1(E \otimes_F A)\Big]$$

(where $i(x) = 1 \otimes x$*) has exponent dividing* n.

Proof By Proposition 1.18, the composite

$$trf \circ K_1(i) : K_1(A) \longrightarrow K_1(E \otimes_F A) \longrightarrow K_1(A)$$

is induced by tensoring with $E \otimes_F A$, regarded as an (A,A)-bimodule. Since F is central in A, $E \otimes_F A \cong A^n$ as (A,A)-bimodules, and so $trf \circ K_1(i)$ is multiplication by n. The result is now immediate. □

If A is any simple $\mathbb{Q}$-algebra with center K, then a real valuation v of K (i. e., an embedding $v: K \hookrightarrow \mathbb{R}$) is called *ramified* in A if $\mathbb{R} \otimes_{vK} A$ is a matrix algebra over $\mathbb{H}$.

<u>Theorem 2.3</u> *Let* A *be a simple* $\mathbb{Q}$- *or* $\hat{\mathbb{Q}}_p$*-algebra with center* $F = Z(A)$, *let* $\mathfrak{M} \subseteq A$ *be a maximal* $\mathbb{Z}$- *or* $\hat{\mathbb{Z}}_p$*-order, and let* $R \subseteq F$ *be the ring of integers. Then*

$$nr_A : K_1(A) \longrightarrow F^*$$

is injective; and $\mathrm{Im}(nr_A)$ *and* $\mathrm{Im}(nr_{\mathfrak{M}})$ *are described as follows:*

(i) If A *is a* $\hat{\mathbb{Q}}_p$*-algebra, then*

$$nr_A(K_1(A)) = F^* \quad \text{and} \quad nr_{\mathfrak{M}}(K_1(\mathfrak{M})) = R^*.$$

(ii) If A *is a* $\mathbb{Q}$*-algebra, then set*

$$F^*_+ = \left\{u \in F^* : v(u) > 0 \ \text{for all ramified} \ v: F \hookrightarrow \mathbb{R}\right\}; \qquad R^*_+ = F^*_+ \cap R^*.$$

Then

$$nr_A(K_1(A)) = F^*_+ \quad \text{and} \quad nr_{\mathfrak{M}}(K_1(\mathfrak{M})) = R^*_+.$$

<u>Proof</u> Recall that the index of $A \cong M_r(D)$ (D a division algebra) is defined by $\mathrm{ind}(A) = \mathrm{ind}(D) = [D:F]^{1/2}$.

<u>Step 1</u> We first consider the formulas for $\mathrm{Im}(nr_A)$ and $\mathrm{Im}(nr_{\mathfrak{M}})$. Set $n = \mathrm{ind}(A)$. Note first that $nr_{\mathfrak{M}}(K_1(\mathfrak{M})) \subseteq R^*$: for any $u \in GL_k(\mathfrak{M})$,

$$nr_{A/F}(u)^n = \det_F(A^k \xrightarrow{u\cdot} A^k) = \det_R(\mathfrak{M}^k \xrightarrow{u\cdot} \mathfrak{M}^k) \in R^*$$

(Lemma 2.1(i)), and so $nr_{A/F}(u) \in R^*$.

If F is a finite extension of $\hat{\mathbb{Q}}_p$, then by Theorem 1.9, we can write $A = M_r(D)$ and $\mathfrak{M} = M_r(\Delta)$, where D is a division algebra and $\Delta \subseteq D$ a maximal order. Also by Theorem 1.9, there is a maximal subfield $E \subseteq D$ and an element $\pi \in D$ such that E/F is unramified, such that

$\pi E\pi^{-1} = E$, and such that π^n generates the maximal ideal in R. By Lemma 2.1(iv), $nr_D|E = N_{E/F}$, the usual norm. If $S \subseteq E$ is the ring of integers, then $N_{E/F}(S^*) = R^*$ by Proposition 1.8(iii); and so

$$nr_{\mathfrak{M}}(K_1(\mathfrak{M})) \supseteq \langle N_{E/F}(S^*), nr_{D/F}(\pi)\rangle = \langle R^*, (-1)^{n-1}\pi^n\rangle = F^*.$$

If F is a finite extension of $\mathbb{Q}$, then the formula for $nr_{A/F}(A^*)$ is the Hasse-Schilling-Maass norm theorem (see, e. g., Reiner [1, Theorem 33.15]). To see that $nr_{\mathfrak{M}}(K_1(\mathfrak{M})) = R^*_+$, fix some $u \in R^*_+$, and choose $M \in GL(A)$ such that $nr_{A/F}(M) = u$. Let n be the product of the distinct primes at which M is not invertible; and set $\hat{A}_n = \prod_{p|n}\hat{A}_p$, $\hat{R}_n = \prod_{p|n}\hat{R}_p$, etc. Then $nr_{\hat{A}_n}([M]) \in (\hat{R}_n)^* = nr_{\hat{\mathfrak{M}}_n}(K_1(\hat{\mathfrak{M}}_n))$. So assuming the injectivity of $nr_{\hat{A}_n} = \prod_{p|n} nr_{\hat{A}_p}$ (shown in Step 2 below), there exist elementary matrices $e_{i_1j_1}(r_1),\ldots,e_{i_kj_k}(r_k) \in E(\hat{A}_n)$ such that

$$M\cdot e_{i_1j_1}(r_1)\cdots e_{i_kj_k}(r_k) \in GL(\hat{\mathfrak{M}}_n).$$

Choose elements $\hat{r}_1,\ldots,\hat{r}_k \in \mathfrak{M}[\frac{1}{n}]$ such that $\hat{r}_t \equiv r_t \pmod{\hat{\mathfrak{M}}_n}$ for all t. Note that it suffices to do this on the individual coordinates (in $\hat{\mathbb{Q}}_n$) of the r_t with respect to some fixed $\mathbb{Z}$-basis of $\mathfrak{M}$. If we now set $\hat{M} = M\cdot e_{i_1j_1}(\hat{r}_1)\cdots e_{i_kj_k}(\hat{r}_k)$, then $nr_{A/F}(\hat{M}) = nr_{A/F}(M) = u$ and $\hat{M} \in GL(\mathfrak{M})$.

<u>Step 2</u> The injectivity of nr_A was first shown by Nakayama & Matsushima [1] in the p-adic case, and for $\mathbb{Q}$-algebras by Wang [1]. The following combined proof, using induction on $[A:F]$, is modelled on that in Draxl [1].

<u>Step 2a</u> Assume first that $E \subseteq A$ is a subfield such that E/F is a cyclic Galois extension of degree $n > 1$, and let B denote the centralizer of E in A. We claim that $[u] = 1$ in $K_1(A)$ for any $u \in B^*$ such that $nr_{A/F}(u) = 1$.

Note first that B is a simple algebra with center E (see Reiner

[1, Theorem 7.11]), and that $[B:E] < [A:F]$. So nr_B is injective by the induction hypothesis. Furthermore, by Lemma 2.1(iv),

$$nr_{A/F}(u) = N_{E/F}(nr_{B/E}(u)) = 1. \tag{2}$$

Set $G = Gal(E/F) \cong \mathbb{Z}/n$, and consider the exact sequence in cohomology

$$\hat{H}^{-2}(G;E^*/nr_B(B^*)) \longrightarrow \hat{H}^{-1}(G;nr_B(B^*)) \longrightarrow \hat{H}^{-1}(G;E^*).$$

Here, if $\psi \in G$ is a generator, we identify for any G-module M:

$$\hat{H}^{-1}(G;M) = \{x \in M : N_G(x) = x+\psi(x)+\ldots+\psi^{p-1}(x) = 0\}/\{\psi(x)-x : x \in M\}.$$

In particular, $\hat{H}^{-1}(G;E^*) = 1$ by Hilbert's Theorem 90 (see Janusz [1, Appendix A]). Also, by Step 1, $\hat{H}^{-2}(G;E^*/nr_B(B^*)) = 1$: in the p-adic case since $E^* = nr_B(B^*)$; and in the $\mathbb{Q}$-algebra case since $E^*/nr_B(B^*)$ is a product of copies of $\{\pm 1\}$ for certain real embeddings $E \hookrightarrow \mathbb{R}$, and these real embeddings are permuted freely by G.

Thus, $\hat{H}^{-1}(G;nr_B(B^*)) = 1$. So by (2), there is an element $v \in B^*$ such that

$$\psi(nr_{B/E}(v))\cdot(nr_{B/E}(v))^{-1} = nr_{B/E}(u).$$

Furthermore, by the Skolem-Noether theorem (Theorem 1.1(iv)), there is an element $\alpha \in A$ such that $\alpha x\alpha^{-1} = \psi(x)$ for all $x \in E$. Then $\alpha v\alpha^{-1} \in B$ (B is the centralizer of E), and

$$nr_{B/E}([\alpha,v]) = nr_{B/E}(\alpha v\alpha^{-1})/nr_{B/E}(v) = \psi(nr_{B/E}(v))/(nr_{B/E}(v)) = nr_{B/E}(u).$$

Thus, $nr_{B/E}(u\cdot[\alpha,v]^{-1}) = 1$, so $u\cdot[\alpha,v]^{-1} = 1$ in $K_1(B)$ by the induction hypothesis; and hence $[u] = 1 \in K_1(A)$.

<u>Step 2b</u> The rest of the proof consists of manipulations, using Lemma 2.2, to reduce the general case to that handled in Step 2a. Fix a prime

p, and any $u \in A^*$ such that $nr_{A/F}(u) = 1$. We will show that $[u] \in K_1(A)$ has finite order prime to p.

Let $\hat{F} \supseteq F$ be a splitting field such that $\hat{F}/F$ is Galois, and let $F' \subseteq \hat{F}$ be the fixed field of a p-Sylow subgroup of $Gal(\hat{F}/F)$. Then $p\nmid[F':F]$, and $[\hat{F}:F']$ is a p-power. Write

$$F' \otimes_F A \cong M_r(D) \qquad \text{and} \qquad [1\otimes u] = [v] \in K_1(D), \tag{3}$$

where D is a division algebra, $F' = Z(D)$, $v \in D^*$, and $nr_{D/F'}(v) = 1$.

Now let $E \supseteq F'(v)$ be any maximal subfield containing v, and let $\hat{E} \supseteq E$ be any normal closure of E over F'. Let $K \subseteq \hat{E}$ be the fixed field of some p-Sylow subgroup of $Gal(\hat{E}/F')$. Thus, $p\nmid[K:F']$; and $ind(D) = [E:F'] \mid [\hat{F}:F']$ is a p-power by Proposition 1.3.

Set $B = K\otimes_{F'} D$, and identify $v \in D$ with $1\otimes v \in B$. Then $nr_{B/K}(v) = 1$, and $v \in \hat{L} = K\otimes_{F'} E$. Also, $\hat{L}$ is a field, since $[K:F']$ and $[E:F']$ are relatively prime; $K \subseteq \hat{L} \subseteq \hat{E}$, and $\hat{E}/K$ is a Galois extension of p-power degree. If $\hat{L} = K$, then $[B:K] = [\hat{L}:K]^2 = 1$, and so $v = 1$. Otherwise, there is a subfield $L \subseteq \hat{L}$ such that L/K is a degree p Galois extension, and v centralizes L. In this case, since $[B:K] = [A:F]$, Step 2a applies to show that $[v] = 1 \in K_1(B) = K_1(K\otimes_{F'} D)$.

Lemma 2.2 now applies to show that $[v]^{[K:F']} = 1$ in $K_1(D)$. Hence $[1\otimes u]^{[K:F']} = 1 \in K_1(F'\otimes_F A)$ by (3), and a second application of Lemma 2.2 shows that $[u]^{[K:F]} = 1$ in $K_1(A)$. But $p\nmid[K:F]$ by construction, and so $[u]$ has order prime to p in $K_1(A)$. □

The following lemma, due to Swan [2], makes it possible to compare $K_1(\mathfrak{A})$ with $K_1(\mathfrak{B})$, when $\mathfrak{A} \subseteq \mathfrak{B}$ is any pair of orders in the same algebra. It will also be used in the next chapter when constructing localization sequences.

<u>Lemma 2.4</u> *Let* $R \subseteq S$ *be any pair of rings, and let* I *be any* S*-ideal contained in* R. *Then*

$$E(S,I^2) \subseteq E(R,I) \subseteq E(R).$$

If, furthermore, S/I^2 *is finite, then the induced map* $K_1(R) \longrightarrow K_1(S)$ *has finite kernel and cokernel.*

<u>Proof</u> If e^r_{ij} denotes the elementary matrix with single off-diagonal entry r in (i,j)-position, then $e^{rs}_{ij} = [e^r_{ik}, e^s_{kj}]$ for any $r,s \in I$ and any distinct i,j,k. Hence, since by definition $E(S,I^2)$ is the smallest normal subgroup in $GL(S)$ containing all such e^{rs}_{ij},

$$E(S,I^2) \subseteq [E(S,I),\ E(S,I)] \subseteq [GL(S,I),\ GL(S,I)]$$

$$= [GL(R,I),\ GL(R,I)] \subseteq [GL(R), GL(R,I)] = E(R,I) \qquad (1)$$

(see Theorem 1.13).

Now consider the following diagram, with exact rows and column:

$$\begin{array}{ccccccc}
 & & E(S,I^2)/E(R,I^2) & & & & \\
 & & \downarrow & & & & \\
K_2(R/I^2) & \longrightarrow & K_1(R,I^2) & \longrightarrow & K_1(R) & \longrightarrow & K_1(R/I^2) \\
\downarrow & & \downarrow & (2) & \downarrow & & \downarrow \\
K_2(S/I^2) & \longrightarrow & K_1(S,I^2) & \longrightarrow & K_1(S) & \longrightarrow & K_1(S/I^2).
\end{array}$$

By Theorem 1.16, $K_i(R/I^2)$ and $K_i(S/I^2)$ $(i = 1,2)$ are all finite. Also, $E(S,I^2)/E(R,I^2)$ is finite since by (1),

$$E(S,I^2)/E(R,I^2) \subseteq (E(R) \cap GL(R,I^2))/E(R,I^2) = \operatorname{Ker}[K_1(R,I^2) \longrightarrow K_1(R)].$$

This shows that three of the maps in square (2) have finite kernel and cokernel, and so the same holds for $K_1(R) \longrightarrow K_1(S)$. □

We are now ready to apply reduced norm homomorphisms to describe the structure of $K_1(\mathfrak{A},I)$ — modulo finite groups, at least — when $\mathfrak{A}$ is a $\mathbb{Z}$- or $\hat{\mathbb{Z}}_p$-order and $I \subseteq \mathfrak{A}$ is an ideal of finite index.

__Theorem 2.5__ *Let* A *be a semisimple* $\mathbb{Q}$- *or* $\hat{\mathbb{Q}}_p$*-algebra, let* $\mathfrak{A}$ *be any* $\mathbb{Z}$- *or* $\hat{\mathbb{Z}}_p$*-order in* A, *and let* $I \subseteq \mathfrak{A}$ *be an ideal of finite index. Then*

(i) $SK_1(\mathfrak{A}) = \mathrm{Ker}(\mathrm{nr}_{\mathfrak{A}})$ *and is finite.*

(ii) $\mathrm{nr}_{\mathfrak{A},I} : K_1(\mathfrak{A},I) \longrightarrow R^*$ *has finite kernel and cokernel, where* R^* *is the product of the rings of integers in the field components of the center* Z(A).

(iii) If A *is a* $\mathbb{Q}$*-algebra, then* $K_1(\mathfrak{A},I)$ *is a finitely generated abelian group. If*

q = *number of simple summands of* A, *and*

r = *number of simple summands of* $\mathbb{R} \otimes_{\mathbb{Q}} A$,

then $rk_{\mathbb{Z}}(K_1(\mathfrak{A},I)) = r - q$.

__Proof__ (i) The equality $SK_1(\mathfrak{A}) = \mathrm{Ker}(\mathrm{nr}_{\mathfrak{A}})$ is immediate from the injectivity of nr_A. If $\mathfrak{M} \supseteq \mathfrak{A}$ is a maximal order, then Lemma 2.4 shows that $SK_1(\mathfrak{A})$ is finite if and only if $SK_1(\mathfrak{M})$ is. By the localization sequences of Theorem 1.17, $SK_1(\mathfrak{M})$ is torsion, and is finite if $\mathfrak{M}$ is a $\hat{\mathbb{Z}}_p$-order. (A proof of this which does not use Quillen's localization sequence is given by Swan in [3, Chapter 8].)

When $\mathfrak{A}$ is a $\mathbb{Z}$-order, then by a theorem of Bass [1, Proposition 11.2], every element of $K_1(\mathfrak{A})$ is represented by a 2x2 matrix. Also, Siegel [1] has shown that $GL_2(\mathfrak{A})$ is finitely generated. So $SK_1(\mathfrak{A})$ is finitely generated, and hence finite, in this case. Alternatively, the finiteness of $SK_1(\mathfrak{M})$ follows from Theorem 4.16(i) below.

(ii) Let $\mathfrak{M} \supseteq \mathfrak{A}$ be a maximal order. Then the maps

$$K_1(\mathfrak{A},I) \longrightarrow K_1(\mathfrak{A}) \longrightarrow K_1(\mathfrak{M}) \xrightarrow{nr_{\mathfrak{M}}} R^*$$

all have finite kernel and cokernel: the first since $K_2(\mathfrak{A}/I)$ and $K_1(\mathfrak{A}/I)$ are finite (Theorem 1.16), the second by Lemma 2.4, and $nr_{\mathfrak{M}}$ by Theorem 2.3 and (i) above.

(iii) If A is a $\mathbb{Q}$-algebra, then write $F = Z(A) = \prod F_i$, where the F_i are fields, and let $R_i \subseteq F_i$ be the ring of integers. By the Dirichlet unit theorem (see Janusz [1, Theorem I.11.19]), $R^* = \prod (R_i)^*$ is finitely generated and

$$rk_{\mathbb{Z}}(R^*) = \sum_{i=1}^{q} rk_{\mathbb{Z}}(R_i^*) = \sum_{i=1}^{q} [(\text{no. field summands of } \mathbb{R} \otimes_{\mathbb{Q}} F_i) - 1]$$

$$= (\text{no. field summands of } \mathbb{R} \otimes_{\mathbb{Q}} F) - q = r - q.$$

By (ii), the same holds for $K_1(\mathfrak{A},I)$. □

In the case of an integral group ring, the formula for $rk(K_1(\mathbb{Z}[G]))$ can be given a still nicer form, using the concept of "K-conjugacy" defined in Section 1a. Note that in any finite group G, two elements $g,h \in G$ are $\mathbb{R}$-conjugate if g is conjugate to h or h^{-1}; and are $\mathbb{Q}$-conjugate if the subgroups $\langle g\rangle$ and $\langle h\rangle$ are conjugate.

<u>Theorem 2.6</u> *Fix a finite group* G, *and set*

r = *no. of* $\mathbb{R}$-*conjugacy classes in* G,

q = *no. of* $\mathbb{Q}$-*conjugacy classes in* G.

Then $rk(Wh(G)) = rk(K_1(\mathbb{Z}[G])) = r - q$.

<u>Proof</u> By the Witt-Berman theorem (Theorem 1.6), for any $K \subseteq \mathbb{C}$,

(no. K-conjugacy classes in G) = (no. irred. K[G]-modules)

= (no. simple summands in K[G]).

The result is now immediate from Theorem 2.5(iii). □

2b. Logarithmic and exponential maps in p-adic orders

In the last section, reduced norms were used to compare $K_1(\mathfrak{A})$, for any $\hat{\mathbb{Z}}_p$-order $\mathfrak{A}$, with the group of units in the center of the maximal order. Now, p-adic logarithms will be used to get more information about the structure of $K_1(\mathfrak{A})$.

Throughout this section, p will be a fixed prime, and the term "p-adic order" will be used to mean any $\hat{\mathbb{Z}}_p$-algebra which is finitely generated and free as a $\hat{\mathbb{Z}}_p$-module. The results here are shown for arbitrary p-adic orders, to emphasize their independence of the more specialized properties of orders in semisimple $\hat{\mathbb{Q}}_p$-algebras. Any p-adic order R is semilocal, since $R/J(R)$ is finite. So by Theorem 1.14(i), $K_1(R)$ is generated by units in R; and $K_1(R,I)$ is generated by units in $1+I$ for any ideal $I \subseteq R$.

For any p-adic order R and any $x \in R$, define

$$\mathrm{Log}(1+x) = x - \frac{x^2}{2} + \frac{x^3}{3} - \cdots \quad \text{and} \quad \mathrm{Exp}(x) = 1 + x + \frac{x^2}{2!} + \frac{x^3}{3!} + \cdots$$

whenever these series converge (in $\mathbb{Q} \otimes_{\mathbb{Z}} R$, at least). Just as is the case with the usual logarithm on $\mathbb{R}$, p-adic logarithms can be used to translate certain multiplicative problems involving units in a p-adic order to additive problems — which usually are much simpler to study. The main results of this section are, for any p-adic order R and any ideal $I \subseteq R$, that Log induces a homomorphism

$$\log_I : K_1(R,I) \longrightarrow \mathbb{Q} \otimes_{\mathbb{Z}} (I/[R,I]),$$

that $\text{Ker}(\log_I)$ is finite and $\text{Im}(\log_I)$ is a $\hat{\mathbb{Z}}_p$-lattice, and that $E(R,I)\cap GL_n(R,I)$ is closed in $GL_n(R,I)$ for all n.

Throughout this section Log and Exp are used to denote set maps between subgroups of R^* and R, while log and exp denote induced group homomorphisms. For any pair of ideals $I_1,I_2 \subseteq R$, $[I_1,I_2]$ denotes the subgroup of R generated by elements $[a,b] = ab - ba$ for all $a \in I_1$ and $b \in I_2$. Recall (Theorem 1.11) that for any radical ideal $I \subseteq J(R)$, every element in $1+I$ is invertible. The following lemma collects most of the technical details which will be needed throughout the section.

Lemma 2.7 *Let R be any p-adic order, let $J = J(R)$ denote the Jacobson radical, and let $I \subseteq J$ be any radical ideal.*

(i) Set $R_{\mathbb{Q}} = \mathbb{Q}\otimes_{\mathbb{Z}} R = R[\frac{1}{p}]$ *and* $I_{\mathbb{Q}} = \mathbb{Q}\otimes_{\mathbb{Z}} I = I[\frac{1}{p}]$. *Then for all* $u,v \in 1+I$, Log(u) *and* Log(v) *converge in* $I_{\mathbb{Q}}$, *and*

$$\text{Log}(uv) \equiv \text{Log}(u) + \text{Log}(v) \pmod{[R_{\mathbb{Q}},I_{\mathbb{Q}}]}. \tag{1}$$

(ii) Assume $I \subseteq \xi R$ *for some central element* $\xi \in Z(R)$ *such that* $\xi^p \in p\xi R$. *Then for all* $u,v \in 1+I$, $\text{Log}(u), \text{Log}(v) \in I$ *and*

$$\text{Log}(uv) \equiv \text{Log}(u) + \text{Log}(v) \pmod{[R,I]}. \tag{2}$$

(iii) Assume $I \subseteq \xi R$ *for some* $\xi \in Z(R)$ *such that* $\xi^p \in p\xi R$, *and also that* $I^p \subseteq pIJ$. *Then* Exp(x) *converges in* $1+I$ *for all* $x \in I$; *and* Exp *and* Log *are inverse bijections between* I *and* $1+I$. *In addition,* $\text{Exp}([R,I]) \subseteq E(R,I)$, *and for any* $x,y \in I$:

$$\text{Exp}(x+y) \equiv \text{Exp}(x)\cdot\text{Exp}(y) \pmod{E(R,I)}. \tag{3}$$

Proof The proof will be carried out in three steps. The convergence of Log(u) or Exp(x) in all three cases will be shown in Step 1. The congruences (1) and (2) will then be shown in Step 2, and congruence (3) in Step 3.

Step 1 For any $n \geq 1$, J/p^nR is nilpotent in R/p^nR. Hence, for any $x \in I \subseteq J$, $\lim_{n\to\infty}(x^n) = 0$, and $\lim_{n\to\infty}(x^n/n) = 0$. The series for $\mathrm{Log}(1+x)$ thus converges in $I_{\mathbb{Q}}$.

Under the hypotheses of (ii), $I^p \subseteq pI$, and so $I^n \subseteq nI$ for all $n \geq 1$ (all rational primes except p are inverted in $\hat{\mathbb{Z}}_p \subseteq R$). So for any $x \in I$, $x^n/n \in I$ for all n, and hence $\mathrm{Log}(1+x) \in I$.

To see that $\mathrm{Exp}(x)$ converges when $I^p \subseteq pIJ$, note first that for any $n \geq 1$,

$$n!\cdot p^{-([n/p]+[n/p^2]+[n/p^3]+\dots)} \in (\hat{\mathbb{Z}}_p)^*$$

where $[\cdot]$ denotes greatest integer. For any $n \geq p$,

$$I^n \subseteq (I^p)^{[n/p]} \subseteq p^{[n/p]}\cdot I^{[n/p]}\cdot J.$$

Similarly, if $n \geq p^2$, then $I^n \subseteq p^{[n/p]}\cdot(p^{[n/p^2]}\cdot I^{[n/p^2]}\cdot J)\cdot J$; and by induction, for any $n \geq 1$,

$$I^n \subseteq p^{([n/p]+[n/p^2]+\dots+[n/p^k])}\cdot I\cdot J^k = n!\cdot IJ^k \quad (\text{if } p^k \leq n < p^{k+1}). \tag{4}$$

Thus, $\frac{1}{n!}\cdot I^n \subseteq I$ for all n, $\lim_{n\to\infty} \frac{1}{n!}\cdot I^n = 0$; and so $\mathrm{Exp}(x)$ converges in $1+I$ for any $x \in I$. The relations $\mathrm{Log}\circ\mathrm{Exp}(x) = x$ and $\mathrm{Exp}\circ\mathrm{Log}(1+x) = 1+x$, for $x \in I$, follow from (4) and the corresponding relations for power series.

<u>Step 2</u> For any radical ideal $I \subseteq R$, set

$$U(I) = \sum_{m,n\geq 1} \frac{1}{m+n}\cdot[I^m, I^n] \subseteq [R_{\mathbb{Q}}, I_{\mathbb{Q}}], \tag{5}$$

a $\hat{\mathbb{Z}}_p$-submodule of R. If $I \subseteq \xi R$, where $\xi \in Z(R)$ and $\xi^p \in p\xi R$, then $\xi^n \in n\xi R$ for all n, and

$$U(I) = \left\langle [r, \frac{\xi^{m+n}}{m+n}\cdot s] : m,n \geq 1,\ \xi^m r \in I^m,\ \xi^n s \in I^n,\ \xi r, \xi s \in I \right\rangle \subseteq [R, I].$$

So congruences (1) and (2) will both follow, once we have shown the relation

$$\mathrm{Log}((1+x)(1+y)) \equiv \mathrm{Log}(1+x) + \mathrm{Log}(1+y) \pmod{U(I)} \tag{6}$$

for any I and any $x,y \in I$.

For each $n \geq 1$, let W_n be the set of formal (ordered) monomials of length n in two variables a, b. For $w \in W_n$, set

$C(w)$ = orbit of w in W_n under cyclic permatations

$k(w)$ = number of occurrences of ab in w

$r(w)$ = coefficient of w in $\mathrm{Log}(1+a+b+ab)$

$$= \sum_{i=0}^{k(w)} (-1)^{n-i-1} \cdot \frac{1}{n-i} \cdot \binom{k(w)}{i}.$$

To see the formula for $r(w)$, note that for each i, w can be written in $\binom{k(w)}{i}$ ways as a product of i (ab)'s and $n-2i$ a's or b's.

Fix an ideal $I \subseteq J$ and elements $x,y \in I$. For any $n \geq 1$, any $w \in W_n$, and any $w' \in C(w)$, w' is a cyclic permutation of w, and so

$$w'(x,y) \equiv w(x,y) \pmod{[I^i, I^j]}.$$

for some i,j such that $i+j=n$. It follows that

$$\begin{aligned} \mathrm{Log}(1+x+y+xy) &= \sum_{n=1}^{\infty} \sum_{w \in W_n} r(w) \cdot w(x,y) \\ &\equiv \sum_{n=1}^{\infty} \sum_{w \in W_n/C} \left(\sum_{w' \in Cw} r(w') \right) \cdot w(x,y) \pmod{U(I)}. \end{aligned} \tag{7}$$

For fixed $w \in W_n$, if $|C(w)| = n/t$ (i. e., w has cyclic symmetry of order t), and if

$$k = \max\left\{k(w') : w' \in C(w)\right\},$$

then $C(w)$ contains k/t elements with $k-1$ (ab)'s (i. e., those of the form $b\cdots a$) and $(n-k)/t$ elements with k (ab)'s. So

$$\sum_{w'\in Cw} r(w') = \frac{1}{t}\cdot\sum_{i=0}^{k}(-1)^{n-i-1}\cdot\frac{1}{n-i}\cdot\left[(n-k)\binom{k}{i} + k\binom{k-1}{i}\right]$$

$$= \frac{1}{t}\cdot\sum_{i=0}^{k}(-1)^{n-i-1}\cdot\frac{1}{n-i}\cdot\left[(n-k)\binom{k}{i} + (k-i)\binom{k}{i}\right]$$

$$= \frac{1}{t}\cdot\sum_{i=0}^{k}(-1)^{n-i-1}\cdot\binom{k}{i} = \begin{cases} 0 & \text{if } k>0 \\ (-1)^{n-1}\cdot\frac{1}{n} & \text{if } k=0 \quad (\text{so } t=n). \end{cases}$$

Formula (7) now takes the form, for any $x,y\in I$,

$$\mathrm{Log}((1+x)(1+y)) \equiv \sum_{n=1}^{\infty}(-1)^{n-1}\cdot\left(\frac{x^n}{n} + \frac{y^n}{n}\right) \qquad (\mathrm{mod}\ \ U(I))$$

$$= \mathrm{Log}(1+x) + \mathrm{Log}(1+y);$$

and this finishes the proof of (6).

<u>Step 3</u> Now assume that $I \subseteq \xi R$ for some central $\xi \in R$ such that $\xi^p \in p\xi R$, and that $I^p \subseteq pIJ$. In particular, by Step 1, Exp and Log are inverse bijections between I and $1+I$. So for any $x,y \in I$,

$$\mathrm{Log}(\mathrm{Exp}(x)\cdot\mathrm{Exp}(y)) \equiv x+y \quad (\mathrm{mod}\ \ U(I))$$

by (6). It follows that

$$\mathrm{Exp}(x)\cdot\mathrm{Exp}(y) \in \mathrm{Exp}(x+y+U(I)) \tag{8}$$

for $x,y \in I$; and hence that

$$\begin{aligned}\mathrm{Exp}(x)\cdot\mathrm{Exp}(y)\cdot\mathrm{Exp}(x+y)^{-1} &\in \mathrm{Exp}(x+y+U(I))\cdot\mathrm{Exp}(-x-y) \\ &\subseteq \mathrm{Exp}(U(I)) \subseteq \mathrm{Exp}([R,I]).\end{aligned} \tag{9}$$

So it remains only to show that $\mathrm{Exp}([R,I]) \subseteq E(R,I)$. Note that for all $r \in R$ and $x \in I$,

$$\mathrm{Exp}(rx)\cdot\mathrm{Exp}(xr)^{-1} = \Big(1+r\big(\sum_{n=1}^{\infty}\frac{x(rx)^{n-1}}{n!}\big)\Big)\Big(1+\big(\sum_{n=1}^{\infty}\frac{x(rx)^{n-1}}{n!}\big)r\Big)^{-1} \in E(R,I) \tag{10}$$

by Vaserstein's identity (Theorem 1.15).

Fix some $\hat{\mathbb{Z}}_p$-basis $[r_1,v_1],\ldots,[r_m,v_m]$ for $[R,I]$, where $r_i \in R$ and $v_i \in I$. Define

$$\psi : [R,I] \longrightarrow \mathrm{Exp}([R,I])$$

by setting, for any $x = \sum_{i=1}^m a_i[r_i,v_i] \in [R,I]$ $(a_i \in \hat{\mathbb{Z}}_p)$:

$$\psi(x) = \prod_{i=1}^{m}\Big(\mathrm{Exp}(a_ir_iv_i)\cdot\mathrm{Exp}(a_iv_ir_i)^{-1}\Big).$$

Then $\mathrm{Im}(\psi) \subseteq E(R,I)$ by (10). For any $k \geq 1$ and any $x,y \in p^kI$,

$$\mathrm{Exp}(x)\cdot\mathrm{Exp}(y) \equiv \mathrm{Exp}(x+y) \qquad (\mathrm{mod}\ \ U(p^kI) \subseteq p^{2k}U(I) \subseteq p^{2k}[R,I])$$

by (9). Also, for any $k,\ell \geq 1$ and any $x \in p^kI$, $y \in p^\ell I$,

$$\mathrm{Exp}(x)\cdot\mathrm{Exp}(y) \equiv \mathrm{Exp}(y)\cdot\mathrm{Exp}(x) \qquad (\mathrm{mod}\ \ [p^kI,p^\ell I] \subseteq p^{k+\ell}[R,I]).$$

So for any $\ell \geq k \geq 1$, and any $x \in p^k[R,I]$ and $y \in p^\ell[R,I]$,

$$\begin{aligned} &\psi(x) \equiv \mathrm{Exp}(x) \qquad (\mathrm{mod}\ \ p^{2k}[R,I]) \\ &\psi(x+y) \equiv \psi(x)\cdot\psi(y) \equiv \psi(x)\cdot\mathrm{Exp}(y) \qquad (\mathrm{mod}\ \ p^{k+\ell}[R,I]). \end{aligned} \tag{11}$$

For arbitrary $u \in \mathrm{Exp}(p[R,I])$, define a sequence $x_0,x_1,x_2,\ldots$ in $[R,I]$ by setting

$$x_0 = \mathrm{Log}(u) \in p[R,I]; \qquad x_{i+1} = x_i + \mathrm{Log}(\psi(x_i)^{-1}\cdot u).$$

By (11), applied inductively for all $i \geq 0$,

$$\psi(x_i) \equiv u, \quad x_{i+1} \equiv x_i \qquad (\mathrm{mod}\ \ p^{2+i}[R,I]).$$

So $\{x_i\}$ converges, and $u = \psi(\lim_{i\to\infty} x_i)$. This shows that

$$\mathrm{Exp}(p[R,I]) \subseteq \mathrm{Im}(\psi) \subseteq E(R,I). \tag{12}$$

Now define subgroups D_k, for all $k \geq 0$, by setting

$$D_k = \left\langle rx - xr \;:\; x\in I,\ \ r\in R,\ \ rx,xr \in IJ^k \right\rangle \subseteq [R,I] \cap IJ^k.$$

Recall the hypotheses on I: $I \subseteq \xi R$, where $\xi \in Z(R)$ and $\xi^p \in p\xi R$, and $I^p \subseteq pIJ$ (so $I^n \subseteq nIJ$ for all n). Then for all $k \geq 0$,

$$U(IJ^k) = \sum_{m,n\geq 1} \frac{1}{m+n}\cdot[(IJ^k)^m,(IJ^k)^n] \qquad \text{(by (5))}$$

$$\subseteq \left\langle [r,\tfrac{\xi^n}{n}\cdot s] \;:\; n\geq 2,\ \xi r,\xi s \in IJ^k,\ \ \xi^n rs,\xi^n sr \in (IJ^k)^n \subseteq nIJ^{k+1} \right\rangle \subseteq D_{k+1}.$$

Together with (8), this shows that $\mathrm{Exp}(D_k) \subseteq \mathrm{Exp}([R,I])$ are both (normal) subgroups of R^*. Also, by (9), for any $x,y \in IJ^k$,

$$\mathrm{Exp}(x)\cdot\mathrm{Exp}(y) \equiv \mathrm{Exp}(x+y) \quad (\mathrm{mod}\ \ \mathrm{Exp}(U(IJ^k)) \subseteq \mathrm{Exp}(D_{k+1})) \tag{13}$$

For any $k \geq 0$ and any $x \in D_k$, if we write $x = \sum(r_i x_i - x_i r_i)$ (where $r_i \in R$, $x_i \in I$; $r_i x_i$, $x_i r_i \in IJ^k$), then

$$\mathrm{Exp}(x) \equiv \prod\left(\mathrm{Exp}(r_i x_i)\cdot\mathrm{Exp}(x_i r_i)^{-1}\right) \quad (\mathrm{mod}\ \ \mathrm{Exp}(D_{k+1})) \qquad \text{(by (13))}$$

$$\equiv 1 \quad (\mathrm{mod}\ \ E(R,I)). \qquad \text{(by (10))}$$

In other words, $\mathrm{Exp}(D_k) \subseteq E(R,I)\cdot\mathrm{Exp}(D_{k+1})$ for all $k\geq 0$. But $D_k \subseteq p[R,I]$ for k large enough ($D_k \subseteq [R,I]\cap IJ^k$); and so using (12):

$$\mathrm{Exp}([R,I]) = \mathrm{Exp}(D_0) \subseteq E(R,I)\cdot\mathrm{Exp}(p[R,I]) \subseteq E(R,I). \qquad \square$$

Constructing a homomorphism induced by logarithms is now straightforward.

Theorem 2.8 *For any p-adic order* R *with Jacobson radical* $J \subseteq R$, *and any 2-sided ideal* $I \subseteq R$, *the p-adic logarithm* $\mathrm{Log}(1+x)$ *(for* $x \in I \cap J$*) induces a unique homomorphism*

$$\log_I : K_1(R,I) \longrightarrow \mathbb{Q} \otimes_{\mathbb{Z}} (I/[R,I]).$$

If, furthermore, $I \subseteq \xi R$ *for some central* $\xi \in Z(R)$ *such that* $\xi^p \in p\xi R$, *then the logarithm induces a homomorphism*

$$\log^I : K_1(R,I) \longrightarrow I/[R,I];$$

and $\log^I$ *is an isomorphism if* $I^p \subseteq pIJ$.

Proof Write $R_{\mathbb{Q}} = \mathbb{Q} \otimes_{\mathbb{Z}} R$ and $I_{\mathbb{Q}} = \mathbb{Q} \otimes_{\mathbb{Z}} I$, for short, and let J be the Jacobson radical of R. Assume first that $I \subseteq J$. By Lemma 2.7(i), the composite

$$L : 1+I \xrightarrow{\mathrm{Log}} I_{\mathbb{Q}} \xrightarrow{\mathrm{proj}} I_{\mathbb{Q}}/[R_{\mathbb{Q}}, I_{\mathbb{Q}}] \tag{1}$$

is a homomorphism.

For each $n \geq 1$, let

$$\mathrm{Tr}_n : M_n(I_{\mathbb{Q}})/[M_n(R_{\mathbb{Q}}), M_n(I_{\mathbb{Q}})] \longrightarrow I_{\mathbb{Q}}/[R_{\mathbb{Q}}, I_{\mathbb{Q}}] \tag{2}$$

be the homomorphism induced by the trace map. Then (1), applied to the ideal $M_n(I) \subseteq M_n(R)$, induces a homomorphism

$$L_n : 1+M_n(I) = GL_n(R,I) \xrightarrow{\mathrm{Log}} M_n(I_{\mathbb{Q}})/[M_n(R_{\mathbb{Q}}), M_n(I_{\mathbb{Q}})] \xrightarrow{\mathrm{Tr}_n} I_{\mathbb{Q}}/[R_{\mathbb{Q}}, I_{\mathbb{Q}}].$$

For any n, and any $u \in 1+M_n(I)$ and $r \in GL_n(R)$,

$$L_n([r,u]) = L_n(rur^{-1}) - L_n(u) = \mathrm{Tr}_n(r\cdot\mathrm{Log}(u)\cdot r^{-1}) - \mathrm{Tr}_n(\mathrm{Log}(u)) = 0;$$

and so $L_\infty = \bigcup(L_n)$ factors through a homomorphism

$$\log_I : K_1(R,I) = GL(R,I)/[GL(R),GL(R,I)] \longrightarrow I_{\mathbb{Q}}/[R_{\mathbb{Q}},I_{\mathbb{Q}}].$$

Now assume that I is arbitrary, and set $I_0 = I \cap J$. Consider the relative exact sequence

$$K_2(R/I_0,I/I_0) \longrightarrow K_1(R,I_0) \longrightarrow K_1(R,I) \longrightarrow K_1(R/I_0,I/I_0)$$

(see Milnor [2, Remark 6.6]). The surjection $R/I_0 \twoheadrightarrow R/J$ sends I/I_0 isomorphically to $(I+J)/J$, which is a 2-sided ideal and hence a ring summand of R/J (R/J is semisimple). In particular, I/I_0 is a semisimple ring summand of R/I_0, and by Theorem 1.16,

$$K_2(R/I_0,I/I_0) \cong K_2(I/I_0) = 1 \quad \text{and} \quad p \nmid |K_1(R/I_0,I/I_0)| = |K_1(I/I_0)|.$$

So $\log_{I_0}$ ($I_0 \subseteq J$) extends uniquely to a homomorphism

$$\log_I : K_1(R,I) \longrightarrow I_{\mathbb{Q}}/[R_{\mathbb{Q}},I_{\mathbb{Q}}].$$

If $I \subseteq \xi R$ for some central $\xi \in R$ such that $\xi^p \in p\xi R$, then by Lemma 2.7(ii), $\text{Log}(1+I) \subseteq I$, and the composite

$$L : 1+I \xrightarrow{\text{Log}} I \xrightarrow{\text{proj}} I/[R,I]$$

is a homomorphism. The same argument as before then shows that L factors through a homomorphism $\log^I$ defined on $K_1(R,I)$. If $I^p \subseteq pIJ$, then Log is bijective and $\text{Log}^{-1}([R,I]) \subseteq E(R,I)$ by Lemma 2.7(iii); and so $\log^I$ is an isomorphism. □

The next result is based on a theorem of Carl Riehm [1]. Roughly, it says that for any p-adic order R, the p-adic topology on R^* makes $K_1(R)$ into a Hausdorff group.

<u>**Theorem 2.9**</u> *For any p-adic order* R *and any 2-sided ideal* $I \subseteq R$, $\text{Ker}(\log_I)$ *is finite; and for all* n *the group*

$$\bar{E}_n(R,I) = GL_n(R,I) \cap E(R,I) = \mathrm{Ker}\left[GL_n(R,I) \longrightarrow K_1(R,I)\right]$$

is closed (in $GL_n(R,I)$*) in the p-adic topology.*

<u>Proof</u> Set $R_{\mathbb{Q}} = \mathbb{Q} \otimes_{\mathbb{Z}} R$ and $I_{\mathbb{Q}} = \mathbb{Q} \otimes_{\mathbb{Z}} I$ as before, and write

$$L = \log_I \circ \mathrm{proj} : GL_1(R,I) \twoheadrightarrow K_1(R,I) \longrightarrow I_{\mathbb{Q}}/[R_{\mathbb{Q}}, I_{\mathbb{Q}}].$$

By Lemma 2.7(iii) and Theorem 2.8, $\mathrm{Log}: 1+p^2I \longrightarrow p^2I$ is a homeomorphism, and factors through an isomorphism

$$\log^{p^2I} : K_1(R,p^2I) \cong (1+p^2I)/\bar{E}_1(R,p^2I) \xrightarrow{\cong} p^2I/[R,p^2I].$$

In particular, since $[R,p^2I] = \mathrm{Log}(\bar{E}_1(R,p^2I))$ is open in $[R_{\mathbb{Q}}, I_{\mathbb{Q}}]$, $\bar{E}_1(R,p^2I) \subseteq \bar{E}_1(R,I)$ are open subgroups of $\mathrm{Ker}(L)$.

Now, $GL_1(R,I)$ is compact: it is the inverse limit of the finite groups $GL_1(R/p^nR, (I+p^nR)/p^nR)$. So $\mathrm{Ker}(L)$ is compact, and any open subgroup of $\mathrm{Ker}(L)$ has finite index. It follows that

$$\mathrm{Ker}(\log_I) = \mathrm{Ker}(L)/\bar{E}_1(R,I)$$

is finite.

Any open subgroup of a topological group is also closed (its complement is a union of open cosets). In particular, $\bar{E}_1(R,I)$ is closed in $\mathrm{Ker}(L)$ and hence also in $GL_1(R,I)$. To see that $\bar{E}_n(R,I)$ is closed in $GL_n(R,I)$ for all n, just note that by definition, $\bar{E}_n(R,I) = \bar{E}_1(M_n(R), M_n(I))$. □

The following description of the structure of $K_1(R,I)$ is an easy consequence of Theorem 2.9.

Theorem 2.10 *For any p-adic order R with Jacobson radical $J \subseteq R$,*

$$K_1(R) \cong K_1(R/J) \oplus K_1(R,J),$$

where $K_1(R/J)$ is finite of order prime to p, and $K_1(R,J)$ is a finitely generated $\hat{\mathbb{Z}}_p$-module. If $I \subseteq R$ is any (2-sided) ideal, then

(i) $K_1(R,I)$ is the product of a finite group with a finitely generated $\hat{\mathbb{Z}}_p$-module and

$$\mathrm{rk}_{\hat{\mathbb{Z}}_p} K_1(R,I) = \mathrm{rk}_{\hat{\mathbb{Z}}_p} I/[R,I];$$

(ii) $K_1(R,I)$ is a $\hat{\mathbb{Z}}_p$-module (i. e., contains no torsion prime to p) if $I \subseteq J$; and

(iii) $K_1(R,I) \cong \varprojlim_n K_1(R/p^nR,(I+p^nR)/p^nR)$.

Proof Note first that $\mathrm{Log}(1+p^2I) = p^2I$ by Lemma 2.7(iii). Hence, since $1+p^2I$ has finite index in $1+I$, the image of

$$\log_I : K_1(R,I) \longrightarrow \hat{\mathbb{Q}}_p \otimes_{\hat{\mathbb{Z}}_p} (I/[R,I])$$

is a $\hat{\mathbb{Z}}_p$-lattice. Since $\mathrm{Ker}(\log_I)$ is finite by Theorem 2.9, $K_1(R,I)$ is now seen to be a product of a finite group with a $\hat{\mathbb{Z}}_p$-module, and

$$\mathrm{rk}_{\hat{\mathbb{Z}}_p} K_1(R,I) = \mathrm{rk}_{\hat{\mathbb{Z}}_p} I/[R,I].$$

To prove (iii), note first that

$$GL_1(R,I) = \varprojlim_n GL_1(R/p^nR,(I+p^nR)/p^nR).$$

Since $E(R,I) \cap GL_1(R,I)$ is closed in $GL_1(R,I)$ (Theorem 2.9), it is also an inverse limit of groups of elementary matrices over R/p^nR. The description of $K_1(R,I)$ as an inverse limit then follows since $\varprojlim$ preserves exact sequences of finite groups.

If $I \subseteq J$, then

$$GL_1(R/p^nR, (I+p^nR)/p^nR) = 1 + (I+p^nR)/p^nR$$

is a p-group for all n. So $K_1(R,I)$ is a pro-p-group, and hence a $\hat{\mathbb{Z}}_p$-module, by (iii). Since $K_2(R/J) = 1$ (Theorem 1.16), the sequence

$$1 \longrightarrow K_1(R,J) \longrightarrow K_1(R) \longrightarrow K_1(R/J) \longrightarrow 1$$

is exact; and is split since $K_1(R,J)$ is a $\hat{\mathbb{Z}}_p$-module and $K_1(R/J)$ is finite of order prime to p (Theorem 1.16 again). □

Theorem 2.10 will be the most important application of these results needed in the next three chapters. P-adic logarithms will again be used directly in Chapters 6 and 7, but in the form of "integral" logarithms for p-adic group rings, whose image is much more easily identified.

We end the chapter with the following theorem of Kuku [1], which applies results from both Sections 2a and 2b. Note in particular that if $\mathfrak{M}$ is a maximal $\hat{\mathbb{Z}}_p$-order in any semisimple $\hat{\mathbb{Q}}_p$-algebra, then $SK_1(\mathfrak{M}) = 1$ if and only if A is a product of matrix algebras over fields.

<u>Theorem 2.11</u> *Let* A *be a simple* $\hat{\mathbb{Q}}_p$*-algebra with center* F, *and let* $\mathfrak{M} \subseteq A$ *be any maximal order. Then* $SK_1(\mathfrak{M})$ *is cyclic of order* $(q^n-1)/(q-1)$, *where* $n = \text{ind}(A)$, *and where* q *is the order of the residue field of* F.

<u>Proof</u> By Theorem 1.9, it suffices to show this when A is a division algebra: otherwise, if $A \cong M_r(D)$, and $\Delta \subseteq D$ is the maximal order, then $\mathfrak{M} \cong M_r(\Delta)$ by Theorem 1.9. In particular, $[A:F] = n^2$. Let $R \subseteq F$ be the ring of integers, and let $\mathfrak{p} \subseteq R$ and $J \subseteq \mathfrak{M}$ be the maximal ideals. By Hasse's description of $\mathfrak{M}$ (Theorem 1.9), $\mathfrak{M}/J$ is a field and $[\mathfrak{M}/J : R/\mathfrak{p}] = n$. Also, $p \nmid |SK_1(\mathfrak{M})|$ by Theorem 1.17(i). It follows that

$$SK_1(\mathfrak{M}) \cong \mathrm{Ker}\Big[nr_{\mathfrak{M}}: K_1(\mathfrak{M}) \twoheadrightarrow K_1(R)\Big][\tfrac{1}{p}] \qquad \text{(Theorem 2.5)}$$

$$\cong \mathrm{Ker}\Big[(\mathfrak{M}/J)^* \twoheadrightarrow (R/p)^*\Big]; \qquad \text{(Theorem 2.10)}$$

where the reduced norm is onto by Theorem 2.3(i). Since $(\mathfrak{M}/J)^*$ is cyclic, this shows that $SK_1(\mathfrak{M})$ is cyclic of order

$$|(\mathfrak{M}/J)^*|/|(R/p)^*| = (q^n - 1)/(q - 1). \qquad \square$$

Chapter 3 CONTINUOUS K_2 AND LOCALIZATION SEQUENCES

So far, all we have shown about $SK_1(\mathbb{Z}[G])$ is that it is finite. In order to learn more about its structure, except in the simplest cases, exact sequences which connect the functors K_1 and K_2 are necessary. The Mayer-Vietoris sequences of Milnor [2, Theorems 3.3 and 6.4] are sufficient for doing this in some cases (see, e. g., the computation of $SK_1(\mathbb{Z}[Q(8)])$ by Keating [2]). But to get more systematic results, some kind of localization exact sequence is needed which compares the K-theory of $\mathbb{Z}[G]$ with that of $\mathbb{Q}[G]$ or a maximal order, and their p-adic completions.

The results here on localization sequences are contained in Section 3c. The principal sequence to be used (Theorem 3.9) takes the form

$$\bigoplus_p K_2^c(\hat{\mathfrak{A}}_p) \longrightarrow C(A) \longrightarrow SK_1(\mathfrak{A}) \longrightarrow \bigoplus_p SK_1(\hat{\mathfrak{A}}_p) \longrightarrow 1 \qquad (1)$$

for any $\mathbb{Z}$-order $\mathfrak{A}$ in a semisimple $\mathbb{Q}$-algebra A. Here,

$$C(A) = \varprojlim_{I \subseteq \mathfrak{A}} SK_1(\mathfrak{A}, I) \cong \operatorname{Coker}\Big[K_2(A) \longrightarrow \bigoplus_p K_2^c(\hat{A}_p)\Big];$$

where the limit is taken over all ideals $I \subseteq \mathfrak{A}$ of finite index, and where the last isomorphism is constructed in Theorem 3.12. A specialized version of (1) in the p-group case is derived in Theorem 3.15.

As can be seen above, the continuous K_2 of p-adic orders and algebras plays an important role in these sequences. These groups $K_2^c(-)$ are defined in Section 3b, and some of their basic properties are derived there. This, in turn, requires some results about Steinberg symbols and symbol generators for $K_2(R)$: results which are surveyed in Section 3a.

3a. Steinberg symbols in $K_2(R)$

For any ring R, the Steinberg group $St(R)$ is defined to be the free group on generators x_{ij}^r for all $i \neq j$ $(i,j \geq 1)$ and all $r \in R$; modulo the relations

$$x_{ij}^r \cdot x_{ij}^s = x_{ij}^{r+s} \quad \text{for any } r,s \in R \text{ and any } i \neq j$$

$$[x_{ij}^r, x_{k\ell}^s] = \begin{cases} x_{i\ell}^{rs} & \text{if } i \neq \ell,\ j = k \\ 1 & \text{if } i \neq \ell,\ j \neq k. \end{cases}$$

An epimorphism $\phi\colon St(R) \twoheadrightarrow E(R)$ is defined by letting $\phi(x_{ij}^r)$ be the elementary matrix whose single nonzero off-diagonal entry is r in the (i,j)-position. Then $St(R)$ is the "universal central extension" of $E(R)$ (in particular, $Ker(\phi) \subseteq Z(St(R))$), and

$$K_2(R) = Ker(\phi) \cong H_2(E(R)).$$

For details, see, e. g., Milnor [2, Chapter 5].

For any pair $u,v \in R^*$ of units, the *Steinberg symbol* $\{u,v\}$ is defined to be the commutator

$$\{u,v\} = \left[\phi^{-1}(diag(u,u^{-1},1))\, ,\, \phi^{-1}(diag(v,1,v^{-1}))\right] \in St(R).$$

Since $Ker(\phi)$ is central in $St(R)$, this is independent of the choice of liftings. We are mostly interested in the case where $uv = uv$, and hence where $\{u,v\} \in Ker(\phi) = K_2(R)$. However, it will occasionally be necessary to work with the $\{u,v\}$ for noncommuting u and v; for example, in Lemma 4.10 and Proposition 13.3 below.

The next theorem lists some of the basic relations between Steinberg symbols.

Theorem 3.1 *For any ring* R, *the following relations hold in* $K_2(R)$ *or* $St(R)$:

(i) For any $u \in R^*$, $\{u,-u\} = 1$; *and* $\{u,1-u\} = 1$ *if* $1-u \in R^*$.

(ii) For any $u,v,w \in R^*$ *such that* $uv = vu$ *and* $uw = wu$,

$$\{u,vw\} = \{u,v\}\cdot\{u,w\}, \quad \textit{and} \quad \{v,u\} = \{u,v\}^{-1}.$$

(iii) For any $x,y,r,s \in R$ *such that* $0 = xy = xry = yx = ysx$,

$$\{1+xr,1+y\} = \{1+x,1+ry\} \quad \textit{and} \quad \{1+sx,1+y\} = \{1+x,\ 1+ys\}.$$

(iv) If $X,Y \in St(R)$ *are such that* $\phi(X) = \mathrm{diag}(u_1,\ldots,u_n)$ *and* $\phi(Y) = \mathrm{diag}(v_1,\ldots,v_n)$, *where* $u_i,v_i \in R^*$, *and* $u_iv_i = v_iu_i$ *for each* $i \geq 2$, *then* $[X,Y] = \prod_{i=1}^{n}\{u_i,v_i\}$.

(v) For any $S \subseteq R$ *such that* R *is finitely generated and projective as an* S*-module, and any commuting units* $u \in R^*$ *and* $v \in S^*$,

$$\mathrm{trf}_S^R(\{u,v\}) = \{\mathrm{trf}_S^R(u),v\}.$$

Here, $\mathrm{trf}_S^R\colon K_n(R) \longrightarrow K_n(S)$ *denotes the transfer homomorphism.*

__Proof__ Point (i), and the relation $\{v,u\} = \{u,v\}^{-1}$, are shown in Milnor [2, Lemmas 9.8 and 8.2] and Silvester [1, Propositions 80 and 79] (it clearly suffices to prove these for commutative R). The relations in (iii) are shown by Dennis & Stein [1, Lemma 1.4(b)] in the commutative case, and follow in the noncommutative case by the same proof. Alternatively, using Dennis-Stein symbols, (iii) follows from the relation: $\{1+x,1+y\} = \langle x,y\rangle$ whenever $xy = 0 = yx$ (see Silvester [1, Propositions 96 and 97]). The formula in (v) is shown in Milnor [2, Theorem 14.1].

When proving (ii) and (iv), it will be convenient to adopt Milnor's notation: $A*B = [\phi^{-1}(A),\phi^{-1}(B)] \in St(R)$ for any $A,B \in E(R)$. This is uniquely defined since $K_2(R) = \mathrm{Ker}(\phi)$ is central in $St(R)$. Also, for any $u \in R^*$ and any $i \neq j$, $d_{ij}(u)$ will denote the diagonal matrix with entries u, u^{-1} in positions i and j (and 1's elsewhere). Note the following two points:

(1) $A*B = 1$ if $A \in E_I(R)$, $B \in E_J(R)$, where I and J are disjoint subsets of $\{1,2,3,\ldots\}$. This follows easily from the defining relation: $[x^r_{ij}, x^s_{k\ell}] = 1$ whenever $i \neq \ell$ and $j \neq k$.

(2) $MAM^{-1} * MBM^{-1} = A*B$ for any $A,B,M \in E(R)$ such that $[A,B]*M = 1$; and in particular whenever $[A,B] = 1$. This is immediate from the obvious relations among commutators.

Now, fix X,Y as in (iv), and set $A = \phi(X) = \mathrm{diag}(u_1,\ldots,u_n)$ and $B = \phi(Y) = \mathrm{diag}(v_1,\ldots,v_n)$ (where $[u_i,v_i] = 1$ for all $i \geq 2$). Then

$$[X,Y] = A*B = \mathrm{diag}(A,A^{-1},1) * \mathrm{diag}(B,1,B^{-1}) \qquad \text{(by (1))}$$

$$= \left(d_{1,n+1}(u_1) * d_{1,2n+1}(v_1)\right)\cdots\left(d_{n,2n-1}(u_n) * d_{n,3n-1}(v_n)\right) \qquad \text{(by (1))}$$

$$= \left(d_{12}(u_1) * d_{13}(v_1)\right)\cdots\left(d_{12}(u_n) * d_{13}(v_n)\right) \qquad \text{(by (2))}$$

$$= \{u_1,v_1\}\cdots\{u_n,v_n\}.$$

To prove (ii), fix units $u,v,w \in R^*$ such that $[u,v] = 1 = [u,w]$. Then, using the relation $[a,bc] = [a,b]\cdot[a,c]\cdot[[c,a],b]$, we get

$$\{u,vw\} = d_{12}(u)*d_{13}(vw) = d_{12}(u) * \left(d_{13}(v)\cdot\mathrm{diag}(w,1,vw^{-1}v^{-1})\right)$$

$$= \left(d_{12}(u)*d_{13}(v)\right)\cdot\left(d_{12}(u)*\mathrm{diag}(w,1,vw^{-1}v^{-1})\right)\cdot\left(1*d_{13}(v)\right)$$

$$= \{u,v\}\cdot\{u,w\}. \quad \text{(by (iv))} \quad \square$$

We now consider relative K_2-groups. Keune [1] has defined groups $K_2(R,I)$ which fit into a long exact sequence involving K_2 and K_3 (note that this is *not* the case with the $K_2(R,I)$ defined by Milnor in [2, Section 6]). In this book, however, K_3 never appears; and it is most convenient to take as definition

$$K_2(R,I) = \operatorname{Ker}\Big[K_2(R) \longrightarrow K_2(R/I)\Big],$$

for any ring R and any (2-sided) ideal $I \subseteq R$. In particular, symbol relations which hold in $K_2(R)$ will automatically hold in $K_2(R,I)$ here.

The following lemma is frequently useful. It will also be needed in the next section when defining continuous K_2.

<u>Lemma 3.2</u> *For any pair* $S \subseteq R$ *of rings, and any R-ideal* $I \subseteq S$,

$$K_2(R,I^4) \subseteq \operatorname{Im}\Big[K_2(S,I) \longrightarrow K_2(R)\Big].$$

<u>Proof</u> Let $\bar{I} \subseteq R$ be any ideal, and consider the pullback square

$$\begin{array}{ccc} D & \xrightarrow{p_2} & R \\ \downarrow{\scriptstyle p_1} & & \downarrow \\ R & \longrightarrow & R/\bar{I} \end{array} \qquad (D = \{(r_1,r_2) \in R\times R \; ; \; r_1 - r_2 \in \bar{I}\}).$$

We identify $\operatorname{Ker}(p_2)$ with $\bar{I}$. By the Mayer-Vietoris sequence for the above square (see Milnor [2, Theorem 6.4]), p_1 induces a surjection of $K_2(D,\bar{I})$ onto $K_2(R,\bar{I})$. Also, $E(D,\bar{I}) \cong E(R,\bar{I})$ by Milnor [2, Lemma 6.3]. Since p_2 is split by the diagonal map $\Delta\colon R \longrightarrow D$, there is a split extension

$$1 \longrightarrow E(R,\bar{I}) \longrightarrow E(D) \longrightarrow E(R) \longrightarrow 1. \qquad (1)$$

The Hochschild-Serre spectral sequence for (1) (see Brown [1, Theorem VII.6.3]) then induces a surjection

$$H_1(E(R);E(R,\bar{I})^{ab}) \xrightarrow{\cong} \frac{\operatorname{Ker}[H_2(E(D)) \to H_2(E(R))]}{\operatorname{Im}[H_2(E(R,\bar{I})) \to H_2(E(D))]}$$

$$\xrightarrow{\cong} \operatorname{Coker}\Big[H_2(E(R,\bar{I})) \to K_2(D,\bar{I})\Big] \twoheadrightarrow \operatorname{Coker}\Big[H_2(E(R,\bar{I})) \to K_2(R,\bar{I})\Big]. \qquad (2)$$

This will now be applied to the ideals $I^4 \subseteq I^2 \subseteq R$. Note first that $E(R,I^4) \subseteq [E(R,I^2),E(R,I^2)]$, since $E(R,I^4)$ is the smallest normal subgroup in $GL(R)$ containing all elementary matrices $e_{ij}^{xy} = [e_{ik}^{x}, e_{kj}^{y}]$, for $x,y \in I^2$ and distinct indices i,j,k. Thus, $E(R,I^4)^{ab}$ maps trivially to $E(R,I^2)^{ab}$, and so by (2),

$$K_2(R,I^4) \subseteq \mathrm{Im}\Big[H_2(E(R,I^2)) \longrightarrow K_2(R,I^2)\Big].$$

Furthermore, $E(R,I^2) \subseteq E(S,I)$ by Lemma 2.4, and hence

$$K_2(R,I^4) \subseteq \mathrm{Im}\Big[H_2(E(R,I^2)) \longrightarrow K_2(R)\Big] \subseteq \mathrm{Im}\Big[H_2(E(S,I)) \longrightarrow K_2(R)\Big]$$

$$\subseteq \mathrm{Im}\Big[K_2(S,I) \longrightarrow K_2(R)\Big]. \qquad \square$$

The next theorem lists some generating sets for the relative groups $K_2(R,I)$. For the purposes in this book, Steinberg symbols are the simplest elements to use as generators. However, in many situations, the Dennis-Stein symbols $\langle a,b\rangle \in K_2(R)$ (defined for any commuting pair $a,b \in R$ with $1+ab \in R^*$) are the most useful. We refer to Stein & Dennis [1], and to Silvester [1, pp. 214-217], for their definition and relations.

__Theorem 3.3__ *Fix a noetherian ring* R, *let* $J = J(R)$ *be its Jacobson radical, and let* $I \subseteq R$ *be a radical ideal of finite index such that* $[J,I] = 0$. *Then*

$$K_2(R,I) = \langle\{1+x,1+y\} : x\in J,\ y\in I\rangle = \langle\langle x,y\rangle : x\in J,\ y\in I\rangle.$$

Moreover, if R *is finite, and if either*

(i) $J = \langle a_1,\ldots,a_k\rangle_R$, *or*

(ii) $J = \langle p,a_1,\ldots,a_k\rangle_R$ *for some prime* p, *where* p *is odd or* $I \subseteq \langle a_1,\ldots,a_k\rangle_R$,

then

$$K_2(R,I) = \langle\{1+a_i,1+x\} : 1 \leq i \leq k, \quad x \in I\rangle.$$

Proof Set $S = \mathbb{Z} + I$, a subring of R. Then by Lemma 3.2,

$$\mathrm{Im}\Big[K_2(S,I) \longrightarrow K_2(R)\Big] \supseteq K_2(R,I^4) = \mathrm{Ker}\Big[K_2(R,I) \longrightarrow K_2(R/I^4,I/I^4)\Big].$$

Also, since $[J,I] = 0$, any symbol $\{1+x,1+y\}$, for $x \in J/I^4$ and $y \in I/I^4$, can be lifted to $K_2(R,I)$. Since R/I^4 is finite by assumption, this shows that we need prove the theorem only when either $R = \mathbb{Z} + I$, or R is finite.

Case 1 Assume $R = \mathbb{Z} + I$. Then R is commutative. By Stein & Dennis [1, Theorem 2.1] or Silvester [1, Corollary 104], $K_2(R,I)$ is generated by symbols $\langle r,x\rangle$ for $r \in R$ and $x \in I$. For any such r and x, $\langle r,x\rangle = \langle r,x\rangle\cdot\langle -1,x\rangle = \langle r-1-rx,x\rangle$ by relations shown in Silvester [1, Propositions 96 and 97]. This procedure can be repeated until $r \in J$ or $r \in -1 + J \subseteq R^*$; and in the latter case

$$\langle r,x\rangle = \{-r,1+rx\} \in \{1+J,1+I\}$$

by Silvester [1, Proposition 96(iv)].

Case 2 If R is finite, then J is nilpotent; and there is a sequence $I = I_0 \supseteq I_1 \supseteq \dots \supseteq I_n = 0$ such that $JI_k + I_kJ \subseteq I_{k+1}$ for all k. By using this to filter $K_2(R,I)$, all claims are reduced to the case where $IJ = 0 = JI$. If this holds, then $K_2(R,I) = \{1+J,1+I\}$ by Oliver [3, Proposition 2.3]; and $\{1+J,1+I\} = \langle J,I\rangle$ since $\{1+x,1+y\} = \langle x,y\rangle$ whenever $xy = 0 = yx$. Point (i) is now an easy consequence of relation (iii) in Theorem 3.1. The refinement in (ii) is shown in Oliver [7, Lemma 1.1], using an argument involving Dennis-Stein symbols similar to that used in Case 1 above. □

Even when Theorem 3.3 does not apply directly to $K_2(R,I)$, one can often filter I by a sequence $I = I_0 \supseteq I_1 \supseteq \dots$ of ideals such that Theorem 3.3 applies to each of the groups $K_2(R/I_k,I_{k-1}/I_k)$, and obtain generators for $K_2(R,I)$ from that. This technique is the basis of the proofs of Theorem 4.11, Proposition 9.4, and Lemma 13.1 below.

The last part of Theorem 3.3 is especially useful in the case of group rings of p-groups.

<u>Corollary 3.4</u> *Fix a prime* p *and a p-group* G, *and let* R *be the ring of integers in some finite unramified extension of* $\hat{\mathbb{Q}}_p$. *Then, for any pair* $I_0 \subseteq I \subseteq R[G]$ *of ideals of finite index such that* $gx - xg \in I_0$ *for all* $g \in G$ *and* $x \in I$,

$$K_2(R[G]/I_0, I/I_0) = \langle \{g, 1+x\} : g \in G,\ x \in I/I_0 \rangle$$

if p *is odd, or if* $p=2$ *and* $I \subseteq \langle 4, g-1 : g \in G \rangle_{RG}$

$$K_2(R[G]/I_0, I/I_0) = \langle \{-1,-1\}, \{g, 1+x\} : g \in G,\ x \in I/I_0 \rangle \quad \textit{otherwise.}$$

<u>Proof</u> By Example 1.12, the Jacobson radical $J \subseteq R[G]$ is generated by p, together with elements $g-1$ for $g \in G$. So the result is immediate from Theorem 3.3. □

Other theorems giving sets of generators for $K_2(R)$ or $K_2(R,I)$ are shown in, for example, Stein & Dennis [1] and Silvester [1]. There are also some much deeper theorems, which give presentations for $K_2(R)$ in terms of Steinberg symbols or Dennis-Stein symbols. The first such result was Matsumoto's presentation for K_2 of a field (see Theorem 4.1 below). Other examples of presentations of $K_2(R)$ or $K_2(R,I)$ have been given by Maazen & Stienstra [1] and Keune [1] for radical ideals in commutative rings, by Rehmann [1] for division algebras, and by Kolster [1] for noncommutative local rings.

<u>3b.</u> Continuous K_2 of p-adic orders and algebras

As mentioned above, the goal is to describe $SK_1(\mathfrak{A})$, for a $\mathbb{Z}$-order $\mathfrak{A}$ in a semisimple $\mathbb{Q}$-algebra A, in terms of $K_i(A)$, $K_i(\hat{\mathfrak{A}}_p)$, and

$K_i(\hat{A}_p)$ (for $i = 1,2$). However, the groups $K_2(\hat{\mathfrak{A}}_p)$ and $K_2(\hat{A}_p)$ are huge: for example, $K_2(F)$ has uncountable rank for any finite extension $F \supseteq \hat{\mathbb{Q}}_p$ (see Bass & Tate [1, Proposition 5.10]). In contrast, we will see in Theorem 4.4 that $K_2^c(F)$ — the continuous K_2 — is finite for such F.

Several different definitions have been used for a "continuous" functor $K_2^c(R)$ for a topological ring R, especially when R is an algebra over $\hat{\mathbb{Z}}_p$ or $\hat{\mathbb{Q}}_p$. Definitions involving continuous universal central extensions of E(R) have been used by Moore [1] and Rehmann [1, Section 5]; and Wagoner [1] has defined $K_n^c(R)$ in all dimensions as a limit of homotopy groups of certain simplicial complexes. But for the purposes here, the following definition is the simplest and most convenient.

For any prime p, any semisimple $\hat{\mathbb{Q}}_p$-algebra A, and any $\hat{\mathbb{Z}}_p$-order $\mathfrak{A}$ in A, set

$$K_2^c(\mathfrak{A}) = \varprojlim_k \operatorname{Coker}\left[K_2(\mathfrak{A}, p^k\mathfrak{A}) \longrightarrow K_2(\mathfrak{A})\right]$$

and

$$K_2^c(A) = \varprojlim_k \operatorname{Coker}\left[K_2(\mathfrak{A}, p^k\mathfrak{A}) \longrightarrow K_2(A)\right].$$

By Lemma 3.2, for any pair $\mathfrak{A} \subseteq \mathfrak{B}$ of orders in A and any $k > 0$,

$$\operatorname{Im}\left[K_2(\mathfrak{B}, p^{4k}\mathfrak{B}) \longrightarrow K_2(A)\right] \subseteq \operatorname{Im}\left[K_2(\mathfrak{A}, p^k\mathfrak{A}) \longrightarrow K_2(A)\right].$$

So $K_2^c(A)$ is well defined, independently of the choice of order $\mathfrak{A} \subseteq A$.

Quillen's localization sequence for maximal $\hat{\mathbb{Z}}_p$-orders (Theorem 1.17) can easily be reformulated in terms of K_2^c.

<u>Theorem 3.5</u> *Fix a prime p, let $\mathfrak{M}$ be a maximal $\hat{\mathbb{Z}}_p$-order in a semisimple $\hat{\mathbb{Q}}_p$-algebra A, and let $J \subseteq \mathfrak{M}$ be the Jacobson radical. Then there is an exact sequence*

$$1 \to K_2^c(\mathfrak{M}) \longrightarrow K_2^c(A) \longrightarrow K_1(\mathfrak{M}/J) \longrightarrow K_1(\mathfrak{M}) \longrightarrow K_1(A) \longrightarrow \ldots.$$

Proof This is almost an immediate consequence of the localization sequence in Theorem 1.17(i). Since $\mathfrak{M}/J$ is semisimple of p-power order, $K_2(\mathfrak{M}/J) = 1$ and $p \nmid |K_1(\mathfrak{M}/J)|$ by Theorem 1.16. Hence $K_2(\mathfrak{M})$ injects into $K_2(A)$; and so $K_2^c(\mathfrak{M})$ injects into $K_2^c(A)$ by definition of K_2^c. □

The formula in the next proposition could just as easily have been taken as the definition of $K_2^c(\mathfrak{A})$. Recall that a *pro-p-group* is a group which is the inverse limit of some system of finite p-groups.

Proposition 3.6 *Fix a prime* p, *and let* $\mathfrak{A}$ *be a* $\hat{\mathbb{Z}}_p$*-order in some semisimple* $\hat{\mathbb{Q}}_p$*-algebra* A. *Then*

$$K_2^c(\mathfrak{A}) \cong \varprojlim_k K_2(\mathfrak{A}/p^k\mathfrak{A}).$$

In particular, $K_2^c(\mathfrak{A})$ *is a pro-p-group, and* $K_2^c(A)$ *is the product of a finite group and a pro-p-group.*

Proof By definition,

$$K_2^c(\mathfrak{A}) = \varprojlim_k \operatorname{Coker}\Big[K_2(\mathfrak{A},p^k\mathfrak{A}) \longrightarrow K_2(\mathfrak{A})\Big].$$

The sequence

$$1 \longrightarrow \operatorname{Coker}\Big[K_2(\mathfrak{A},p^k\mathfrak{A}) \longrightarrow K_2(\mathfrak{A})\Big] \longrightarrow K_2(\mathfrak{A}/p^k\mathfrak{A}) \longrightarrow K_1(\mathfrak{A},p^k\mathfrak{A})$$

is exact for all k, and hence is still exact after taking inverse limits. But by Theorem 2.10(iii),

$$\varprojlim_k K_1(\mathfrak{A},p^k\mathfrak{A}) \cong \varprojlim_{k\le n} K_1(\mathfrak{A}/p^n\mathfrak{A},p^k\mathfrak{A}/p^n\mathfrak{A})$$

$$\cong \varprojlim_n K_1(\mathfrak{A}/p^n\mathfrak{A},p^n\mathfrak{A}/p^n\mathfrak{A}) = 1.$$

In particular, $K_2^c(\mathfrak{A})$ is a pro-p-group by Theorem 1.16(ii). If $\mathfrak{M} \subseteq A$ is any maximal order, then $[K_2^c(A) : K_2^c(\mathfrak{M})]$ is finite by Theorem 3.5, and so $K_2^c(A)$ is pro-finite since $K_2^c(\mathfrak{M})$ is. □

In fact, in Chapter 4, $K_2^c(A)$ will be shown to always be finite.

3c. Localization sequences for torsion in Whitehead groups

We now want to describe $SK_1(\mathfrak{A})$, when $\mathfrak{A}$ is a $\mathbb{Z}$-order in a semisimple $\mathbb{Q}$-algebra A, in terms of K_1 and K_2 of A, $\hat{\mathfrak{A}}_p$ and $\hat{A}_p$. The usual way of doing this is via Mayer-Vietoris exact sequences based on "arithmetic squares", and one example of such sequences is given at the end of the section (Theorem 3.16). But for the purposes here, it has been convenient to make a different approach, using the relative exact sequences for ideals $I \subseteq \mathfrak{A}$ of finite index. This will be based on the following definitions.

Definition 3.7 *For any semisimple $\mathbb{Q}$-algebra A, and any $\mathbb{Z}$-order $\mathfrak{A} \subseteq A$, define*

$$Cl_1(\mathfrak{A}) = \operatorname{Ker}\Big[SK_1(\mathfrak{A}) \xrightarrow{\ \ell\ } \bigoplus_p SK_1(\hat{\mathfrak{A}}_p)\Big].$$

More generally, for any (2-sided) ideal $I \subseteq \mathfrak{A}$, set

$$SK_1(\mathfrak{A},I) = \operatorname{Ker}\Big[K_1(\mathfrak{A},I) \longrightarrow K_1(A)\Big] \quad \textit{and}$$

$$Cl_1(\mathfrak{A},I) = \operatorname{Ker}\Big[SK_1(\mathfrak{A},I) \xrightarrow{\ \ell\ } \bigoplus_p SK_1(\hat{\mathfrak{A}}_p,\hat{I}_p)\Big].$$

Then define

$$C(A) = \varprojlim_I SK_1(\mathfrak{A},I)$$

where the limit is taken over all ideals $I \subseteq \mathfrak{A}$ of finite index.

The subgroup $Cl_1(\mathfrak{A})$ can be thought of as the part of $SK_1(\mathfrak{A})$ which is hit from behind (i. e., detected by K_2) in the localization sequences. Recall (Theorem 1.14(ii)) that $SK_1(\hat{\mathfrak{A}}_p) = 1$ for all p if $\mathfrak{A}$ is commutative, so that $Cl_1(\mathfrak{A}) = SK_1(\mathfrak{A})$ in this case.

As is suggested by the notation, $C(A) = \varprojlim SK_1(\mathfrak{A},I)$ is independent of the choice of order $\mathfrak{A}$ in A. This is an easy consequence of Lemma 2.4; and will be shown explicitly in Theorem 3.9. The C(A) can be characterized in several ways:

$$C(A) = \varprojlim_I SK_1(\mathfrak{A},I) \cong \varprojlim_I Cl_1(\mathfrak{A},I) \qquad \text{(Theorem 3.9)}$$

$$\cong \varprojlim_{\mathfrak{A}} Cl_1(\mathfrak{A}) \qquad \text{(taken over all } \mathbb{Z}\text{-orders } \mathfrak{A} \text{ in } A)$$

$$\cong \operatorname{Coker}\Big[K_2(A) \longrightarrow \bigoplus_p K_2^c(\hat{A}_p)\Big]. \qquad \text{(Theorem 3.12)}$$

It is the last description of C(A), in terms of $K_2(-)$, which will be used to calculate these groups in Section 4c below.

The appearance of C(A) in the localization sequence for $SK_1(\mathfrak{A})$ helps to explain the close connection between computations of $SK_1(\mathfrak{A})$ and the congruence subgroup problem. In fact, the original conjecture would have implied that $SK_1(R,I) = 1$ whenever R is the ring of integers in a number field K and $I \subseteq R$ is an ideal of finite index. The computation of the groups $C(K) \cong \varprojlim SK_1(R,I)$ follows as a special case of results of Bass, Milnor, and Serre [1, Theorem 4.1 and Corollary 4.3] in their solution to the problem.

One difficulty which always occurs in localization sequences based on comparing $K_i(\mathfrak{A})$ with $K_i(A)$ (when $\mathfrak{A} \subseteq A$ is a $\mathbb{Z}$-order) is dealing with the infinite products which arise in the p-adic completions of $\mathfrak{A}$ and A. The next lemma says that in the case of $\oplus_p SK_1(\hat{\mathfrak{A}}_p)$ and $\oplus_p K_2^c(\hat{\mathfrak{M}}_p)$, at least, there is no problem — both of these are finite products.

__Lemma 3.8__ *For any semisimple* $\mathbb{Q}$*-algebra* A *and any* $\mathbb{Z}$*-order* $\mathfrak{A} \subseteq A$, $K_2^c(\hat{\mathfrak{A}}_p) = 1$ *and* $SK_1(\hat{\mathfrak{A}}_p) = 1$ *for almost all primes* p.

__Proof__ Let $\mathfrak{M} \supseteq \mathfrak{A}$ be a maximal order. Then $\hat{\mathfrak{A}}_p = \hat{\mathfrak{M}}_p$ for all $p \nmid [\mathfrak{M}:\mathfrak{A}]$, and $\hat{\mathfrak{M}}_p$ is a maximal order in $\hat{A}_p$ by Theorem 1.7(iv). So we can assume that $\mathfrak{A} = \mathfrak{M}$. Also, since $\mathfrak{M}$ factors as a product of maximal orders in the simple summands of A, we can assume that A is simple.

Let $K = Z(A)$ be the center, let $R \subseteq K$ be the ring of integers, and set $n = [A:K]^{1/2}$. Then $\hat{A}_p \cong M_n(\hat{K}_p)$ for almost all p (Theorem 1.7(iii)); and $\hat{\mathfrak{M}}_p \cong M_n(\hat{R}_p)$ for such p by Theorem 1.9. In particular, $SK_1(\hat{\mathfrak{M}}_p) \cong SK_1(\hat{R}_p) = 1$ for almost all p by Theorem 1.14(ii).

It remains to consider the case of $K_2^c(\hat{\mathfrak{M}}_p) \cong K_2^c(\hat{R}_p)$; it suffices to do this when $A = K$ and $\mathfrak{M} = R$. Recall that for each p, $\hat{R}_p \cong \prod_{\mathfrak{p}|p} \hat{R}_\mathfrak{p}$ and $\hat{K}_p \cong \prod_{\mathfrak{p}|p} \hat{K}_\mathfrak{p}$, where the products are taken over all maximal ideals $\mathfrak{p} \subseteq R$ containing p. We claim that $K_2^c(\hat{R}_\mathfrak{p}) = 1$ for any $\mathfrak{p}$ such that (i) $p = \mathrm{char}(R/\mathfrak{p})$ is odd, and (ii) $\hat{K}_\mathfrak{p}$ is unramified over $\hat{\mathbb{Q}}_p$. By Theorem 1.7(ii), this is the case for all but finitely many $\mathfrak{p}$.

Fix such a $\mathfrak{p}$. Then, $\hat{R}_\mathfrak{p}$ is the ring of integers in $\hat{K}_\mathfrak{p}$, and $p\hat{R}_\mathfrak{p}$ is its maximal ideal by (ii). Thus, $(\hat{R}_\mathfrak{p}/p\hat{R}_\mathfrak{p})^*$ has order prime to p. Furthermore, $(1 + p^2\hat{R}_\mathfrak{p})$ consists of p-th powers in $\hat{R}_\mathfrak{p}$: p is odd, and so the binomial series for $(1 + p^2x)^{1/p}$ converges for any $x \in \hat{R}_\mathfrak{p}$. In particular, by Proposition 1.8(i),

$$(\hat{K}_\mathfrak{p})^* = \langle u^p,\ 1-pu^p,\ p : u \in (\hat{R}_\mathfrak{p})^* \rangle.$$

This, together with identities of the form $\{a,1-a\} = 1 = \{a,-a\}$ (Theorem 3.1(i)), shows that $K_2^c(\hat{K}_\mathfrak{p})$ is generated by symbols of the form

$$\{u^p,v\} \qquad (\text{for } u \in (\hat{R}_\mathfrak{p})^*,\ v \in (\hat{K}_\mathfrak{p})^*)$$

$$\{p,p\} = \{-1,p\} = \{(-1)^p,p\}$$

$$\{p,1-pu^p\} = \{u^p,1-pu^p\}^{-1} \qquad (\text{for } u \in (\hat{R}_p)^*)$$

$$\{1-pu^p,1-pv^p\} = \{pv^p(1-pu^p),1-pv^p\} \qquad (\text{for } u,v \in (\hat{R}_p)^*)$$

$$= \{pv^p(1-pu^p),(1-pv^p)(1-pv^p+p^2u^pv^p)^{-1}\} \in \{(\hat{R}_p)^*,1+p^2\hat{R}_p\}.$$

In other words, every element of $K_2^c(\hat{R}_p)$ is a p-th power. But the localization sequence of Theorem 3.5 takes the form

$$1 \longrightarrow K_2^c(\hat{R}_p) \longrightarrow K_2^c(\hat{R}_p) \longrightarrow (R/p)^*;$$

$K_2^c(\hat{R}_p)$ is a pro-p-group and $(R/p)^*$ is finite, and so $K_2^c(\hat{R}_p) = 1$. □

We are now ready to derive the main localization sequence for describing $SK_1(\mathfrak{A})$. At the same time, we show that $C(A)$ is well defined, independently of the choice of order $\mathfrak{A} \subseteq A$.

<u>Theorem 3.9</u> *For any $\mathbb{Z}$-order $\mathfrak{A}$ in a semisimple $\mathbb{Q}$-algebra A, there is an exact sequence*

$$\bigoplus_p K_2^c(\hat{\mathfrak{A}}_p) \xrightarrow{\varphi} C(A) \xrightarrow{\partial} SK_1(\mathfrak{A}) \xrightarrow{\ell} \bigoplus_p SK_1(\hat{\mathfrak{A}}_p) \longrightarrow 1; \qquad (1)$$

where ℓ is induced by the inclusions $\mathfrak{A} \subseteq \hat{\mathfrak{A}}_p$, ∂ by the inclusions $(\mathfrak{A},I) \subseteq \mathfrak{A}$, and φ by the composites $K_2^c(\hat{\mathfrak{A}}_p) \longrightarrow K_2(\hat{\mathfrak{A}}_p/\hat{I}_p) \longrightarrow K_1(\mathfrak{A},I)$. Furthermore,

$$C(A) = \varprojlim_I SK_1(\mathfrak{A},I) \cong \varprojlim_I Cl_1(\mathfrak{A},I); \qquad (I \subseteq \mathfrak{A} \textit{ of finite index})$$

and is independent of the choice of order $\mathfrak{A}$ in A.

<u>Proof</u> For each ideal $I \subseteq \mathfrak{A}$ of finite index, the relative exact sequence for the pair $(\mathfrak{A},I)$ (Theorem 1.13) restricts to an exact sequence

$$K_2(\mathfrak{A}/I) \longrightarrow SK_1(\mathfrak{A},I) \longrightarrow SK_1(\mathfrak{A}) \longrightarrow K_1(\mathfrak{A}/I).$$

The first term is finite (Theorem 1.16), and so taking the inverse limit over all I gives a new exact sequence

$$\varprojlim_I K_2(\mathfrak{A}/I) \longrightarrow C(A) \longrightarrow SK_1(\mathfrak{A}) \longrightarrow \varprojlim_I K_1(\mathfrak{A}/I). \tag{2}$$

The first term in (2) is isomorphic to $\oplus_p K_2^c(\hat{\mathfrak{A}}_p)$ by Proposition 3.6 (and Lemma 3.8), and the last term is isomorphic to $\prod_p K_1(\hat{\mathfrak{A}}_p)$ by Theorem 2.10(iii). This shows that sequence (1) is defined, and is exact except possibly for the surjectivity of ℓ.

For each prime p, $SK_1(\hat{\mathfrak{A}}_p)$ is finite (Theorem 2.5(i)), and $K_1(\hat{\mathfrak{A}}_p) \cong \varprojlim K_1(\mathfrak{A}/p^k\mathfrak{A})$. Since $SK_1(\hat{\mathfrak{A}}_p) = 1$ for almost all p, this shows that we can choose $n \geq 1$ such that for all primes p,

$$SK_1(\hat{\mathfrak{A}}_p) \rightarrowtail K_1(\hat{\mathfrak{A}}_p/n\hat{\mathfrak{A}}_p) \tag{3}$$

is injective.

Fix a prime p and an element $[M] \in SK_1(\hat{\mathfrak{A}}_p)$. In other words, $M \in GL(\hat{\mathfrak{A}}_p) \cap E(\hat{A}_p)$. Write

$$M = e_{i_1j_1}(r_1)\cdots e_{i_kj_k}(r_k), \qquad (r_i \in \hat{A}_p)$$

a product of elementary matrices. Write $n = p^a \cdot m$, where $p \nmid m$. For each r_i, choose a global approximation $\tilde{r}_i \in \mathfrak{A}[\frac{1}{p}]$ such that

$$\tilde{r}_i \equiv r_i \pmod{p^a\hat{\mathfrak{A}}_p}, \text{ and}$$

$$\tilde{r}_i \equiv 0 \pmod{m\mathfrak{A}}.$$

Note that it suffices to do this on the individual coordinates (in $\hat{\mathbb{Q}}_p$) of r_i with respect to some fixed $\mathbb{Z}$-basis of $\mathfrak{A}$. If we now set

$$\tilde{M} = e_{i_1j_1}(\tilde{r}_1)\cdots e_{i_kj_k}(\tilde{r}_k) \in GL(\mathfrak{A}) \cap E(A),$$

then $\tilde{M} \equiv M \pmod{p^a\hat{\mathfrak{A}}_p}$, $\tilde{M} \equiv I \pmod{m\mathfrak{A}}$, and

$$[\tilde{M}] \in SK_1(\mathfrak{A}) = \mathrm{Ker}\Big[K_1(\mathfrak{A}) \longrightarrow K_1(A)\Big].$$

By (3), the congruences guarantee that $\ell([\tilde{M}]) = [M]$. So ℓ is onto.

It remains to prove the last statement. First let $\mathfrak{B} \subseteq \mathfrak{A}$ be any pair of $\mathbb{Z}$-orders in A. For any $\mathfrak{A}$-ideal $I \subseteq \mathfrak{B}$ of finite index, there is a short exact sequence

$$1 \longrightarrow E(\mathfrak{A},I)/E(\mathfrak{B},I) \longrightarrow SK_1(\mathfrak{B},I) \xrightarrow{f_I} SK_1(\mathfrak{A},I) \longrightarrow 1.$$

By Lemma 2.4, $E(\mathfrak{A},I^2) \subseteq E(\mathfrak{B},I)$ for any such I, so that $\mathrm{Ker}(f_{I^2})$ maps trivially to $\mathrm{Ker}(f_I)$. In particular, the inclusion $\mathfrak{B} \subseteq \mathfrak{A}$ induces an isomorphism $\varprojlim_I SK_1(\mathfrak{B},I) \cong \varprojlim_I SK_1(\mathfrak{A},I)$, so that $C(A)$ is well defined. Also, by definition of $Cl_1(\mathfrak{A},I)$, there is an exact sequence

$$1 \longrightarrow \varprojlim_I Cl_1(\mathfrak{A},I) \longrightarrow \varprojlim_I SK_1(\mathfrak{A},I) \longrightarrow \varprojlim_I \bigoplus_p K_1(\hat{\mathfrak{A}}_p,\hat{I}_p);$$

and the last term vanishes by Theorem 2.10(iii). □

Note in particular that for any $\mathfrak{A}$, $Cl_1(\mathfrak{A}) = \mathrm{Im}[C(A) \xrightarrow{\partial} SK_1(\mathfrak{A})]$ in the localization sequence above; and that $SK_1(\mathfrak{A})$ sits in an extension

$$1 \longrightarrow Cl_1(\mathfrak{A}) \longrightarrow SK_1(\mathfrak{A}) \xrightarrow{\ell} \bigoplus_p SK_1(\hat{\mathfrak{A}}_p) \longrightarrow 1.$$

One easy consequence of Theorem 3.9 is the following:

<u>Corollary 3.10</u> *Let $f\colon A \twoheadrightarrow B$ be a surjection of semisimple $\mathbb{Q}$-algebras, and let $\mathfrak{A} \subseteq A$ and $\mathfrak{B} \subseteq B$ be $\mathbb{Z}$-orders such that $f(\mathfrak{A}) \subseteq \mathfrak{B}$. Then the induced map*

$$Cl_1(f) : Cl_1(\mathfrak{A}) \twoheadrightarrow Cl_1(\mathfrak{B})$$

is surjective. If $A = B$ *and* $\mathfrak{A} \subseteq \mathfrak{B}$, *then* $Ker(Cl_1(f))$ *has torsion only for primes* $p|[\mathfrak{B}:\mathfrak{A}]$.

<u>Proof</u> Consider the following diagram of localization sequences from Theorem 3.9:

$$\begin{array}{ccccccc}
\bigoplus_p K_2^c(\hat{\mathfrak{A}}_p) & \longrightarrow & C(A) & \longrightarrow & Cl_1(\mathfrak{A}) & \longrightarrow & 1 \\
\downarrow{\scriptstyle K_2^c(\hat{f})} & & \downarrow{\scriptstyle C(f)} & & \downarrow{\scriptstyle Cl_1(f)} & & \\
\bigoplus_p K_2^c(\hat{\mathfrak{B}}_p) & \longrightarrow & C(B) & \longrightarrow & Cl_1(\mathfrak{B}) & \longrightarrow & 1.
\end{array}$$

Since $f: A \longrightarrow B$ is projection onto a direct summand, $C(f)$ is onto. Hence $Cl_1(f)$ is also onto. If $A = B$, then $C(f)$ is an isomorphism, so $Coker(K_2^c(\hat{f}))$ surjects onto $Ker(Cl_1(f))$. But $K_2^c(\hat{\mathfrak{B}}_p) = K_2^c(\hat{\mathfrak{A}}_p)$ whenever $p\nmid[\mathfrak{B}:\mathfrak{A}]$, $K_2^c(\hat{\mathfrak{B}}_p)$ is a pro-p-group for all p, and so $Coker(K_2^c(\hat{f}))$ and $Ker(Cl_1(f))$ have torsion only for primes $p|[\mathfrak{B}:\mathfrak{A}]$. □

We next want to prove an alternate description of $C(A)$, in terms of $K_2(-)$. The key problem when doing this is to define and compare certain boundary maps for localization squares. In fact, given any commutative square

$$\begin{array}{ccc}
R & \xrightarrow{\alpha} & R' \\
\downarrow{\scriptstyle f} & & \downarrow{\scriptstyle f'} \\
S & \xrightarrow{\beta} & S'
\end{array}$$

of rings, *inverse* boundary maps

$$\delta: Ker(K_n(\alpha) \oplus K_n(f)) \longrightarrow Coker(K_{n+1}(f') \oplus K_{n+1}(\beta))$$

can always be defined (and the problem is to determine when δ is an isomorphism). When $n = 1$, there are two obvious ways of defining δ:

(i) For any $M \in GL(R)$ such that $[M] \in \mathrm{Ker}(K_1(\alpha) \oplus K_1(f))$, choose elements $x \in St(R')$ and $y \in St(S)$ such that $\phi(x) = \alpha(M)$, $\phi(y) = f(M)$. Then

$$\delta_1([M]) = [x^{-1}y] \in K_2(S') \pmod{\mathrm{Im}(K_2(\beta) \oplus K_2(f'))}.$$

(ii) Define

$$K_1(\alpha) = \pi_1\Big(\text{homotopy fiber of } BGL(R)^+ \longrightarrow BGL(R')^+\Big),$$

and similarly for $K_1(\beta)$. Consider the following diagram:

$$\begin{array}{ccccccccc}
K_2(R) & \xrightarrow{K_2(\alpha)} & K_2(R') & \xrightarrow{\partial_\alpha} & K_1(\alpha) & \xrightarrow{K_1(i_\alpha)} & K_1(R) & \xrightarrow{K_1(\alpha)} & K_1(R') \\
\downarrow{\scriptstyle K_2(f)} & & \downarrow{\scriptstyle K_2(f')} & & \downarrow{\scriptstyle K_1(f_o)} & & \downarrow{\scriptstyle K_1(f)} & & \downarrow{\scriptstyle K_1(f')} \\
K_2(S) & \xrightarrow{K_2(\beta)} & K_2(S') & \xrightarrow{\partial_\beta} & K_1(\beta) & \xrightarrow{K_1(i_\beta)} & K_1(R) & \xrightarrow{K_1(\beta)} & K_1(S');
\end{array}$$

where the rows are induced by the homotopy exact sequence for a fibration; and let δ_2 be the composite

$$\delta_2 = \partial_\beta^{-1} \circ K_1(f_o) \circ K_1(i_\alpha)^{-1} : \mathrm{Ker}(K_1(\alpha) \oplus K_1(f)) \longrightarrow \mathrm{Coker}(K_2(\beta) \oplus K_2(f')).$$

Note that any boundary homomorphisms constructed using Quillen's localization sequences (Theorems 1.17 and 3.5) will be of type δ_2.

<u>Lemma 3.11</u> *Let*

$$\begin{array}{ccc}
R & \xrightarrow{\alpha} & R' \\
\downarrow{\scriptstyle f} & & \downarrow{\scriptstyle f'} \\
S & \xrightarrow{\beta} & S'
\end{array}$$

be any commutative square of rings. Then

$$\delta_1 = \delta_2 : \mathrm{Ker}(K_1(\alpha) \oplus K_1(f)) \longrightarrow \mathrm{Coker}(K_2(\beta) \oplus K_2(f')).$$

<u>Proof</u> Note first the following more direct description of δ_2. Fix any $\sigma: S^1 \longrightarrow BGL(R)^+$ such that $[\sigma] \in \mathrm{Ker}(\pi_1(\alpha_+) \oplus \pi_1(f_+))$, and extend $\alpha_+ o\sigma$ and $f_+ o\sigma$ to maps

$$\tilde{\sigma}_\alpha : D^2 \longrightarrow BGL(R')^+, \qquad \tilde{\sigma}_f : D^2 \longrightarrow BGL(S)^+.$$

Then $\delta_2([\sigma])$ is the homotopy class of the map

$$(f'_+ o\tilde{\sigma}_\alpha) \cup (\beta_+ o\tilde{\sigma}_f) : S^2 = D^2 \cup_{S^1} D^2 \longrightarrow BGL(S')^+.$$

Regard $BGL(S)^+$ as a CW complex whose 2-skeleton consists of one vertex, a 1-cell $\langle A \rangle$ for each element $A \in GL(S)$, a 2-cell for each relation among the elements in $GL(S)$, and a 2-cell $[x^s_{ij}]$ for each elementary matrix $e^s_{ij} \in E(S)$. Then, given any $A \in E(S)$, a lifting of A to some $X \in St(S)$ induces a null-homotopy of the loop $\langle A \rangle$. The same argument applies to $BGL(R')^+$ and $BGL(S')^+$, and shows that $\delta_1 = \delta_2$. □

We are now ready to reinterpret $C(A)$ in terms of $K_2(-)$. The description of $C(A)$ in the following theorem will be the basis of its computation in the next chapter.

<u>Theorem 3.12</u> *For any semisimple $\mathbb{Q}$-algebra A, there is a natural isomorphism*

$$C(A) \cong \mathrm{Coker}\Big[K_2(A) \longrightarrow \bigoplus_p K^c_2(\hat{A}_p)\Big].$$

Under this identification, in the localization sequence

$$\bigoplus_p K^c_2(\hat{\mathfrak{A}}_p) \xrightarrow{\ \varphi\ } C(A) \xrightarrow{\ \partial\ } Cl_1(\mathfrak{A}) \longrightarrow 1$$

for a $\mathbb{Z}$-order $\mathfrak{A} \subseteq A$, φ is induced by the inclusions $\hat{\mathfrak{A}}_p \subseteq \hat{A}_p$, and ∂ is described as follows. Given any $M \in GL(\mathfrak{A})$ such that

$$[M] \in Cl_1(\mathfrak{A}) = \mathrm{Ker}\Big[K_1(\mathfrak{A}) \longrightarrow K_1(A) \oplus \prod_p K_1(\hat{\mathfrak{A}}_p)\Big],$$

lift M *to* $x \in St(A)$ *and to* $y = (y_p) \in \prod_p St(\hat{\mathfrak{A}}_p)$ *such that* $x = y_p$ *in* $St(\hat{A}_p)$ *for almost all* p. *Then* $[M] = \partial([x^{-1}y]) = \partial([yx^{-1}])$, *where*

$$x^{-1}y = yx^{-1} \in \mathrm{Coker}[K_2(A) \longrightarrow \bigoplus_p K_2^c(\hat{A}_p)] = C(A).$$

<u>Proof</u> Fix a maximal $\mathbb{Z}$-order $\mathfrak{M} \subseteq A$. For any $x \in K_2(A)$, with localizations $\ell_p(x) \in K_2^c(\hat{A}_p)$, $\ell_p(x) \in \mathrm{Im}[K_2^c(\hat{\mathfrak{M}}_p) \longrightarrow K_2^c(\hat{A}_p)]$ for almost all p (this holds for each generator $x_{ij}^r \in St(A)$), and $K_2^c(\hat{\mathfrak{M}}_p) = 1$ for almost all p by Lemma 3.8. This shows that $K_2(A)$ maps into the direct sum $\oplus_p K_2^c(\hat{A}_p)$.

For each ideal $I \subseteq \mathfrak{M}$ of finite index, consider the commutative square

$$\begin{array}{ccc} (\mathfrak{M}, I) & \longrightarrow & A \\ \downarrow & & \downarrow \\ \prod_p (\hat{\mathfrak{M}}_p, \hat{I}_p) & \longrightarrow & \prod_p \hat{A}_p. \end{array} \qquad (1)$$

Recall (Milnor [2, Chapters 4 and 6]) that the $K_i(\mathfrak{M}, I)$ can be regarded as direct summands of $K_i(D)$, where $D = \{(r,s) \in \mathfrak{M} \times \mathfrak{M}\colon r \equiv s \pmod I\}$; and similarly for $K_i(\hat{\mathfrak{M}}_p, \hat{I}_p)$. So Lemma 3.11 can also be applied to this relative case. The inverse boundary maps $\delta_1 = \delta_2$ for (1) then take the form

$$\delta_I : Cl_1(\mathfrak{M}, I) \longrightarrow \mathrm{Coker}\Big[K_2(A) \oplus \bigoplus_p K_2^c(\hat{\mathfrak{M}}_p, \hat{I}_p) \longrightarrow \bigoplus_p K_2^c(\hat{A}_p)\Big].$$

To see that δ_I actually maps into the direct sum (as opposed to the direct product), note that for any $[M] \in Cl_1(\mathfrak{A}, I)$, and any explicit decomposition of M as a product of elementary matrices over A, this decomposition will have coefficients in $\hat{I}_p$ for almost all p (so that $\delta_I([M])$ can be taken to be trivial in $K_2^c(\hat{A}_p)$ for such p).

For each p, $\varprojlim_I \operatorname{Coker}[K_2^c(\hat{\mathfrak{A}}_p, \hat{I}_p) \longrightarrow K_2^c(\hat{A}_p)] \cong K_2^c(\hat{A}_p)$ by definition of $K_2^c(-)$. Since, in addition, $K_2^c(\hat{\mathfrak{M}}_p) = 1$ for almost all p, the inverse limit of the δ_I takes the form

$$\hat{\delta} : C(A) \cong \varprojlim_I Cl_1(\mathfrak{M}, I) \longrightarrow \operatorname{Coker}\Big[K_2(A) \longrightarrow \bigoplus_p K_2^c(\hat{A}_p)\Big].$$

Now consider the localization sequences

$$\begin{array}{ccccccccc}
1 \longrightarrow K_2(\mathfrak{M}) & \longrightarrow & K_2(A) & \xrightarrow{\partial_A} & \bigoplus_p K_1(\hat{\mathfrak{M}}_p/J_p) & \longrightarrow & SK_1(\mathfrak{M}) \longrightarrow 1 \\
\downarrow & & \downarrow & & \cong\downarrow \mathrm{Id} & & \downarrow \\
1 \longrightarrow \bigoplus_p K_2^c(\hat{\mathfrak{M}}_p) & \longrightarrow & \bigoplus_p K_2^c(\hat{A}_p) & \xrightarrow{\oplus\partial_p} & \bigoplus_p K_1(\hat{\mathfrak{M}}_p/J_p) & \longrightarrow & \bigoplus_p SK_1(\hat{\mathfrak{M}}_p) \longrightarrow 1.
\end{array}$$

of Theorems 1.17 and 3.5 (where $J_p \subseteq \hat{\mathfrak{M}}_p$ is the Jacobson radical). This diagram, together with the localization sequence of Theorem 3.9, induces the following commutative diagram with exact rows:

$$\begin{array}{ccccc}
\operatorname{Coker}[K_2(\mathfrak{M}) \to \bigoplus_p K_2^c(\hat{\mathfrak{M}}_p)] & \xrightarrow{\varphi} & C(A) & \xrightarrow{\partial} & Cl_1(\mathfrak{M}) \to 1 \\
\cong\downarrow \mathrm{Id} & (2a) & \downarrow \hat{\delta} & (2b) & \cong\downarrow \mathrm{Id} \\
1 \to \operatorname{Coker}[K_2(\mathfrak{M}) \to \bigoplus_p K_2^c(\hat{\mathfrak{M}}_p)] & \longrightarrow & \operatorname{Coker}[K_2(A) \to \bigoplus_p K_2^c(\hat{A}_p)] & \xrightarrow{\partial'} & Cl_1(\mathfrak{M}) \to 1
\end{array} \qquad (2)$$

Square (2b) commutes since $(\partial')^{-1} = \delta_2$ in the notation of Lemma 3.11. To see that (2a) commutes (up to sign), fix $x \in \oplus_p K_2^c(\hat{\mathfrak{M}}_p)$, and for each I let $x_I \in St(\mathfrak{M})$ be a mod I approximation to x (i. e., replace each x_{ij}^r in x by some mod I approximation to r). Then

$$\varphi(x) = ([\phi(x_I)])_{I \subseteq \mathfrak{M}} \in \varprojlim_I SK_1(\mathfrak{M}, I) = C(A).$$

Each $\phi(x_I)$ lifts to $x_I \in St(A)$ and $x^{-1}x_I \in \oplus_p St(\hat{\mathfrak{A}}_p,\hat{I}_p)$; so that $\hat{\delta}\circ\varphi(x) = (x^{-1}x_I)\cdot x_I^{-1} = x^{-1}$.

It now follows from (2) that $\hat{\delta}$ is an isomorphism. The descriptions of $\hat{\delta}\circ\varphi$ and $\partial\circ\hat{\delta}^{-1}$ are immediate. □

The description of $\partial\colon C(A) \twoheadrightarrow Cl_1(\mathfrak{A})$ in Theorem 3.12 is in itself of only limited use when working with concrete matrices. No matter how well $K_2^c(\hat{A}_p)$ is understood, it is difficult to deal with an element which is presented only as a product of generators $x_{ij}^r \in St(\hat{A}_p)$. In contrast, the formula in the following proposition, while complicated to state, is easily applied in many concrete calculations.

Proposition 3.13 *Let $\mathfrak{A}$ be a $\mathbb{Z}$-order in a semisimple $\mathbb{Q}$-algebra A, and let*

$$\partial : \mathrm{Coker}\Big[K_2(A) \longrightarrow \oplus K_2^c(\hat{A}_p)\Big] \cong C(A) \longrightarrow\!\!\!\!\!\rightarrow Cl_1(\mathfrak{A})$$

be the boundary map of Theorems 3.9 and 3.12. Fix $n \geq 1$, and fix factorizations $\mathfrak{A}[\frac{1}{n}] = \mathfrak{B} \times \mathfrak{B}'$, $A = B \times B'$; where $\mathfrak{B} \subseteq B$ and $\mathfrak{B}' \subseteq B'$ are $\mathbb{Z}[\frac{1}{n}]$-orders. Let $\begin{pmatrix} a & b \\ c & d \end{pmatrix} \in GL_2(\mathfrak{A})$ be any matrix such that $ac = ca$ and $ad - cb = 1$, and such that

$$c \in B^* \times B', \quad a \in \mathfrak{B} \times (\mathfrak{B}')^*, \quad \textit{and} \quad a \in (\hat{\mathfrak{A}}_p)^* \ \textit{for all}\ p|n.$$

Then $\begin{bmatrix} a & b \\ c & d \end{bmatrix} = \partial(X) \in Cl_1(\mathfrak{A})$, *where*

$$X = (\{a,c\}^{-1},1) \in \mathrm{Im}\Big[\Big(\bigoplus_{p|n} K_2^c(\hat{B}_p) \times \bigoplus_{p|n} K_2^c(\hat{B}'_p)\Big) \longrightarrow C(A)\Big]; \quad \text{and}$$

$$X = (\{a,c\},1) \in \mathrm{Im}\Big[\Big(\bigoplus_{p\nmid n} K_2^c(\hat{B}_p) \times \bigoplus_{p\nmid n} K_2^c(\hat{B}'_p)\Big) \longrightarrow C(A)\Big].$$

Proof Note first that these two definitions of X are equivalent:

$$C(A) = \mathrm{Coker}\Big[K_2(A) \longrightarrow \bigoplus_p K_2^c(\hat{A}_p)\Big],$$

and $(\{a,c\},1) \in K_2(B)\times K_2(B') = K_2(A)$.

Consider the matrix decompositions

$$\begin{pmatrix} a & b \\ c & d\end{pmatrix} = \begin{pmatrix} 1 & 0 \\ ca^{-1} & 1\end{pmatrix}\begin{pmatrix} a & 0 \\ 0 & a^{-1}\end{pmatrix}\begin{pmatrix} 1 & a^{-1}b \\ 0 & 1\end{pmatrix} \quad \text{in } E_2(\mathcal{B}') \text{ and } E_2(\hat{\mathfrak{A}}_p) \quad (p|n)$$

$$\begin{pmatrix} a & b \\ c & d\end{pmatrix} = \begin{pmatrix} 1 & ac^{-1} \\ 0 & 1\end{pmatrix}\begin{pmatrix} 0 & -c^{-1} \\ c & 0\end{pmatrix}\begin{pmatrix} 1 & c^{-1}d \\ 0 & 1\end{pmatrix} \quad \text{in } E_2(\mathcal{B}).$$

These give liftings of $\begin{pmatrix} a & b \\ c & d\end{pmatrix}$ to elements

$$x_{21}^{ca^{-1}}\cdot h_{21}(a)^{-1}\cdot x_{12}^{a^{-1}b} \quad \text{in } St(B'),\ St(\hat{B}'_p)\ (p\nmid n) \text{ and } St(\hat{\mathfrak{A}}_p)\ (p|n)$$

$$x_{12}^{ac^{-1}}\cdot w_{21}(c)\cdot x_{12}^{c^{-1}d} \quad \text{in } St(B),\ St(\hat{B}_p)\ (p\nmid n).$$

Here,

$$h_{21}(a) = x_{21}^{a}x_{12}^{-a^{-1}}x_{21}^{a}x_{21}^{-1}x_{12}^{1}x_{21}^{-1} \quad \text{and} \quad w_{21}(c) = x_{21}^{c}x_{12}^{-c^{-1}}x_{21}^{c}$$

are liftings of $\begin{pmatrix} a^{-1} & 0 \\ 0 & a\end{pmatrix}$ and $\begin{pmatrix} 0 & -c^{-1} \\ c & 0\end{pmatrix}$, respectively (see Milnor [2, Chapter 9] for more details). The description of $\partial^{-1}\left(\begin{bmatrix} a & b \\ c & d\end{bmatrix}\right)$ now follows from the following computation in $St(\hat{B}_p)$ for $p|n$, based on relations in Milnor [2, Corollary 9.4 and Lemma 9.6]:

$$\left(x_{21}^{ca^{-1}}\cdot h_{21}(a)^{-1}\cdot x_{12}^{a^{-1}b}\right)\left(x_{12}^{ac^{-1}}\cdot w_{21}(c)\cdot x_{12}^{c^{-1}d}\right)^{-1}$$

$$= x_{21}^{ca^{-1}}\cdot h_{21}(a)^{-1}\cdot x_{12}^{-a^{-1}c^{-1}}\cdot w_{21}(c)^{-1}\cdot x_{12}^{-ac^{-1}} \quad (c^{-1}d - a^{-1}b = a^{-1}c^{-1})$$

$$= h_{21}(a)^{-1}\cdot x_{21}^{ac}\cdot x_{12}^{-a^{-1}c^{-1}}\cdot x_{21}^{ac}\cdot w_{21}(c)^{-1}$$

$$= h_{21}(a)^{-1}\cdot w_{21}(ac)\cdot w_{21}(c)^{-1} = w_{21}(ac)\cdot w_{21}(c)^{-1}\cdot h_{21}(a)^{-1}$$

$$= h_{21}(ac)\cdot h_{21}(c)^{-1}\cdot h_{21}(a)^{-1} = \{c,a\} = \{a,c\}^{-1}. \quad \square$$

The use of this formula for detecting whether or not an explicit matrix vanishes in $Cl_1(\mathbb{Z}[G])$ will be illustrated in Example 5.1 (Step 3). Note that when $\mathfrak{A}$ is commutative, any element of $SK_1(\mathfrak{A})$ can be reduced to a 2×2 matrix $\begin{pmatrix} a & b \\ c & d \end{pmatrix}$ where $ad - bc = 1$ (see Bass [1, Proposition 11.2]). So in principle Proposition 3.13 (or some variant) can always be applied in this case. Another example where this formula is used can be seen in Oliver [5, Proposition 2.6].

We now focus attention on group rings. The following theorem shows that $SK_1(\hat{\mathbb{Z}}_p[G])$ is a p-group for *any* finite G, not only when G is a p-group. This does not, of course, hold for arbitrary $\hat{\mathbb{Z}}_p$-orders.

<u>Theorem 3.14</u> (Wall) *Fix a prime* p, *let* $F/\hat{\mathbb{Q}}_p$ *be any finite extension, and let* $R \subseteq F$ *be the ring of integers. Then for any finite group* G, $SK_1(R[G])$ *is a p-group.*

<u>Proof</u> Let $J \subseteq R[G]$ be the Jacobson radical. Then $SK_1(R[G])$ is finite by Theorem 2.5(i), and $\mathrm{Ker}\big[K_1(R[G]) \longrightarrow K_1(R[G]/J)\big]$ is a pro-p-group by Theorem 2.10(ii). So it will suffice to show that $SK_1(R[G])$ maps trivially to $K_1(R[G]/J)$.

Fix a finite extension $E \supseteq F$ with ring of integers $S \subseteq E$, such that the residue field $\bar{S}$ of S is a splitting field for G; i. e., such that $S[G]/J$ is a product of matrix rings over $\bar{S}$. Let $u \in (R[G])^*$ be such that $[u] \in SK_1(R[G])$. If V is any finitely generated E[G]-module, and if M is any S[G]-lattice in V, then

$$\det_S(M \xrightarrow{u\cdot} M) = \det_E(V \xrightarrow{u\cdot} V) = 1$$

since $[u] = 1$ in $K_1(F[G])$. Hence, if we set $\bar{M} = \bar{S} \otimes_S M$, then

$$\det_{\bar{S}}(\bar{M} \xrightarrow{u\cdot} \bar{M}) = 1. \tag{1}$$

By the surjectivity of the decomposition map for modular representations (see Serre [2, §16.1, Theorem 33] or Curtis & Reiner [1, Corollary

18.14]), the representation ring $R_{\bar{S}}(G) \cong K_0(S[G]/J)$ is generated by modules of the form $\bar{M} = \bar{S} \otimes_S M$. So (1) extends to show that $\det_{\bar{S}}(T \xrightarrow{u\cdot} T) = 1$ for any irreducible $S[G]/J$-module T. Since $S[G]/J \cong \bar{S} \otimes_{\bar{R}} (R[G]/J)$, and is a product of matrix algebras over $\bar{S}$, it follows that

$$[u] \in \operatorname{Ker}\Big[SK_1(R[G]) \longrightarrow K_1(R[G]/J) \rightarrowtail K_1(S[G]/J)\Big]. \quad \square$$

The localization sequence of Theorem 3.9, in the case of group rings, at least, can now be split up in a very simple fashion into their p-primary components. For any semisimple $\mathbb{Q}$-algebra A, we define $C_p(A)$ to be the p-localization of $C(A)$. Note that $C(A)$ splits as a direct sum $C(A) = \oplus_p C_p(A)$ — since $C(A)$ is a quotient of $\oplus_p K_2^c(\hat{A}_p)$, and each $K_2^c(\hat{A}_p)$ is a product of a finite group and a pro-p-group by Proposition 3.6. In fact, $C(A)$ will be seen in Chapter 4 to be finite for all A.

<u>Theorem 3.15</u> *Fix a number field K, and let R be its ring of integers. Then, for any finite group G and each prime p, there are exact sequences*

$$K_2^c(\hat{R}_p[G]) \xrightarrow{\varphi_G^p} C_p(K[G]) \xrightarrow{\partial_G^p} Cl_1(R[G])_{(p)} \longrightarrow 1, \qquad (1)$$

and

$$1 \longrightarrow Cl_1(R[G])_{(p)} \longrightarrow SK_1(R[G])_{(p)} \longrightarrow SK_1(\hat{R}_p[G]) \longrightarrow 1. \qquad (2)$$

These sequences, together with the isomorphism

$$\hat{\delta} : C(K[G]) \xrightarrow{\cong} \operatorname{Coker}\Big[K_2(K[G]) \longrightarrow \bigoplus_p K_2^c(\hat{R}_p[G])\Big],$$

are natural with respect to homomorphisms of group rings, as well as transfer (restriction) maps for inclusions of groups or of base rings.

<u>Proof</u> The sequences follow immediately from Theorems 3.9 and 3.14; since $K_2^c(\hat{R}_p[G])$ is a pro-p-group for each p by Proposition 3.6.

Naturality with respect to homomorphisms of group rings is immediate. For any inclusion $S[H] \subseteq R[G]$, where $H \subseteq G$ and S is the ring of integers in a subfield of K, the transfer maps for the terms in (1) and (2) are all induced by some fixed inclusion $R[G] \subseteq M_k(S[H])$, together with the usual isomorphisms $K_i(M_k(S[H])) \cong K_i(S[H])$, etc. Sequence (2) is clearly natural with respect to these last isomorphisms, and the naturality of (1) and $\hat{\delta}$ follow from the descriptions of φ and ∂ in Theorems 3.9 and 3.12. □

It has been simplest to derive the localization sequences used here by indirect means. The usual way to regard localization sequences is as Mayer-Vietoris sequences for certain "arithmetic" pullback squares. We end the chapter with an example of such sequences, due to Bak [2] in dimensions up to 2, and to Quillen (Grayson [1]) for arbitrary dimensions.

__Theorem 3.16__ *Let* $\mathfrak{A}$ *be any* $\mathbb{Z}$*-order in a semisimple* $\mathbb{Q}$*-algebra* A, *and fix a set* $\mathcal{P}$ *of (rational) primes. Define*

$$\mathfrak{A}[\tfrac{1}{\mathcal{P}}] = \mathfrak{A}[\tfrac{1}{p}: p \in \mathcal{P}], \qquad \hat{\mathfrak{A}}_{\mathcal{P}} = \prod_{p\in\mathcal{P}} \hat{\mathfrak{A}}_p, \qquad \hat{A}_{\mathcal{P}} = \hat{\mathfrak{A}}_{\mathcal{P}}[\tfrac{1}{\mathcal{P}}].$$

Then the pullback square

$$\begin{array}{ccc} \mathfrak{A} & \longrightarrow & \mathfrak{A}[\frac{1}{\mathcal{P}}] \\ \big\downarrow & & \big\downarrow \\ \hat{\mathfrak{A}}_{\mathcal{P}} & \longrightarrow & \hat{A}_{\mathcal{P}} \end{array}$$

induces a Mayer-Vietoris exact sequence

$$\begin{aligned} \cdots \longrightarrow K_i(\mathfrak{A}) \longrightarrow K_i(\mathfrak{A}[\tfrac{1}{\mathcal{P}}]) \oplus K_i(\hat{\mathfrak{A}}_{\mathcal{P}}) \longrightarrow K_i(\hat{A}_{\mathcal{P}}) \longrightarrow K_{i-1}(\mathfrak{A}) \longrightarrow \cdots \\ \cdots \longrightarrow K_0(\mathfrak{A}) \longrightarrow K_0(\mathfrak{A}[\tfrac{1}{\mathcal{P}}]) \oplus K_0(\hat{\mathfrak{A}}_{\mathcal{P}}) \longrightarrow K_0(\hat{A}_{\mathcal{P}}). \end{aligned} \tag{1}$$

__Proof__ Let $\underline{P}^t(\mathfrak{A},\mathcal{P})$ and $\underline{P}^t(\hat{\mathfrak{A}}_{\mathcal{P}},\mathcal{P})$ denote the categories of finitely generated $\mathcal{P}$-torsion $\mathfrak{A}$- and $\hat{\mathfrak{A}}_{\mathcal{P}}$-modules of projective dimension one.

These categories are equivalent: note, for example, that a finitely generated $\mathcal{P}$-torsion module over either $\mathfrak{A}$ or $\hat{\mathfrak{A}}_{\mathcal{P}}$ must be finite. So the localization sequences of Quillen for nonabelian categories (see Grayson [1]) induce the following commutative diagram with exact rows:

$$\begin{array}{ccccccccc}
\cdots \longrightarrow & K_i(\underline{P}^t(\mathfrak{A},\mathcal{P})) & \longrightarrow & K_i(\mathfrak{A}) & \longrightarrow & K_i(\mathfrak{A}[\tfrac{1}{\mathcal{P}}]) & \longrightarrow & K_{i-1}(\underline{P}^t(\mathfrak{A},\mathcal{P})) & \longrightarrow \cdots \\
 & \cong\downarrow & & \downarrow & & \downarrow & & \cong\downarrow & \\
\cdots \longrightarrow & K_i(\underline{P}^t(\hat{\mathfrak{A}}_{\mathcal{P}},\mathcal{P})) & \longrightarrow & K_i(\hat{\mathfrak{A}}_{\mathcal{P}}) & \longrightarrow & K_i(\hat{A}_{\mathcal{P}}) & \longrightarrow & K_{i-1}(\underline{P}^t(\hat{\mathfrak{A}}_{\mathcal{P}},\mathcal{P})) & \longrightarrow \cdots .
\end{array} \tag{2}$$

The snake lemma applied to (2) now gives sequence (1), except for exactess at $K_0(\mathfrak{A}[\frac{1}{\mathcal{P}}]) \oplus K_0(\hat{\mathfrak{A}}_{\mathcal{P}})$; and this last point is easily checked. □

Chapter 4 THE CONGRUENCE SUBGROUP PROBLEM

The central result in this chapter is the computation in Theorem 4.13 of

$$C(A) = \varprojlim_{I} SK_1(\mathfrak{A}, I) \cong \operatorname{Coker}\Big[K_2(A) \longrightarrow \bigoplus_p K_2^c(\hat{A}_p)\Big]$$

for a simple $\mathbb{Q}$-algebra A: a complete computation when A is a summand of any group ring $K[G]$ for finite G, but only up to a factor $\{\pm 1\}$ in the general case. This computation is closely related to the solution of the congruence subgroup problem by Bass, Milnor, and Serre [1]. The groups $C(A)$ have already been seen (Theorems 3.9 and 3.15) to be important for computing $Cl_1(\mathfrak{A})$ for $\mathbb{Z}$-orders $\mathfrak{A} \subseteq A$. In fact, Theorem 4.13 is needed when computing $SK_1(\mathbb{Z}[G])$ in all but the most elementary cases.

It is the second formula for $C(A)$ (involving $K_2(A)$ and $K_2^c(\hat{A}_p)$) which is used as the basis for the results here. This is the approach originally taken by C. Moore in [1]. The idea is to construct isomorphisms between $C(A)$ and $K_2^c(\hat{A}_p)$ and certain groups of roots of unity.

Norm residue symbols are defined in Section 4a, and applied there to prove Moore's theorem (Theorem 4.4) that $K_2^c(F) \cong \mu_F$ (the group of roots of unity in F) for any finite field extension $F \supseteq \hat{\mathbb{Q}}_p$. In Section 4b, this is extended to the case of a simple $\hat{\mathbb{Q}}_p$-algebra A: the computation of $K_2^c(A)$ is not complete but does at least include all simple summands of p-adic group rings.

The final computation of $\oplus_p K_2^c(\hat{A}_p)/\operatorname{Im}(K_2(A))$ is then carried out in Section 4c, based on Moore's reciprocity law (Theorem 4.12), and results of Suslin needed to handle certain division algebras. A few simple applications are then listed: for example, that $Cl_1(\mathfrak{A}) = 1$ whenever $\mathfrak{A}$ is a maximal $\mathbb{Z}$-order, or an arbitrary Λ-order when $\mathbb{Z} \subsetneq \Lambda \subseteq \mathbb{Q}$.

4a. Symbols in K_2 of p-adic fields

By a *symbol* on a field F is meant a bimultiplicative function $\chi: F^* \times F^* \longrightarrow G$, where G is any abelian group, such that $\chi(u,1-u) = 1$ for any $1 \neq u \in F^*$. The importance of symbols when working in K-theory comes from the following theorem of Matsumoto, which says that the Steinberg symbol with values in $K_2(F)$ is the "universal symbol" for F.

Theorem 4.1 (Matsumoto) *For any field F, the Steinberg symbol*

$$\{,\} : F^* \otimes F^* \longrightarrow K_2(F)$$

is surjective, and its kernel is the subgroup generated by all elements $u \otimes (1-u)$ for units $1 \neq u \in F^*$. In particular, any symbol $\chi: F^* \times F^* \longrightarrow G$ factors through a unique homomorphism $\hat{\chi}: K_2(F) \longrightarrow G$.

Proof See, for example, Milnor [2, Theorem 11.1]. □

It is an easy exercise to show that the relations $\{u,-u\} = 1$ and $\{u,v\}\cdot\{v,u\} = 1$ in $K_2(F)$ follow as a formal consequence of the identity $\{u,1-u\} = 1$ (when F is a field, at least).

When constructing symbols, the hardest part is usually to check the relation $\chi(u,1-u) = 1$. The following general result is very often useful when doing this.

Lemma 4.2 *Fix a field F and an abelian group G. Let*

$$\chi_E : E^* \times E^* \longrightarrow G$$

be bimultiplicative maps, defined for each finite extension $E \supseteq F$, and which satisfy the relations

$$\chi_E(u,v) = \chi_F(N_{E/F}(u),v) \qquad (\textit{all } u \in E^*,\ v \in F^*)$$

for all E. Then, for any $n \geq 1$ and any $1 \neq u \in F^*$,

$$\chi_F(u,1-u) \in \left\langle \chi_E(v,1-v)^n : E/F \textit{ finite extension}, \quad 1 \neq v \in E^* \right\rangle.$$

In particular, χ_F *is a symbol if* G *contains no nontrivial infinitely divisible elements.*

<u>Proof</u> Fix $u \in F^* \smallsetminus 1$, and let

$$x^n - u = \prod_{i=1}^{k} f_i(x)^{e_i} \in F[x]$$

be the factorization as a product of powers of distinct irreducible polynomials. In some algebraic closure of F, fix roots u_i of f_i, and set $F_i = F(u_i)$. Then $u_i^n = u$ for all i, and

$$1 - u = \prod_{i=1}^{k} f_i(1)^{e_i} = \prod_{i=1}^{k} N_{F_i/F}(1-u_i)^{e_i}.$$

It follows that

$$\chi_F(u,1-u) = \prod_{i=1}^{k} \chi_F(u, N_{F_i/F}(1-u_i)^{e_i}) = \prod_{i=1}^{k} \chi_{F_i}(u_i^n, (1-u_i)^{e_i})$$

$$= \prod_{i=1}^{k} \chi_{F_i}(u_i, 1-u_i)^{ne_i}. \qquad \square$$

We now consider a more concrete example. Fix a prime p, let F be any finite extension of $\hat{\mathbb{Q}}_p$, and let μ_F be the group of roots of unity in F. For any $\mu \subseteq \mu_F$, the *norm residue symbol*

$$(\,,\,)_\mu : F^* \otimes F^* \longrightarrow \mu$$

is defined by setting $(u,v)_\mu = s(v)(\alpha)/\alpha$; where $F(\alpha)/F$ is some extension such that $\alpha^n = u$ $(n = |\mu|)$, and where

$$s : F^*/N_{F(\alpha)/F}(F(\alpha)^*) \xrightarrow{\cong} \mathrm{Gal}(F(\alpha)/F)$$

is the reciprocity map (see Proposition 1.8(ii)).

Theorem 4.3 *Let* F *be any finite extension of* $\hat{\mathbb{Q}}_p$, *and fix some group* $\mu \subseteq \mu_F$ *of roots of unity in* F. *Then*

(i) $(\,,\,)_\mu : F^* \times F^* \longrightarrow \mu$ *is a symbol.*

(ii) If $E \supseteq F$ *is any finite extension, and if* $(\,,\,)_{\mu,E}$ *denotes the symbol on* E *with values in* μ, *then for any* $u \in F^*$ *and any* $v \in E^*$,

$$(u,v)_{\mu,E} = (u, N_{E/F}(v))_\mu .$$

(iii) For any $\mu_0 \subseteq \mu$, *and any* $u, v \in F^*$,

$$(u,v)_{\mu_0} = \Big((u,v)_\mu\Big)^{[\mu:\mu_0]}.$$

(iv) For any $n \,|\, |\mu|$, *and any* $u \in F^*$ *such that* $u^{1/q} \notin F$ *for all primes* $q|n$, *there exists* $v \in F^*$ *such that* $(u,v)_\mu$ *generates the n-power torsion in* μ.

Proof (ii,iii) Set $n = |\mu|$ and $m = |\mu_0|$, fix $u \in F^*$, and let $E(\alpha)/E$ be an extension such that $\alpha^n = u$. The diagrams

$$\begin{array}{ccc} E^* & \xrightarrow{\ s_E\ } & \mathrm{Gal}(E(\alpha)/E) \\ \big\downarrow{\scriptstyle N_{E/F}} & & \big\downarrow{\scriptstyle \mathrm{res}} \\ F^* & \xrightarrow{\ s\ } & \mathrm{Gal}(F(\alpha)/F) \end{array} \qquad \text{and} \qquad \begin{array}{ccc} F^* & \xrightarrow{\ s\ } & \mathrm{Gal}(F(\alpha)/F) \\ \big\downarrow{\scriptstyle \mathrm{Id}} & & \big\downarrow{\scriptstyle \mathrm{res}} \\ F^* & \xrightarrow{\ s_0\ } & \mathrm{Gal}(F(\alpha^{n/m})/F) \end{array}$$

commute by Serre [1, Section XI.3], where s, s_E, and s_0 are the reciprocity maps, and where res denotes restriction maps. By the definition of $(\,,\,)_\mu$, for any $v \in E^*$,

$$(u, N_{E/F}(v))_\mu = s(N_{E/F}(v))(\alpha)/\alpha = s_E(v)(\alpha)/\alpha = (u,v)_{\mu,E};$$

and the proof of (iii) is similar.

(i) The relation $(u,1-u)_\mu = 1$ is immediate from (ii) and Lemma 4.2.

(iv) It suffices to show this when $n = q$ is prime. Fix $u \in F^*$ such that $u^{1/q} \notin F$, set $E = F(u^{1/q})$, and let μ_q be the group of q-th roots of unity in F. Then the reciprocity map for E/F takes the form

$$F^*/N_{E/F}(E^*) \xrightarrow[\cong]{s} Gal(E/F) \cong \mathbb{Z}/q.$$

So for any $v \in F^* \smallsetminus N_{E/F}(E^*)$, $(u,v)_{\mu_q} = s(v)(u^{1/q})/u^{1/q}$ generates μ_q, and $(u,v)_\mu$ generates the q-power torsion in μ by (iii). □

Now, for any prime p and any finite extension F of $\hat{\mathbb{Q}}_p$, $(,)_F$ will denote the norm residue symbol for F with values in μ_F: the group of roots of unity of F. We can now prove the main theorem in this section, which says that $(,)_F$ is the universal continuous symbol for F.

__Theorem 4.4__ (C. Moore [1]) *Let p be any prime, and let F be any finite extension of $\hat{\mathbb{Q}}_p$. Then the norm residue symbol $(,)_F$ induces an isomorphism*

$$K_2^c(F) \xrightarrow[\cong]{\sigma_F} \mu_F.$$

Furthermore, if $R \subseteq F$ is the ring of integers, then $K_2^c(R) \cong K_2^c(F)_{(p)} \cong (\mu_F)_p$: the group of p-th power roots of unity.

__Proof__ Let $\mathfrak{p} \subseteq R$ be the maximal ideal. The relation $K_2^c(R) \cong K_2^c(F)_{(p)}$ is clear from Theorem 3.5: $K_2^c(R)$ is a pro-p-group by Proposition 3.6, and $p \nmid |K_1(R/\mathfrak{p})|$ by Theorem 1.16.

By Matsumoto's theorem (Theorem 4.1), the norm residue symbol induces a homomorphism $\tilde{\sigma}_F\colon K_2(F) \longrightarrow \mu_F$. If $n = |\mu_F|$, then $K_2(R,p^2nR) =$

$\{R^*, 1+p^2nR\}$ (Theorem 3.3). Also, $\{R^*, 1+p^2nR\} \subseteq \mathrm{Ker}(\tilde{\sigma}_F)$: all elements of $1+p^2nR$ are n-th powers, since the Taylor series for $(1+p^2nx)^{1/n}$ converges for $x \in R$. So $\tilde{\sigma}_F$ factors through $\sigma_F\colon K_2^c(F) \longrightarrow \mu_F$.

Set $n = |\mu_F|$, and let $\zeta \in \mu_F$ be a generator. By Theorem 4.3(iv), there exists $u \in F^*$ such that $\sigma_F(\{\zeta,u\}) = (\zeta,u)_F$ generates μ_F. Since $\{\zeta,u\}$ has order at most n $(\zeta^n = 1)$, this shows that σ_F is split surjective. Also, in the localization sequence

$$1 \longrightarrow K_2^c(R) \longrightarrow K_2^c(F) \longrightarrow K_1(R/p)$$

of Theorem 3.5, $K_2^c(R)$ is a pro-p-group, and $K_1(R/p) \cong (\mu_F)[\frac{1}{p}]$ by Proposition 1.8(i). Thus, σ_F is an isomorphism of non-p-torsion; and we will be done if we can show that

$$K_2^c(F) \otimes \mathbb{Z}/p \cong \begin{cases} \mathbb{Z}/p & \text{if } p \mid |\mu_F| \\ 1 & \text{if } p \nmid |\mu_F|. \end{cases} \tag{1}$$

Fix any $\pi \in p \smallsetminus p^2$. Then

$$F^* = \langle\pi\rangle \times R^* = \langle\pi\rangle \times \mu \times (1+p) \tag{2}$$

by Proposition 1.8(i). Let e be the ramification index of F (i. e., $pR = p^e$), and set $e_0 = e/(p-1)$. For any $n \geq 1$ and any $r \in R$,

$$\begin{aligned} (1+\pi^n r)^p &\equiv 1 + \pi^{pn} r^p && (\mathrm{mod}\ p^{pn+1}) && \text{if } n < e_0 \\ &\equiv 1 + p\pi^n r + \pi^{pn} r^p && (\mathrm{mod}\ p^{n+e+1}) && \text{if } n = e_0 \quad (n+e = pn) \\ &\equiv 1 + p\pi^n r && (\mathrm{mod}\ p^{n+e+1}) && \text{if } n > e_0. \end{aligned}$$

In particular,

$$\begin{aligned} (1 + p^n) &\subseteq (1 + p^{n+1})\cdot(F^*)^p && \text{if } p \mid n \text{ and } n < pe_0, \\ (1 + p^n) &= (1 + p^{n-e})^p && \text{if } n > pe_0. \end{aligned} \tag{3}$$

If $(p-1)|e$ (so $e_0 \in \mathbb{Z}$), then consider the following diagram

$$\begin{array}{ccccccccc} 1 & \longrightarrow & 1+p^{e_0+1} & \longrightarrow & 1+p^{e_0} & \longrightarrow & (1+p^{e_0})/(1+p^{e_0+1}) & \longrightarrow & 1 \\ & & \cong\downarrow\alpha_1 & & \downarrow\alpha_2 & & \downarrow\alpha_3 & & \\ 1 & \longrightarrow & 1+p^{pe_0+1} & \longrightarrow & 1+p^{pe_0} & \longrightarrow & (1+p^{pe_0})/(1+p^{pe_0+1}) & \longrightarrow & 1, \end{array}$$

where $\alpha_i(u) = u^p$ (and $pe_0 = e_0 + e$). The domain and range of α_3 have the same order (both are isomorphic to R/p), and so

$$|\mathrm{Ker}(\alpha_2)| = |\mathrm{Ker}(\alpha_3)| = |\mathrm{Coker}(\alpha_3)| = |\mathrm{Coker}(\alpha_2)|.$$

Also, $|\mathrm{Ker}(\alpha_2)| = p$ or 1, depending on whether $\zeta_p \in F$. So whether or not $(p-1)|e$, one of the following two cases holds: either

(a) $p\nmid|\mu_F|$, and $(1+p^m) \subseteq (F^*)^p$ for any $m \geq pe_0$; or

(b) $p\,|\,|\mu_F|$, $e_0 \in \mathbb{Z}$, and there exists $\delta \in 1+p^{pe_0}$ such that

$$\delta \notin (1+p^{e_0})^p \qquad \text{and} \qquad (1+p^{pe_0}) = \langle\delta\rangle\cdot(1+p^{e_0})^p.$$

In case (b), for any $u \in R^*$, there exists $x \in p^{e_0}$ such that

$$\delta \equiv 1 - ux^p \pmod{(1+p^{pe_0+1}) \subseteq (F^*)^p}$$

(every element of R is a p-th power mod p). Then

$$\{u,\delta\} \equiv \{u, 1-ux^p\} = \{ux^p, 1-ux^p\}\cdot\{x^p, 1-ux^p\}^{-1} \equiv 1 \pmod{K_2^c(F)^p}.$$

It follows that

$$\{R^*,\delta\} \subseteq K_2^c(F)^p. \tag{4}$$

Now fix any $1 \leq n < pe_0$. If $p\nmid n$, then for any $u \in 1+p^n$ we can write $u = (1-\pi^n\omega y)^n$ for some $\omega \in \mu$ and $y \in 1+p$ ($1+p^n$ is a pro-p-group); and so

$$\{\pi,u\} = \{\pi,1-\pi^n\omega y\}^n \equiv \{\pi^n,1-\pi^n\omega\} = \{\pi^n\omega,1-\pi^n\omega\}\cdot\{\omega,1-\pi^n\omega\}^{-1} \equiv 1 \pmod{\{\pi,1+p^{n+1}\}\cdot K_2^c(F)^p}.$$

If, on the other hand, $n = pm < pe_0$, then (3) applies to show that

$$\{\pi,1+p^n\} \subseteq \{\pi,1+p^{n+1}\}\cdot\{\pi,(F^*)^p\} \subseteq \{\pi,1+p^{n+1}\}\cdot K_2^c(F)^p.$$

This shows that if r is such that $pe_0 \lesssim r < pe_0+1$, then

$$\{\pi,1+p\} \subseteq \{\pi,1+p^r\}\cdot K_2^c(F)^p = \begin{cases} K_2^c(F)^p & \text{if } p\nmid|\mu_F| \\ \langle\{\pi,\delta\}\rangle\cdot K_2^c(F)^p & \text{if } p\,|\,|\mu_F| \end{cases} \tag{5}$$

Recall that π was an arbitrary element of $p\smallsetminus p^2$, and that δ was chosen independently of π. Hence, for any $u \in R^*$, π can be replaced by $u\pi$ in (5) to get

$$\{u\pi,1+p\} \subseteq \begin{cases} K_2^c(F)^p & \text{if } p\nmid|\mu_F| \\ \langle\{u\pi,\delta\}\rangle\cdot K_2^c(F)^p = \langle\{\pi,\delta\}\rangle\cdot K_2^c(F)^p & \text{if } p\,|\,|\mu_F|; \end{cases}$$

where the last step follows from (4). By (2), and since $\{\pi,-\pi\} = 1$,

$$K_2^c(F) = \{\mu,F^*\}\cdot\{\pi,1+p\}\cdot\{R^*,1+p\} = \{\mu,F^*\}\cdot\{R^*\pi,1+p\}$$

$$= \begin{cases} K_2^c(F)^p & \text{if } p\nmid|\mu_F| \\ \langle\{\pi,\delta\}\rangle\cdot K_2^c(F)^p & \text{if } p\,|\,|\mu_F|. \end{cases}$$

This proves (1), and hence the theorem. □

One consequence of Theorem 4.4 is the following lemma, which is often useful when checking naturality properties involving $K_2^c(F)$.

<u>Lemma 4.5</u> *For any pair* $E \supseteq F \supseteq \hat{\mathbb{Q}}_p$ *of finite extensions, there is a sequence*

$$F = F_0 \subseteq F_1 \subseteq \cdots \subseteq F_{k-1} \subseteq F_k = E$$

of intermediate fields such that $K_2^c(F_i) = \{F_i^*, F_{i-1}^*\}$ *for each* $1 \le i \le k$.

<u>Proof</u> It will suffice to show that $K_2^c(E) = \{E^*, F^*\}$ whenever either

(a) E/F is a Galois extension of prime degree, or

(b) there is no intermediate field $\bar{E} \subseteq E$ such that $\bar{E}/F$ is Galois and abelian.

In case (b), $F^* = N_{E/F}(E^*)$ by Proposition 1.8(ii). Fix $u \in F^*$ and $v \in E^*$ such that $K_2^c(F) \cong \mu_F$ is generated by $\{u, N_{E/F}(v)\} = trf_F^E(\{u,v\})$ (see Theorem 3.1(v)). Since $K_2^c(E) \cong K_2^c(F)$ ($\mu_E = \mu_F$ since $F(\mu_E)/F$ is an abelian Galois extension), this shows that $\{u,v\}$ generates $K_2^c(E)$.

Now assume that E/F is Galois of prime degree. Fix any prime $q \,|\, |\mu_E|$. We claim that there exists $u \in F^*$ such that $u^{1/q} \notin E$; then by Theorem 4.3(iv) there exists $v \in E^*$ such that $\{u,v\}$ generates $K_2^c(E)_{(q)}$, and so $K_2^c(E)_{(q)} \subseteq \{F^*, E^*\}$.

If $q \ne [E:F]$, then we can take any $u \in F^*$ with valuation 1 (u has valuation 1 or $[E:F]$ in E, and cannot be a q-th power). If $q = [E:F]$, then in particular, $q \,|\, |\mu_F|$ (otherwise, $E = F(\zeta_q)$, and $[E:F] \,|\, q-1$). Fix any element $\pi \in F^*$ with valuation 1, and any $\xi \in F^*$ which generates the group of q-th power roots of unity. Then ξ and π are linearly independent in $F^*/(F^*)^q$; and so at most one of them can be a q-th power in E (see Janusz [1, Theorem 5.8.1] or Cassels & Fröhlich [1, §III.2, Lemma 3]). □

Lemma 4.5 implies in turn the following description of how the isomorphism $K_2^c(F) \cong \mu_F$ behaves under transfer maps. This can be useful when making concrete calculations in $SK_1(\mathbb{Z}[G])$ for finite G.

<u>Theorem 4.6</u> *Fix a prime* p *and finite extensions* $E \supseteq F \supseteq \hat{\mathbb{Q}}_p$, *and let* $\hat{\mu} \subseteq E^*$ *and* $\mu \subseteq F^* \cap \hat{\mu}$ *be groups of roots of unity. Set*

$r = [\hat{\mu} : \mu]$. *Then the following square commutes:*

$$\begin{array}{ccc} K_2^c(E) & \xrightarrow{(\,,\,)_{\hat{\mu}}} & \hat{\mu} \\ \Big\downarrow \mathrm{trf}_F^E & & \Big\downarrow (\zeta \mapsto \zeta^r) \\ K_2^c(F) & \xrightarrow{(\,,\,)_{\mu}} & \mu. \end{array}$$

In particular, trf_F^E *is onto.*

Proof By Lemma 4.5, it suffices to show this when $K_2^c(E) = \{E^*, F^*\}$. But for any $u \in E^*$ and any $v \in F^*$,

$$(\,,\,)_\mu \circ \mathrm{trf}_F^E(\{u,v\}) = (N_{E/F}(u), v)_{\mu,F} \qquad \text{(by Theorem 3.1(v))}$$

$$= (u,v)_{\mu,E} = \left((u,v)_{\hat{\mu},E}\right)^r \qquad \text{(by Theorem 4.3(ii,iii)).}$$

The surjectivity of trf_F^E now follows from Moore's theorem. □

We finish the section by listing some explicit symbol formulas. These are often useful when making computations: for example, in Example 5.1 below, when constructing matrices to represent nonvanishing elements in $SK_1(\mathbb{Z}[C_4 \times C_2 \times C_2])$; and in Chapter 9, when deriving the formula for $Cl_1(\mathbb{Z}[G])$ when G is a p-group for odd p.

Theorem 4.7 (i) *Let* F *be any finite extension of* $\hat{\mathbb{Q}}_p$, *and let* $p \subseteq R \subseteq F$ *be the maximal ideal and ring of integers. Let* $\mu \subseteq \mu_F$ *be any group of roots of unity of order prime to* p, *regard* μ *as a subgroup of* $(R/p)^*$, *and set* $m = [(R/p)^* : \mu]$. *Then, for any* $u, v \in F^*$,

$$(u,v)_\mu = \left((-1)^{p(u)p(v)} \cdot u^{p(v)}/v^{p(u)}\right)^m \in (R/p)^*.$$

Here, $p(-)$ *denotes the p-adic valuation* ($p(u) = r$ *if* $u \in p^r \smallsetminus p^{r+1}$).

(ii) Fix any prime power $p^n \geq 2$, *set* $\zeta = \exp(2\pi i/p^n)$, *and let* $K = \hat{\mathbb{Q}}_p(\zeta)$. *Let* $\mathrm{Tr}\colon K \longrightarrow \hat{\mathbb{Q}}_p$ *and* $N\colon K^* \longrightarrow (\hat{\mathbb{Q}}_p)^*$ *be the trace and norm maps, and set* $\mu = \langle\zeta\rangle$. *Then for any* $u \in 1+(1-z)\hat{\mathbb{Z}}_p[\zeta]$, $(\zeta,u)_\mu = \zeta^R$, *where (modulo* p^n*):*

$$R = \frac{N(u)-1}{p^n} \equiv \begin{cases} p^{-n}\cdot\mathrm{Tr}(\log u) & \text{if } p \text{ is odd} \\ (1+2^{n-1})\cdot 2^{-n}\cdot\mathrm{Tr}(\log u) & \text{if } p = 2 \ (\text{and } n \geq 2). \end{cases}$$

<u>Proof</u> See Serre [1, Proposition XIV.8 and Corollary] for the first formula (the *tame symbol*). The formula for $(\zeta,u)_\mu$ is due to Artin & Hasse [1]. □

Note that Artin & Hasse in [1] also derive a formula for symbols of the form $(1-\zeta,u)_\mu$, in the situation of (ii) above.

<u>4b.</u> Continuous K_2 of simple $\hat{\mathbb{Q}}_p$-algebras

We now want to describe $K_2^c(A)$, whenever A is a simple $\hat{\mathbb{Q}}_p$-algebra with center F, by comparing it with $K_2^c(F)$. This will be based on a homomorphism $\psi_A^c\colon K_2^c(F) \longrightarrow K_2^c(A)$, which is a special case of a construction by Rehmann & Stuhler [1].

<u>Proposition 4.8</u> *If* A *is any simple* $\hat{\mathbb{Q}}_p$*-algebra with center* F, *then there are unique homomorphisms*

$$\psi_A : K_2(F) \longrightarrow K_2(A) \qquad \text{and} \qquad \psi_A^c : K_2^c(F) \longrightarrow K_2^c(A)$$

such that $\psi_A(\{u, \mathrm{nr}_{A/F}(v)\}) = \{u,v\}$ *(and similarly for* ψ_A^c*) for any* $u \in F^*$ *and* $v \in A^*$. *Furthermore, the following naturality relations hold:*

(i) If $E \subseteq A$ *is any self-centralizing subfield (e. g., if* A *is a division algebra and* E *is a maximal subfield), then the following triangle commutes:*

$$\begin{array}{ccc} K_2^c(E) & \xrightarrow{\text{trf}} & K_2^c(F) \\ & {}_{\text{incl}}\searrow & \downarrow{\scriptstyle \psi_A^c} \\ & & K_2^c(A) \end{array}$$

(ii) If $E \supseteq F$ *is any finite extension, then the following squares commute:*

$$\begin{array}{ccccc} K_2^c(F) & \xrightarrow{\text{incl}} & K_2^c(E) & \xrightarrow{\text{trf}} & K_2^c(F) \\ \downarrow{\scriptstyle \psi_A^c} & & \downarrow{\scriptstyle \psi_{E\otimes A}^c} & & \downarrow{\scriptstyle \psi_A^c} \\ K_2^c(A) & \xrightarrow{1\otimes} & K_2^c(E\otimes_F A) & \xrightarrow{\text{trf}} & K_2^c(A). \end{array}$$

(iii) If $E \supseteq F$ *is any splitting field — i. e.,* $E \otimes_F A \cong M_r(E)$ *for some* r *— then the following square commutes:*

$$\begin{array}{ccc} K_2(F) & \xrightarrow{\text{incl}} & K_2(E) \\ \downarrow{\scriptstyle \psi_A} & & \downarrow{\scriptstyle \delta} \\ K_2(A) & \xrightarrow{1\otimes} & K_2(E\otimes_F A) \end{array}$$

where δ *is induced by the identification* $GL_k(M_r(E)) \cong GL_{rk}(E)$.

(iv) For any $r > 1$, *the triangle*

$$\begin{array}{ccc} K_2^c(F) & \xrightarrow{\psi_A} & K_2^c(A) \\ & {}_{\psi_{M_r(A)}}\searrow & \cong\downarrow{\scriptstyle \delta} \\ & & K_2^c(M_r(A)) \end{array}$$

commutes; where δ *is again induced by* $GL_k(M_r(A)) \cong GL_{kr}(A)$.

<u>Proof</u> Let $\tilde{\psi}$ denote the composite

$$\tilde{\psi} = \tilde{\psi}_A : F^* \times F^* \xrightarrow{1 \times (nr_A)^{-1}} F^* \times K_1(A) \xrightarrow{\{,\}} K_2(A).$$

(nr_A is an isomorphism by Theorem 2.3). By Matsumoto's theorem (Theorem 4.1), showing that $\tilde{\psi}_A$ factors through $K_2(F)$ is equivalent to checking that $\tilde{\psi}(u,1-u) = 1$ for all $u \in F^* \setminus \{1\}$.

This will be done using Lemma 4.2. For any finite extension E/F, define $\chi_E: E^* \times E^* \longrightarrow K_2(A)$ by setting

$$\chi_E(u,v) = trf_A^{E\otimes A}(\tilde{\psi}_{E\otimes A}(u,v)).$$

For any $u \in E^*$, any $v \in F^*$, and any $\eta \in A^*$ such that $nr_{A/F}(\eta) = v$,

$$\chi_E(u,v) = trf_A^{E\otimes A}(\{u,1\otimes\eta\}) = \{N_{E/F}(u),\eta\} \qquad \text{(Theorem 3.1(v))}$$

$$= \chi_F(N_{E/F}(u),v).$$

If $n = [A:F]^{1/2}$; then Lemma 4.2 now shows that for any $u \in F^* \setminus \{1\}$, $\chi_F(u,1-u) = \tilde{\psi}(u,1-u)$ is a product of elements

$$\chi_E(v,1-v)^n = trf_A^{E\otimes A}(\{v,nr_{E\otimes A/E}^{-1}(1-v)^n\}) = trf_A^{E\otimes A}(\{v,1-v\}) = 1$$

for $v \in E^* \setminus \{1\}$ ($nr_{E\otimes A/E}(1-v) = (1-v)^n$ by Lemma 2.1(ii)).

This shows that ψ_A is well defined on $K_2(F)$. If $R \subseteq F$ is the ring of integers, and if $\mathfrak{M} \subseteq A$ is a maximal order, then for all $k \geq 1$,

$$\psi_A(K_2(R,p^kR)) = \psi_A(\{1+p^kR,R^*\}) \qquad \text{(Theorem 3.3)}$$

$$= \{1+p^kR,\mathfrak{M}^*\} \subseteq K_2(\mathfrak{M},p^k\mathfrak{M}).$$

So ψ_A factors through $\psi_A^c: K_2^c(F) \longrightarrow K_2^c(A)$.

To prove (i), choose intermediate fields $F = F_0 \subseteq F_1 \subseteq \dots \subseteq F_k = E$ such that $K_2^c(F_i) = \{F_i^*,F_{i-1}^*\}$ for all i (use Lemma 4.5). For each i, let $A_i \subseteq A$ denote the centralizer of F_i in A (so $A_k = E$). Consider the following diagram:

$$\begin{array}{ccccccccc} K_2^c(F_k) & \longrightarrow \cdots \longrightarrow & K_2^c(F_i) & \xrightarrow{trf_i} & K_2^c(F_{i-1}) & \longrightarrow \cdots \longrightarrow & K_2^c(F_0) & & \\ \downarrow \psi_k^c = Id & & \downarrow \psi_i^c & & \downarrow \psi_{i-1}^c & & \downarrow \psi_0^c = \psi_A^c & & (1) \\ K_2^c(A_k) & \longrightarrow \cdots \longrightarrow & K_2^c(A_i) & \xrightarrow{inc_i} & K_2^c(A_{i-1}) & \longrightarrow \cdots \longrightarrow & K_2^c(A_0) & & \end{array}$$

(where $\psi_i^c = \psi_{A_i}^c$). For any $u \in F_{i-1}^*$ and $v \in A_i^*$,

$$\psi_{i-1}^c \circ trf_i(\{u, nr_{A_i/F_i}(v)\}) = \psi_{i-1}^c(\{u, N_{F_i/F_{i-1}} \circ nr_{A_i/F_i}(v)\}) \quad \text{(Thm 3.1(v))}$$

$$= \psi_{i-1}^c(\{u, nr_{A_{i-1}/F_{i-1}}(v)\}) \quad \text{(Lemma 2.1(iv))}$$

$$= \{u,v\} = inc_i \circ \psi_i^c(\{u, nr_{A_i/F_i}(v)\}).$$

Since $K_2^c(F_i) = \{F_i^*, F_{i-1}^*\}$ by assumption (and nr_{A_i/F_i} is onto by Theorem 2.3), this shows that each square in (1) commutes. In particular,

$$incl_E^A = \psi_A^c \circ trf_F^E : K_2^c(E) \longrightarrow K_2^c(A).$$

By Lemma 4.5, it suffices to prove point (ii) when $K_2^c(E) = \{E^*, F^*\}$. And this follows easily upon noting that the reduced norm for A/F is the restriction of the reduced norm for $E \otimes_F A/E$ (by definition).

The last two points are immediate, once one notes that for any A and r, the standard isomorphism $\delta: K_2(A) \xrightarrow{\cong} K_2(M_r(A))$ sends $\{u,v\}$, for commuting $u,v \in A^*$, to the symbol $\{diag(u,\ldots,u), diag(v,1,\ldots,1)\}$ (see Theorem 3.1(iv)). □

The goal now throughout the rest of the section is to show, for as many simple $\hat{\mathbb{Q}}_p$-algebras as possible, that ψ_A^c is an isomorphism. The difficult (and still not completely solved) problem is to prove injectivity. The next proposition will be used to do this when p is odd, and in certain cases when $p = 2$.

<u>Proposition 4.9</u> *Fix a prime* p, *and let* A *be a simple* $\hat{\mathbb{Q}}_p$*-algebra with center* F. *Assume that* p *is odd, or that* $p=2$ *and* $\zeta_{2^n}-\zeta_{2^n}^{-1}\in F$ *for some* $n\geq 2$, *or that* $\mathrm{ind}(A)$ *is odd. Then there is a finite extension* $E\supseteq F$ *which splits* A, *and such that the induction map* $K_2^c(F)\rightarrowtail K_2^c(E)$ *is injective.*

<u>Proof</u> We first show that for any $n\,|\,\mathrm{ind}(A)$, there is a cyclotomic extension $E\supseteq F$ of degree n such that the norm homomorphism $N_{E/F}$ restricts to a surjection of μ_E onto μ_F. It suffices to do this when $n=q$ is prime, and to show surjectivity onto the group $(\mu_F)_q$ of q-power roots of unity.

Write $|(\mu_F)_q|=q^r$; we may assume $r\geq 1$. Set $E=F(\zeta)$, where ζ is a primitive q^{r+1}-st root of unity. Then $[E:F]=q$. If $q\neq p$, then E/F is unramified (Theorem 1.10(i)), and so $N_{E/F}$ induces a surjection of $(\mu_E)_q$ onto $(\mu_F)_q$ by Proposition 1.8(iii). If $q^r>2$, then

$$N_{E/F}(\zeta)=(\zeta)\cdot(\zeta^{1+q^r})\cdot(\zeta^{1+2q^r})\cdots(\zeta^{1+(q-1)q^r})=\pm\zeta^q$$

generates $(\mu_F)_q$. If $p=q^r=2$, then $\zeta_{2^m}-\zeta_{2^m}^{-1}\in F$ by assumption (for some $m\geq 3$); so $\zeta_{2^m}\in E$, and $N_{E/F}(\zeta_{2^m})=(\zeta_{2^m})\cdot(-\zeta_{2^m}^{-1})=-1$.

Now set $n=\mathrm{ind}(A)$. Let $E\supseteq F$ be any extension of degree n such that $N_{E/F}(\mu_E)=\mu_F$. The condition $[E:F]=n$ implies that E is a splitting field for F (see Reiner [1, Corollary 31.10]). To see that $K_2^c(F)$ injects into $K_2^c(E)$, consider the following diagram:

$$\begin{array}{ccccc}
K_2^c(E) & \xrightarrow{\ \mathrm{trf}\ } & K_2^c(F) & \xrightarrow{\ \mathrm{incl}\ } & K_2^c(E) \\
\cong\Big\downarrow\sigma_E & & & & \cong\Big\downarrow\sigma_E \\
\mu_E & & \xrightarrow{\qquad N_{E/F}\qquad} & & \mu_F\subseteq\mu_E.
\end{array}$$

This commutes by the naturality of σ_E: $\mathrm{incl}\circ\mathrm{trf}$ is induced by tensoring with the bimodule $E\otimes_F E$ (see Proposition 1.18); and is hence the norm homomorphism for the action of $\mathrm{Gal}(E/F)$ on $K_2^c(E)$. Also, trf is onto by Theorem 4.6, $K_2^c(F) \cong \mu_F$; and so $|\mathrm{Im}(\mathrm{incl})| = |\mathrm{Im}(N_{E/F})| = |\mu_F| = |K_2^c(F)|$. □

The next lemma will be needed when showing that ψ_A^c is injective for any simple $\hat{\mathbb{Q}}_2$-algebra A of index 2.

Lemma 4.10 *Fix a finite extension F of $\hat{\mathbb{Q}}_p$, and let D be a division algebra with center F for which $[D:F] = 4$. Let $\mathfrak{M} \subseteq D$ be the maximal order. Then for any given n, each element in*

$$\mathrm{Ker}\Big[K_2(D) \longrightarrow K_2^c(D)\Big] = \bigcap_{i=1}^{\infty} \mathrm{Im}\Big[K_2(\mathfrak{M},p^i\mathfrak{M}) \longrightarrow K_2(D)\Big]$$

can be represented as a product of symbols $\{1+p^n x, 1+p^n y\}$ for commuting pairs of elements $x,y \in \mathfrak{M}$.

Proof The proof is modelled on the proof by Rehmann & Stuhler [1, Proposition 4.1] that $K_2(D)$ is generated by Steinberg symbols $\{u,v\}$ for commuting $u,v \in D^*$. However, since we have to work modulo $p^n\mathfrak{M}$, the proof is much more delicate in this setting.

Fix $n \geq 2$, and define

$$X_n = \langle \{u,v\} : u,v \in 1+p^n\mathfrak{M}, \;\; uv = vu \rangle \subseteq K_2(D).$$

We must show that $\mathrm{Ker}[K_2(D) \longrightarrow K_2^c(D)] \subseteq X_n$.

Step 1 Recall the symbols $\{u,v\} \in \mathrm{St}(D)$, defined in Section 3a for any pair of units $u,v \in D^*$, and such that $\phi(\{u,v\}) = [u,v]$ $(\in GL_1(D))$. We are particularly interested here in the case where u and v do not commute. By Theorem 3.1(iv), $\{u,v\} = [x,y]$ for any $x,y \in \mathrm{St}(D)$ such that

$$\phi(x) = \mathrm{diag}(u,u_2,\ldots,u_k) \qquad \text{and} \qquad \phi(y) = \mathrm{diag}(v,v_2,\ldots,v_k),$$

and such that $u_i = 1$ or $v_i = 1$ for each $2 \leq i \leq k$. Using this, the following relations among symbols, for arbitrary $u,v,x,y \in D^*$, follow easily from corresponding relations among commutators:

$$\{v,u\} = \{u,v\}^{-1} \tag{1}$$

$$\{u,v\}\cdot\{vuv^{-1},vyv^{-1}\} = \{u,vy\}; \qquad \{uxu^{-1},uvu^{-1}\}\cdot\{u,v\} = \{ux,v\} \tag{2}$$

$$\{u,v\}\cdot\{v,x\} = \{ux^{-1},xvx^{-1}\} \tag{3}$$

$$\{u,v\}\cdot\{x,y\} = \{u\tilde{y}^{-1},\tilde{y}v\tilde{y}^{-1}\}\cdot\{\tilde{y},x^{-1}v\}. \qquad (\tilde{y} = v^{-1}xyx^{-1}v) \tag{4}$$

In particular, the relations in (2) show that for any $u,v,x,y \in 1+p^n\mathfrak{M}$ such that $[u,y] = [v,x] = 1$,

$$\{u,v\} \equiv \{u,vy\} \equiv \{ux,v\} \pmod{X_n}. \tag{5}$$

<u>Step 2</u> Set $\mathfrak{A} = \hat{\mathbb{Z}}_p + p^{2n}\mathfrak{M} \subseteq D$, a $\hat{\mathbb{Z}}_p$-order in D. By definition of $K_2^c(-)$ (and Lemma 3.2),

$$\mathrm{Ker}\Big[K_2(D) \longrightarrow K_2^c(D)\Big] \subseteq \mathrm{Im}\Big[K_2(\mathfrak{A},p^{2n}\mathfrak{M}) \longrightarrow K_2(D)\Big].$$

Also, since $\mathfrak{A}$ is a local ring, results of Kolster [1] apply to show that each element of $K_2(\mathfrak{A})$ is a product of symbols $\{u,v\}$ for (not necessarily commuting) pairs of units $u,v \in \mathfrak{A}^*$. Since $\mathfrak{A}^*$ is generated by $(\hat{\mathbb{Z}}_p)^*$ and $1+p^{2n}\mathfrak{M}$, relations (2) above show that any $\xi \in \mathrm{Ker}[K_2(D) \longrightarrow K_2^c(D)]$ has the form

$$\xi = \xi_0\cdot\xi_1\cdot\{u_1,v_1\}\cdot\{u_2\cdot v_2\}\cdots\{u_k,v_k\};$$

where $\xi_0 \in K_2(\hat{\mathbb{Z}}_p)$, ξ_1 is a product of symbols $\{(\hat{\mathbb{Z}}_p)^*, 1+p^{2n}\mathfrak{M}\}$, and $u_i,v_i \in 1+p^{2n}\mathfrak{M}$. Furthermore, ξ vanishes under projection to $K_2(\mathfrak{A}/p^{2n}\mathfrak{M}) \cong K_2(\mathbb{Z}/p^{2n}\mathbb{Z})$, so $\xi_0 \in K_2(\hat{\mathbb{Z}}_p,p^{2n}\hat{\mathbb{Z}}_p) = \{(\hat{\mathbb{Z}}_p)^*, 1+p^{2n}\hat{\mathbb{Z}}_p\}$ (Theorem 3.3). But for any $\alpha \in (\hat{\mathbb{Z}}_p)^*$ and any $x \in \mathfrak{M}$,

$$\{\alpha,1+p^{2n}x\} = \left\{\alpha^{(p-1)p^n}, \left(1+p^{2n}x\right)^{1/(p-1)p^n}\right\} \in \left\{1+p^{n+1}\hat{\mathbb{Z}}_p, 1+p^n\mathfrak{M}\right\}.$$

Here, the $(p-1)p^n$-th root is taken using the binomial expansion.

We have now shown that

$$\mathrm{Ker}\left[K_2(D) \longrightarrow K_2^c(D)\right] \subseteq \left\langle \{1+p^nx, 1+p^ny\} : x,y \in \mathfrak{M}\right\rangle \quad (\subseteq \mathrm{St}(D)). \qquad (6)$$

<u>Step 3</u> Let R be the ring of integers in $F = Z(D)$, and let $p = \langle\pi\rangle \subseteq R$ be the maximal ideal. We regard $\mathfrak{M}/p\mathfrak{M}$ as a 4-dimensional R/p-vector space. For any $a \in \mathfrak{M}$, $\bar{a}$ denotes its image in $\mathfrak{M}/p\mathfrak{M}$.

Define functions

$$\mu : (1+p^n\mathfrak{M}) \smallsetminus 1 \longrightarrow (\mathfrak{M}/p\mathfrak{M}) \smallsetminus 0 \quad \text{and} \quad \upsilon : (1+p^n\mathfrak{M}) \smallsetminus 1 \longrightarrow \mathbb{Z}_{\geq 0},$$

by setting, for any $k \geq 0$ and any $a \in \mathfrak{M} \smallsetminus p\mathfrak{M}$:

$$\mu(1+p^n\pi^k a) = \bar{a} \in \mathfrak{M}/p\mathfrak{M} \qquad \text{and} \qquad \upsilon(1+p^n\pi^k a) = k.$$

For any sequence $u_1,\ldots,u_k \in 1+p^n\mathfrak{M}$, set

$$\hat{\mu}(u_1,\ldots,u_k) = \langle \mu(u_1),\ldots,\mu(u_k),1\rangle_{R/p} \subseteq \mathfrak{M}/p\mathfrak{M};$$

i. e., the R/p-vector subspace generated by these elements.

These functions will be used as a "bookkeeping system" when manipulating symbols $\{u,v\}$. The following two points will be needed.

(7) For any $u,v \in 1+p^n\mathfrak{M}$, there exist $u_0,v_0 \in 1+p^n\mathfrak{M}$ such that $\upsilon(v_0) = 0$ (alternatively, $\upsilon(u_0) = 0$), $\mu(u_0) = -\mu(v)$, $\mu(v_0) = \mu(u)$, and $\{u_0,v_0\} \equiv \{u,v\} \pmod{X_n}$. To see this, write $u = 1+p^n\pi^k a$ and $v = 1+p^n\pi^\ell b$, where $a,b \in \mathfrak{M} \smallsetminus p\mathfrak{M}$. Then, by (5),

$$\{u,v\} \equiv \{1+p^n\pi^k a, (1+p^n\pi^\ell b)(1+p^n a)\} = \{1+p^n\pi^k a, 1+p^n(a+\pi^\ell b+p^n\pi^\ell ba)\}$$

$$\equiv \left\{(1+p^n\pi^k a)(1+p^n\pi^k(a+\pi^\ell b+p^n\pi^\ell ba))^{-1}, 1+p^n(a+\pi^\ell b+p^n\pi^\ell ba)\right\} \pmod{X_n}$$

$$= \left\{1-p^n\pi^{k+\ell}b(1+p^n a)(1+p^n\pi^k(a+\pi^\ell b+p^n\pi^\ell ba))^{-1}, 1+p^n(a+\pi^\ell b+p^n\pi^\ell ba)\right\}.$$

This proves the claim if $\ell>0$. If $\ell=0$, then a third such operation finishes the proof.

(8) For any $u,v \in 1+p^n\mathfrak{M}$ such that $[u,v] \neq 1$, there exists $u_0,v_0 \in 1+p^n\mathfrak{M}$ such that $\{u_0,v_0\} \equiv \{u,v\}$ (mod X_n), and such that $\dim_{R/p}(\hat{\mu}(u_0,v_0))=3$. Furthermore, we can do this with $u_0=u$ if $\dim_{R/p}(\hat{\mu}(u))=2$; and similarly for v. To see this, again write $u=1+p^n\pi^k a$ and $v=1+p^n\pi^\ell b$ where $a,b \in \mathfrak{M}\smallsetminus p\mathfrak{M}$. The condition $[u,v]\neq 1$ implies that the elements $1,u,v$ (and hence $1,a,b$) are F-linearly independent in D. If $\dim_{R/p}(\hat{\mu}(u))=2$, so that $a \in \mathfrak{M}\smallsetminus(p\mathfrak{M}\cup R)$, then we can write $b=\alpha+\beta a+\pi^m b_0$, where $\alpha,\beta \in R$ and $\bar{b}_0 \notin \langle \bar{a},1\rangle_{R/p}$. So

$$\{u,v\} \equiv \{1+p^n\pi^k a,(1+p^n\pi^\ell b)(1+p^n\pi^\ell(\alpha+\beta a))^{-1}\} \qquad (\text{mod } X_n)$$

$$= \{u,1-p^n\pi^{\ell+m}b_0(1+p^n\pi^\ell(\alpha+\beta a))^{-1}\} = \{u,v_0\};$$

and $\hat{\mu}(u,v_0) = \langle \bar{a},\bar{b}_0,1\rangle_{R/p}$ is 3-dimensional. The proof when $\hat{\mu}(v)$ is 2-dimensional is similar. If both $\mu(u)$ and $\mu(v)$ lie in R/p then an analogous operation replaces u by u_0 such that $\mu(u_0) \notin R/p$.

<u>Step 4</u> Now consider any 4-tuple of elements $u,v,x,y \in 1+p^n\mathfrak{M}$ such that $\hat{\mu}(u,v)$ and $\hat{\mu}(x,y)$ are 3-dimensional. We will show that there are elements $u_0,v_0,x_0,y_0 \in 1+p^n\mathfrak{M}$ such that $\{u_0,v_0\} \equiv \{u,v\}$ and $\{x_0,y_0\} \equiv \{x,y\}$ (mod X_n); such that $\hat{\mu}(u_0,v_0) = \hat{\mu}(u,v)$ and $\hat{\mu}(x_0,y_0) = \hat{\mu}(x,y)$; and such that either $v_0 = x_0$, or $x_0^{-1}v_0 \notin \hat{\mu}(u_0,v_0) = \hat{\mu}(x_0,y_0)$.

Using (7), we may assume that $\upsilon(v) = 0 = \upsilon(x)$. Write

$$u = 1+p^n\pi^k a, \qquad v = 1+p^n b, \qquad x = 1+p^n c, \qquad y = 1+p^n\pi^\ell d,$$

where $a,b,c,d \in \mathfrak{M}\smallsetminus p\mathfrak{M}$. Since $\dim_{R/p}(\mathfrak{M}/p\mathfrak{M}) = 4$, and since the sets $\{\bar{a},\bar{b},1\}$ and $\{\bar{c},\bar{d},1\}$ are linearly independent, there is a relation

$$\bar{\kappa}\cdot\bar{a}+\bar{\lambda}\cdot\bar{b}+\bar{\alpha}\cdot\bar{c}+\bar{\beta}\cdot\bar{d}+\bar{\gamma} = 0 \qquad (\bar{\alpha},\bar{\beta},\bar{\gamma},\bar{\kappa},\bar{\lambda} \in R/p); \tag{9}$$

where $\bar{\kappa}$ or $\bar{\lambda}$ is nonzero and $\bar{\alpha}$ or $\bar{\beta}$ is nonzero. Using (7) if necessary to make some switches, we can arrange that $\bar{\beta} \neq 0 \neq \bar{\kappa}$.

For any $\alpha,\beta,\gamma \in R$ such that $\beta \in R^*$,

$$\{x,y\} \equiv \{1+p^n c\,,\,(1+p^n\pi^\ell d)(1+p^n\pi^\ell\alpha\beta^{-1}c+p^n\pi^\ell\gamma\beta^{-1})\} \pmod{X_n}$$

$$= \{1+p^n c\,,\,1+p^n\pi^\ell(d+\alpha\beta^{-1}yc+\gamma\beta^{-1}y)\}$$

$$\equiv \left\{(1+p^n c)(1+p^n(\beta d+\alpha yc+\gamma y))\,,\,1+p^n\pi^\ell(d+\alpha\beta^{-1}yc+\gamma\beta^{-1}y)\right\}$$

$$= \left\{1+p^n(c+\beta xd+\alpha xyc+\gamma xy)\,,\,1+p^n\pi^\ell(d+\alpha\beta^{-1}yc+\gamma\beta^{-1}y)\right\} = \{x_o,y_o\}.$$

Note in particular that $\nu(x_o) = 0$, and that

$$\mu(x_o) = \bar{c}+\bar{\beta}\cdot\bar{d}+\bar{\alpha}\cdot\bar{c}+\bar{\gamma}, \qquad \mu(y_o) = \bar{d}+\overline{\alpha\beta}^{-1}\cdot\bar{c}+\bar{\gamma}\cdot\bar{\beta}^{-1}.$$

Thus, $\hat{\mu}(x_o,y_o) = \hat{\mu}(x,y)$. Similarly, for any $\kappa \in R^*$ and $\lambda \in R$, if

$$u_o = 1+p^n\pi^k(a+\lambda\kappa^{-1}ub) \quad\text{and}\quad v_o = 1+p^n(b+\kappa va+\lambda vub),$$

then $\{u_o,v_o\} \equiv \{u,v\} \pmod{X_n}$, and $\hat{\mu}(u_o,v_o) = \hat{\mu}(u,v)$.

Now consider the equation

$$\kappa\cdot va+(1+\lambda\cdot vu)\cdot b = \beta\cdot xd+(1+\alpha\cdot xy)\cdot c+\gamma\cdot xy. \tag{10}$$

By (9), we can find $\alpha,\beta,\gamma,\kappa,\lambda \in R$, where $\beta,\kappa \in R^*$, such that (10) holds (mod $p\mathfrak{M} = \pi\mathfrak{M}$). If (10) holds (mod $\pi^\ell\mathfrak{M}$), for some $\ell > 1$, then we can find a solution (mod $\pi^{\ell+1}\mathfrak{M}$) unless

$$\left(\kappa\cdot va+(1+\lambda\cdot vu)\cdot b\right) - \left(\beta\cdot xd+(1+\alpha\cdot xy)\cdot c+\gamma\cdot xy\right) = \pi^\ell r,$$

and $\bar{r} \notin \langle\bar{a},\bar{b},\bar{c},\bar{d},1\rangle_{R/p} = \hat{\mu}(u,v)+\hat{\mu}(x,y)$. If this ever happens, then

$$\mu(x_o^{-1}v_o) = \bar{r} \notin \hat{\mu}(u,v)+\hat{\mu}(x,y) = \hat{\mu}(u_o,v_o)+\hat{\mu}(x_o,y_o);$$

and $\hat{\mu}(u,v) = \hat{\mu}(x,y)$ since each has codimension one. Otherwise, successive approximations yield $\alpha,\beta,\gamma,\kappa,\lambda$ such that (10) holds, and hence such that $v_0 = x_0$.

<u>Step 5</u> We are now ready to prove the lemma. By Step 2, any element $\xi \in \mathrm{Ker}[K_2(D) \longrightarrow K_2^c(D)]$ is a product of symbols $\{u,v\}$ for $u,v \in 1+p^n\mathfrak{M}$. So to show that $\xi \in X_n$, i. e., that ξ is a product of such symbols for *commuting* pairs $u,v \in 1+p^n\mathfrak{M}$, it will suffice to show that any product $\{u,v\}\cdot\{x,y\}$, for $u,v,x,y \in 1+p^n\mathfrak{M}$, is congruent (mod X_n) to another single symbol of the same form.

Fix such u,v,x,y. We may assume that $[u,v] \neq 1 \neq [x,y]$; and hence (using (8)) that $\hat{\mu}(u,v)$ and $\hat{\mu}(x,y)$ are 3-dimensional. By Step 4, there exist u_0,v_0,x_0,y_0 such that $\{u_0,v_0\}\cdot\{x_0,y_0\} \equiv \{u,v\}\cdot\{x,y\}$; and such that either $v_0 = x_0$ or

$$x_0^{-1}v_0 \notin \hat{\mu}(u_0,v_0) = \hat{\mu}(u,v) = \hat{\mu}(x,y) = \hat{\mu}(x_0,y_0). \tag{11}$$

In the first case, we are done by relation (3). In the second case,

$$\{u_0,v_0\}\cdot\{x_0,y_0\} = \{u_0\tilde{y}_0^{-1},\tilde{y}_0v_0\tilde{y}_0^{-1}\}\cdot\{\tilde{y}_0,x_0^{-1}v_0\} = \{u_1,v_1\}\cdot\{x_1,y_1\} \pmod{X_n}$$

by (4), where $\tilde{y}_0 = v_0^{-1}x_0y_0x_0^{-1}v_0$. Then $\mu(v_1) = \mu(v_0) = \mu(x_0)$, $\mu(x_1) = \mu(y_0)$, and $\mu(y_1) = \mu(x_0^{-1}v_0)$. So by (11),

$$\dim_{R/p}\Big(\hat{\mu}(u_1,v_1) + \hat{\mu}(x_1,y_1)\Big) > \dim_{R/p}(\mu(x_0,y_0)) = 3.$$

Step 4 (and (3)) can now be applied again, this time to $\{u_1,v_1\}\cdot\{x_1,y_1\}$, to show that it is congruent mod X_n to a symbol of the same form. □

The next theorem, due mostly to Bak & Rehmann [1], and Prasad & Raghunathan [1], shows that ψ_A^c is an isomorphism for many simple $\hat{\mathbb{Q}}_p$-algebras. Recall that the index of a central simple F-algebra A is defined by setting $\mathrm{ind}(A) = [D:F]^{1/2}$ if $A \cong M_r(D)$ and D is a division algebra.

<u>Theorem 4.11</u> *Fix a simple* $\hat{\mathbb{Q}}_p$*-algebra* A *with center* F. *Then there is a unique isomorphism*

$$\sigma_A : K_2^c(A) \xrightarrow{\cong} \mu_F/T,$$

where $T \subseteq \{\pm 1\}$, *and such that for any* $a \in F^*$ *and any* $u \in A^*$:

$$\sigma_A(\{a,u\}) = (a,\ nr_{A/F}(u))_F.$$

Furthermore, $T = 1$ *if any of the following three conditions hold:*

(*i*) p *is odd, or* $p = 2$ *and* $\zeta_{2^n} - \zeta_{2^n}^{-1} \in F$ *for some* $n \geq 2$; *or*

(*ii*) $4 \nmid \text{ind}(A)$; *or*

(*iii*) A *is a simple summand of* $K[G]$, *for some finite group* G *and some finite extension* $K \supseteq \hat{\mathbb{Q}}_p$.

Also, for any maximal order $\mathfrak{M} \subseteq A$, $K_2^c(\mathfrak{M}) \cong K_2^c(A)_{(p)} \cong (\mu_F)_p/T$.

Proof The last statement, that $K_2^c(\mathfrak{M}) \cong K_2^c(A)_{(p)}$, is immediate from the localization sequence of Theorem 3.5. By Theorem 4.4, it suffices to show that $\psi_A^c : K_2^c(F) \longrightarrow K_2^c(A)$ is surjective with kernel of order at most 2, and an isomorphism if any of conditions (i) to (iii) hold. The proof will be carried out in three steps: torsion prime to p will be dealt with in Step 1, the surjectivity of ψ_A^c will be shown in Step 2, and $\text{Ker}(\psi_A^c)$ will be handled in Step 3. By Proposition 4.8(iv), it suffices to assume that A is a division algebra.

Let $R \subseteq F$ be the ring of integers, and let $J \subseteq \mathfrak{M}$ and $\mathfrak{p} \subseteq R$ be the maximal ideals. Set $n = [A:F]^{1/2}$. By Theorem 1.9, A is generated by a field $E \supseteq F$ and an element π such that

(a) E/F is unramified, $[E:F] = n$, and $\pi E \pi^{-1} = E$

(b) there is a generator $\eta \in \text{Gal}(E/F)$ such that $\pi x \pi^{-1} = \eta(x)$ for all $x \in E$

(c) $\mathfrak{M} = S[\pi]$ (where $S \subseteq E$ is the ring of integers); $J = J(\mathfrak{M}) = \pi\mathfrak{M}$,

$\pi^n \in R$, and π^n generates the maximal ideal $p \in R$.

__Step 1__ By Theorem 2.11, $|SK_1(\mathfrak{M})| = |K_1(\mathfrak{M}/J)|/|K_1(R/p)|$. A comparison of this with the localization sequence

$$1 \longrightarrow K_2^c(\mathfrak{M}) \longrightarrow K_2^c(A) \longrightarrow K_1(\mathfrak{M}/J) \longrightarrow SK_1(\mathfrak{M}) \longrightarrow 1$$

of Theorem 3.5 shows that

$$|K_2^c(A)[\tfrac{1}{p}]| = |K_2^c(A)/K_2^c(\mathfrak{M})| = |K_1(\mathfrak{M}/J)|/|SK_1(\mathfrak{M})| = |K_1(R/p)| = |K_2^c(F)[\tfrac{1}{p}]|.$$

Since E/F is unramified, the commutative diagram

$$\begin{array}{ccccc}
K_1(R/p) & \xleftarrow{\;\cong\;} & K_2^c(F)/K_2^c(R) \cong K_2^c(F)[\tfrac{1}{p}] & \xrightarrow{\;\psi_A^c\;} & K_2^c(A)[\tfrac{1}{p}] \\
\big\downarrow & & \big\downarrow \text{incl} & & \big\downarrow \text{incl} \\
K_1(S/pS) & \xleftarrow{\;\cong\;} & K_2^c(E)/K_2^c(S) \cong K_2^c(E)[\tfrac{1}{p}] & \xrightarrow{\;\cong\;} & K_2^c(E\otimes_F A)[\tfrac{1}{p}]
\end{array}$$

(from Theorem 3.5 and Proposition 4.8(ii)) shows that ψ_A^c induces an injection of $K_2^c(F)[\frac{1}{p}]$ into $K_2^c(A)/[\frac{1}{p}]$, and hence a bijection.

__Step 2__ We next show that $\psi_A^c(K_2^c(R)) = K_2^c(\mathfrak{M})$, by filtering $K_2^c(\mathfrak{M})$ via the subgroups $K_2^c(\mathfrak{M},J^k)$. By Theorem 1.16, $K_2(\mathfrak{M}/J) = 1$, and so $K_2^c(\mathfrak{M}) = K_2^c(\mathfrak{M},J)$. By Theorem 3.3, for each $k \geq 1$,

$$K_2(\mathfrak{M}/J^k, J^{k-1}/J^k) = \langle\{1+\pi, 1+a\pi^{k-1}\} : a \in S\rangle.$$

If $n \nmid k$, then for any $a,b \in S$, the symbol relations in Theorem 3.1 show that in $K_2(\mathfrak{M}/J^k, J^{k-1}/J^k)$:

$$\{1+\pi,\ 1+ab\pi^{k-1}\} = \{1+\pi, 1+ba\pi^{k-1}\} = \{1+\pi b, 1+a\pi^{k-1}\}$$

$$= \{1+\eta(b)\pi, 1+a\pi^{k-1}\} = \{1+\pi,\ 1+a\pi^{k-1}\cdot\eta(b)\}$$

$$= \{1+\pi,\ 1+a\cdot\eta^k(b)\cdot\pi^{k-1}\}.$$

So $\{1+\pi, 1+a(b-\eta^k(b))\pi^{k-1}\} = 1$, and b can be chosen so that $b-\eta^k(b) \in S^*$ ($\eta^k \neq 1$ since $n \nmid k$). Hence $K_2(\mathfrak{M}/J^k, J^{k-1}/J^k) = 1$ in this case.

If $n|k$, then consider the relative exact sequence

$$\begin{array}{ccccccc} K_2^c(\mathfrak{M},J^k) & \longrightarrow & K_2^c(\mathfrak{M},J^{k-1}) & \longrightarrow & K_2(\mathfrak{M}/J^k,J^{k-1}/J^k) & \xrightarrow{\partial} & K_1(\mathfrak{M},J^k) \\ & & & & & \searrow^{\partial'} & \downarrow \\ & & & & & & K_1(\mathfrak{M}/J^{k+1},J^k/J^{k+1}). \end{array}$$

Since $n|k$, $\pi^k \in R$, and hence $[\mathfrak{M},J^k] \subseteq J^{k+1}$. Then by Theorem 1.15,

$$K_1(\mathfrak{M}/J^{k+1},J^k/J^{k+1}) \cong J^k/J^{k+1} = \pi^k\cdot\mathfrak{M}/J.$$

For any $\{1+\pi, 1+a\pi^{k-1}\} \in K_2(\mathfrak{M}/J^k, J^{k-1}/J^k)$ $(a \in S)$,

$$\partial'(\{1+\pi,1+a\pi^{k-1}\}) = [1+\pi,1+a\pi^{k-1}] = 1 + (\eta(a)-a)\pi^k,$$

and this vanishes if and only if $a \in R + pS$.

This shows that $K_2^c(\mathfrak{M})$ is generated by symbols $\{1+\pi, 1+a\pi^k\}$, for $k \geq 1$ and $a \in R$. In particular, using Proposition 4.8(i),

$$K_2^c(\mathfrak{M}) \subseteq \mathrm{Im}[K_2^c(F(\pi)) \xrightarrow{\text{incl}} K_2^c(A)] = \mathrm{Im}[K_2^c(F(\pi)) \xrightarrow{\text{trf}} K_2^c(F) \xrightarrow{\psi_A^c} K_2^c(A)]$$

(note that $F(\pi)$ is its own centralizer in A). So $K_2^c(\mathfrak{M}) \subseteq \mathrm{Im}(\psi_A^c)$, and ψ_A^c is onto.

<u>Step 3</u> If none of conditions (i) to (iii) are fulfilled, then $p = 2$ and $(\mu_F)_2 = \{\pm 1\}$; and so $|\mathrm{Ker}(\psi_A^c)| \leq 2$. It thus remains to prove the injectivity of ψ_A^c in p-torsion, when (i), (ii), or (iii) holds. By Theorem 1.10(ii), any simple summand of a 2-adic group ring has index at most 2; so it suffices to consider the first two conditions.

(i) Assume first that p is odd, or that $p = 2$ and $\zeta_{2^n} - \zeta_{2^n}^{-1} \in F$ for some $n \geq 2$, or that $\mathrm{ind}(A)$ is odd. Then by Proposition 4.9, there is a splitting field $E \supseteq F$ for A such that the induced homomorphism $K_2^c(F) \rightarrowtail K_2^c(E)$ is injective. By Proposition 4.8(ii,iv), there is a commutative square

$$\begin{array}{ccc} K_2^c(F) & \xrightarrow{\text{incl}} & K_2^c(E) \\ \Big\downarrow \psi_A^c & & \cong \Big\downarrow \delta \\ K_2^c(A) & \longrightarrow & K_2^c(E\otimes_F A); \end{array}$$

and so ψ_A^c is also injective.

(ii) Next assume that $p = 2$, and that $\mathrm{ind}(A) = 2$. There is a trancendental extension $E \supseteq F$ (the "Brauer field") such that E splits A and such that F is algebraically closed in E (see, e. g., Roquette [2, Lemma 3 and Proposition 7]). Then $K_2(F)$ injects into $K_2(E)$ by a theorem of Suslin [1, Theorem 3.6]. The following square commutes by Proposition 4.8(iii):

$$\begin{array}{ccc} K_2(F) & \xrightarrow{\text{incl}} & K_2(E) \\ \Big\downarrow \psi_A & & \cong \Big\downarrow \delta \\ K_2(A) & \longrightarrow & K_2(E\otimes_F A); \end{array}$$

(note that we are using *discrete* K_2 here); and so ψ_A is injective. On the other hand, ψ_A is surjective by a theorem of Rehmann & Stuhler [1, Theorem 4.3]. By Lemma 4.10, any $\eta \in \mathrm{Ker}[K_2(A) \longrightarrow K_2^c(A)]$ is an n-th power for arbitrary $n > 1$. Since ψ_A is an isomorphism, and since $K_2^c(F)$ is finite, this implies that $\psi_A^{-1}(\eta) \in \mathrm{Ker}[K_2(F) \longrightarrow K_2^c(F)]$. It follows that $K_2^c(F) \cong K_2^c(A)$.

Now assume that $\mathrm{ind}(A) = 2m$, where m is odd. Let $E \supseteq F$ be any extension of degree m. Then $E\otimes_F A$ is a central simple E-algebra of index 2: this follows from Reiner [1, Theorems 31.4 and 31.9]. Consider the following commutative diagram of Proposition 4.8(ii):

$$\begin{array}{ccccc} K_2^c(F) & \xrightarrow{\text{incl}} & K_2^c(E) & \xrightarrow{\text{trf}} & K_2^c(F) \\ \downarrow{\scriptstyle \psi_A^c} & & \cong\downarrow{\scriptstyle \psi_{E\otimes A}^c} & & \downarrow{\scriptstyle \psi_A^c} \\ K_2^c(A) & \xrightarrow{1\otimes} & K_2^c(E\otimes_F A) & \xrightarrow{\text{trf}} & K_2^c(A). \end{array}$$

The composite $\text{trf} \circ \text{incl}$ in the top row is multiplication by m (use Proposition 1.18), and so incl is injective in 2-power torsion. Hence ψ_A^c is also injective in 2-power torsion; and this finishes the proof. □

It is still unknown whether $K_2^c(A) \cong K_2^c(F)$ for an arbitrary simple $\hat{\mathbb{Q}}_2$-algebra A with center F. The argument in Step 3(ii) (based on Suslin [1, Theorem 3.6]) shows that $\psi_A: K_2(F) \rightarrowtail K_2(A)$ is always injective (using discrete K_2). But we have been unable to extend any of these results to the case of continuous K_2. This difference between $K_2(-)$ and $K_2^c(-)$ is the source of the (erroneous) claim by Rehmann [2] to show that $K_2^c(A) \cong K_2^c(F) \cong \mu_F$ in general.

4c. The calculation of $C(\mathbb{Q}[G])$

If R is the ring of integers in a number field K, then a *congruence subgroup* of $SL_n(R)$ (for any $n \geq 2$) is a subgroup of the form

$$SL_n(R,I) = \{M \in SL_n(R) : M \equiv 1 \pmod{M_n(I)}\}.$$

for any nonzero ideal $I \subseteq R$. The *congruence subgroup problem* as originally stated was to determine whether every subgroup of $SL_n(R)$ of finite index contains a congruence subgroup.

Any subgroup of $SL_n(R)$ of finite index m contains $E_n(R,mR)$: by definition, $E_n(R,mR)$ is generated by m-th powers in $E_n(R)$. Conversely, if $n \geq 3$, the $E_n(R,I)$ all have finite index in $SL_n(R)$ since

$$SK_1(R,I) \cong SL_n(R,I)/E_n(R,I)$$

(see Bass [2, Corollary V.4.5]) is finite. Furthermore, for any pair $J \subseteq I \subseteq R$ of nonzero ideals, $SL_n(R,I)$ is generated by $SL_n(R,J)$ and $E_n(R,I)$ — any matrix in $SL_n(R/J,I/J)$ can be diagonalized. Thus, the conjecture holds for $n \geq 3$ if and only if the groups $SK_1(R,I)$ vanish for all $I \subseteq R$; if and only if the group $C(K) = \varprojlim SK_1(R,I)$ vanishes.

For the original solution to the problem, where the use of Mennicke symbols helps to maintain more clearly the connection with the groups $SL_n(R,I)$, we refer to the paper of Bass et al [1], as well as to the treatment in Bass [2, Chapter VI]. The presentation here is based on the approach of C. Moore, using the isomorphism

$$C(A) = \varprojlim_{I} SK_1(\mathfrak{A},I) \cong \operatorname{Coker}\Big[K_2(A) \longrightarrow \bigoplus_{p} K_2^c(\hat{A}_p)\Big]$$

shown in Theorem 3.12. The groups $K_2^c(\hat{A}_p)$ have already been described in Theorem 4.11; and so it remains only to understand the image of $K_2(A)$. The key to doing this — for fields at least — is Moore's reciprocity law. Norm residue symbols will again play a central role; and the description of $C(A)$ for a simple $\mathbb{Q}$-algebra A (Theorem 4.13 below) will be in terms of roots of unity in the center of A.

Recall that the *valuations*, or *primes*, in an algebraic number field K consist of the prime ideals in the ring of integers (the "finite primes"), and the real and complex embeddings of K.

<u>Theorem 4.12</u> (Moore's reciprocity law) *Let* K *be an algebraic number field, and let* A *be a simple* $\mathbb{Q}$*-algebra with center* K. *Let* Σ *be the set of noncomplex valuations of* K *(i. e., the set of prime ideals and real embeddings) and set*

$$\Sigma_A = \Sigma \smallsetminus \{v\colon K \hookrightarrow \mathbb{R} : \mathbb{R} \otimes_{vK} A \cong M_r(\mathbb{H}),\ \text{some } r\}.$$

Then the sequence

$$K^* \otimes \mathrm{nr}_{A/K}(A^*) \xrightarrow{\prod(\,,\,)_v} \bigoplus_{v\in\Sigma_A} \mu_{\hat{K}_v} \xrightarrow{\ \rho\ } \mu_K \longrightarrow 1$$

is exact. Here, $\mu_{\hat{K}_v}$ *and* μ_K *denote the groups of roots of unity, and for any* $\zeta = (\zeta_v)_{v\in\Sigma}$ $(\zeta_v \in \mu_{\hat{K}_v})$:

$$\rho(\zeta) = \prod_{v\in\Sigma} (\zeta_v)^{m_v/m}. \qquad (m_v = |\mu_{\hat{K}_v}|, \quad m = |\mu_K|)$$

<u>Proof</u> This was proven (at least in the case $A = K$) by C. Moore [1, Theorem 7.4].

Note that $(K^*, \mathrm{nr}_{A/K}(A^*))_v = 1$ for any $v \in \Sigma \smallsetminus \Sigma_A$: since $v(a) > 0$ $(v: K \hookrightarrow \mathbb{R})$ whenever $a \in \mathrm{nr}_{A/K}(A^*)$. It thus suffices to prove the statement $\rho \circ \prod(\,,\,)_v = 1$ when $A = K$ (so $\Sigma_A = \Sigma$). This is just the usual reciprocity law (see, e. g., Cassels & Fröhlich [1, Exercise 2.9]). For example, when $A = K = \mathbb{Q}$, and p and q are odd primes, the relation $\rho \circ \prod(\,,\,)_v(\{p,q\}) = 1$ reduces to classical quadratic reciprocity using the formula in Theorem 4.7(i).

A second, shorter proof of the relation $\mathrm{Ker}(\rho) \subseteq \mathrm{Im}(\prod(\,,\,)_v)$, in the case $A = K$, is given by Chase & Waterhouse in [1]. By the Hasse-Schilling-Maass norm theorem (Theorem 2.3(ii) above),

$$\mathrm{nr}_{A/K}(A^*) = \{x \in K : v(x) > 0, \quad \text{all} \quad v \in \Sigma \smallsetminus \Sigma_A\};$$

and using this the proof in Chase & Waterhouse [1] of the relation $\mathrm{Ker}(\rho) \subseteq \mathrm{Im}(\prod(\,,\,)_v)$ is easily extended to cover arbitrary A. □

We are now ready to present the description of the groups $C(A)$ — up to a factor $\{\pm 1\}$, at least — in terms of norm residue symbols and roots of unity. This is due to Bass, Milnor, and Serre [1] in the case where A is a field; and (mostly) to Bak & Rehmann [1] and Prasad & Raghunathan [1] in the general case.

<u>Theorem 4.13</u> *Let* A *be a simple* $\mathbb{Q}$-*algebra with center* K, *and let* μ_K *denote the group of roots of unity in* K. *Then*

(*i*) $C(A) = 1$ *if* $\mathbb{R}\otimes_{vK} A \cong M_r(\mathbb{R})$ *for some* $v\colon K \hookrightarrow \mathbb{R}$, *some* r

(*ii*) $C(A) \cong \mu_K$ *if no embedding* $v\colon K \hookrightarrow \mathbb{R}$ *splits* A, *and if for each* 2-*adic valuation* v *of* K, *either* $\zeta_{2^n}-\zeta_{2^n}^{-1} \in \hat{K}_v$ *for some* $n \geq 2$, *or* $4 \nmid \mathrm{ind}(\hat{A}_v)$

(*iii*) $C(A) \cong \mu_K$ *or* $\mu_K/\{\pm 1\}$ *otherwise.*

More precisely, if $C(A) \cong \mu_K/T \neq 1$, *then there is an isomorphism*

$$\sigma_A\colon C(A) \cong \mathrm{Coker}\Big[K_2(A) \longrightarrow \bigoplus_p K_2^c(\hat{A}_p)\Big] \xrightarrow{\cong} \mu_K/T$$

such that for each p, *each prime* $\mathfrak{p}|p$ *of* K, *and each* $\{a,u\} \in K_2^c(\hat{A}_\mathfrak{p})$ (*where* $a \in (\hat{K}_\mathfrak{p})^*$ *and* $u \in (\hat{A}_\mathfrak{p})^*$),

$$\sigma_A(\{a,u\}) = (a, \mathrm{nr}_{A/K}(u))_{\mu_K} \in \mu_K.$$

In particular, each summand $K_2^c(\hat{A}_\mathfrak{p})$ *surjects onto* C(A).

<u>Proof</u> Let Σ be the set of all noncomplex valuations of K (i. e., all finite primes and real embeddings). Fix subsets $\Sigma_0 \subseteq \Sigma_A \subseteq \Sigma$: Σ_0 is the set of finite primes of K (i. e., prime ideals in the ring of integers); and as in Theorem 4.12,

$$\Sigma_A = \Sigma \smallsetminus \{v\colon K \hookrightarrow \mathbb{R} : \mathbb{R}\otimes_{vK} A \cong M_r(\mathbb{H}),\ \text{some } r\}.$$

For each (rational) prime p, $\hat{K}_p \cong \prod_{v|p}\hat{K}_v$ and $\hat{A}_p \cong \prod_{v|p}\hat{A}_v$ (see Theorem 1.7(i)). In other words, we can identify $\bigoplus_p K_2^c(\hat{A}_p)$ with $\bigoplus_{v\in\Sigma_0} K_2^c(\hat{A}_v)$.

Consider the following commutative diagram:

$$\begin{array}{ccccccc} K^* \otimes nr_{A/K}(A^*) & \xrightarrow{\prod(\,,\,)_v} & \bigoplus_{v\in\Sigma_o} \mu_{\hat{K}_v} & \xrightarrow{\rho} & \mu_K & \longrightarrow & 1 \\ \Big\downarrow s & & \cong \Big\uparrow \prod(\sigma_{\hat{A}_v}) & & & & \\ K_2(A) & \xrightarrow{f} & \bigoplus_{v\in\Sigma_o} K_2(\hat{A}_v) & \longrightarrow & C(A) & \longrightarrow & 1. \end{array} \qquad (1)$$

Here, ρ is defined as in Theorem 4.12, and s is induced by the symbol map

$$\{,\} : K^* \otimes K_1(A) \longrightarrow K_2(A)$$

(where $K_1(A) \cong nr_{A/K}(A^*) \subseteq K^*$ by Theorem 2.3). Note that by Theorem 4.3(iii), the composite $\rho \circ \prod(\sigma_{\hat{A}_v})$ satisfies the above formula for σ_A.

If $\Sigma_o \subsetneqq \Sigma_A$, then

$$\mu_K = \{\pm 1\} \quad (K \subseteq \mathbb{R}), \qquad \mu_{\hat{K}_v} = \mu_{\mathbb{R}} = \{\pm 1\} \quad \text{for} \quad v \in \Sigma_A \smallsetminus \Sigma_o,$$

and so $\prod_{v\in\Sigma_o}(\,,\,)_v$ is onto by Theorem 4.12. Hence f is onto, and $C(A) = \mathrm{Coker}(f) = 1$ in this case.

If $\Sigma_A = \Sigma_o$, i. e., if A ramifies at all real places of K, then both rows in (1) are exact. In particular, (1) induces a surjection of μ_K onto $C(A)$. If A is a matrix algebra over a field, then s is onto, and so $C(A) \cong \mu_K$.

Otherwise, we use the K_2 reduced norm homomorphism of Suslin [1, Corollary 5.7] to get control on Coker(s). If F is any field and A is any central simple F-algebra, then there is a unique homomorphism $nr^2_A\colon K_2(A) \longrightarrow K_2(F)$ which satisfies the naturality condition:

(2) if $E \supseteq F$ is any splitting field, then the square

$$\begin{array}{ccc} K_2(A) & \xrightarrow{1\otimes} & K_2(E \otimes_F A) \\ \Big\downarrow nr^2_A & & \cong \Big\downarrow \delta \\ K_2(F) & \xrightarrow{incl} & K_2(E) \end{array}$$

commutes, where δ is induced by the isomorphism $E \otimes_F A \cong M_r(E)$.

Also, there is a splitting field $E \supseteq F$ for A such that F is algebraically closed in E (see, e. g., Roquette [2, Lemma 3 and Proposition 7]). So by a theorem of Suslin [1, Theorem 3.6]:

(3) there exists a splitting field $E \supseteq F$ such that the induced map $K_2(F) \xrightarrow{\text{incl}} K_2(E)$ is injective.

Then (2) and (3) (and Proposition 4.8(ii)) combine to imply

(4) for any $u \in F^*$ and any $a \in A^*$, $nr_A^2(\{u,a\}) = \{u, nr_{A/F}(a)\}$.

Assume now that condition (ii) holds; then $\psi^c_{\hat{A}_v} : K_2^c(\hat{K}_v) \xrightarrow{\cong} K_2^c(\hat{A}_v)$ is an isomorphism for all $v \in \Sigma_0$ by Theorem 4.11. Consider the following diagram:

$$\begin{array}{ccccc}
K_2(A) & \xrightarrow{\quad f \quad} & \bigoplus_{v\in\Sigma_0} K_2^c(\hat{A}_v) & \xrightarrow[\cong]{\prod(\sigma_{\hat{A}_v})} & \bigoplus_{v\in\Sigma_0} \mu_{\hat{K}_v} \\
\Big\downarrow nr_A^2 & (5a) & \Big\downarrow \oplus(\psi^c_{\hat{A}_v})^{-1} & (5b) & \Big\downarrow \text{incl} \\
K_2(K) & \xrightarrow{\quad f_K \quad} & \bigoplus_{v\in\Sigma} K_2^c(\hat{K}_v) & \xrightarrow[\cong]{\prod(\sigma_{\hat{K}_v})} & \bigoplus_{v\in\Sigma} \mu_{\hat{K}_v} .
\end{array} \qquad (5)$$

Here, for $v \in \Sigma \smallsetminus \Sigma_0$, we define for convenience, $K_2^c(\mathbb{R}) = \mu_{\mathbb{R}} = \{\pm 1\}$ (and $\sigma_{\mathbb{R}}(\{u,v\}) = -1$ if and only if $u,v < 0$). Square (5b) commutes by the definition of the $\sigma_{\hat{A}_v}$. If square (5a) also commutes, then a comparison of diagrams (5) and (1) shows that

$$\mathrm{Im}\Big(\prod(\sigma_{\hat{A}_v}) \circ f\Big) \subseteq \mathrm{Im}\Big(\prod(\sigma_{\hat{K}_v}) \circ f_K\Big) \cap \Big(\bigoplus_{v\in\Sigma_0} \mu_{\hat{K}_v}\Big)$$

$$= \mathrm{Ker}\Big[\rho : \bigoplus_{v\in\Sigma_0} K_2(\hat{K}_v) \twoheadrightarrow \mu_K\Big];$$

and it follows that $C(A) = \mathrm{Coker}(f) \cong \mu_K$.

It remains to check that (5a) commutes; we do this separately for each $v \in \Sigma$. This splits into three cases.

<u>Case 1</u> Assume first that $\hat{K}_v \supseteq \hat{\mathbb{Q}}_p$, where either p is odd, or $p = 2$ and $\zeta_{2^n} - \zeta_{2^n}^{-1} \in \hat{K}_v$ for some $n \geq 2$, or $p = 2$ and $\operatorname{ind}(\hat{A}_v)$ is odd. Then, by Proposition 4.9, there is a finite extension $E \supseteq \hat{K}_v$ which splits $\hat{A}_v$, and such that $K_2^c(\hat{K}_v)$ injects into $K_2^c(E)$. In the following diagram:

$$\begin{array}{ccccc} K_2(A) & \longrightarrow & K_2^c(\hat{A}_v) & \longrightarrow & K_2^c(E \otimes_K A) \\ \big\downarrow {\scriptstyle nr_A^2} & (6a) & \cong \big\downarrow {\scriptstyle (\psi^c_{\hat{A}_v})^{-1}} & (6b) & \cong \big\downarrow {\scriptstyle \delta} \\ K_2(K) & \longrightarrow & K_2^c(\hat{K}_v) & \rightarrowtail & K_2^c(E) \end{array} \qquad (6)$$

square (6b) commutes by Proposition 4.8(ii,iv), and (6a+6b) commutes by (2) above. So (6a) also commutes.

<u>Case 2</u> Assume now that $\hat{K}_v \supseteq \hat{\mathbb{Q}}_2$, and that $\operatorname{ind}(\hat{A}_v) = 2m$ for some odd m. Using Proposition 4.9 again, choose an extension $E \supseteq \hat{K}_v$ of degree m such that $K_2^c(\hat{K}_v)$ injects into $K_2^c(E)$. Then $E \otimes_K A$ has index 2 (see Reiner [1, Theorems 31.4 and 31.9]). The same argument as in Case 1 shows that square (5a) commutes for $\hat{A}_v$ if it commutes for $E \otimes_K A$; i. e., that we are reduced to the case where $\operatorname{ind}(\hat{A}_v) = 2$.

If $\operatorname{ind}(\hat{A}_v) = 2$, then consider the following diagram:

$$\begin{array}{ccccc} K_2(A) & \longrightarrow & K_2(\hat{A}_v) & \longrightarrow & K_2^c(\hat{A}_v) \\ \big\downarrow {\scriptstyle nr_A^2} & (7a) & \big\downarrow {\scriptstyle nr^2} & (7b) & \cong \big\downarrow {\scriptstyle (\psi^c_{\hat{A}_v})^{-1}} \\ K_2(K) & \longrightarrow & K_2(\hat{K}_v) & \longrightarrow & K_2^c(\hat{K}_v). \end{array} \qquad (7)$$

By Rehmann & Stuhler [1, Theorem 4.3], $K_2(\hat{A}_v)$ is generated by symbols of the form $\{a,u\}$ for $a \in (\hat{K}_v)^*$ and $u \in (\hat{A}_v)^*$; and so (7b) commutes by (4) (and the definition of ψ). Square (7a) commutes by (2) and (3) above; and so (5a) commutes in this case.

Case 3 Finally, assume that $v \in \Sigma \smallsetminus \Sigma_0$; i. e., that $\hat{A}_v \cong M_r(\mathbb{H})$ for some r. Then $K_2(\hat{A}_v) = \{(\hat{K}_v)^*, (\hat{A}_v)^*\}$ by Rehmann & Stuhler [1, Theorem 4.3]. The composite $K_2(\hat{A}_v) \xrightarrow{nr^2} K_2(\hat{K}_v) \longrightarrow K_2^c(\hat{K}_v) \cong K_2^c(\mathbb{R}) \cong \{\pm 1\}$ is thus trivial (use (4) again); and so (5a) also commutes at such v.

This finishes the proof of the theorem when (i) or (ii) holds. If neither of these hold, then $(\mu_K)_2 = \{\pm 1\}$, so we need only check that $C(A)$ is isomorphic to μ_K in odd torsion. The proof of this is identical to that given above. □

Theorem 4.13 immediately suggests the following conjecture.

Conjecture 4.14 *For any simple $\mathbb{Q}$-algebra A with center K,*

$$C(A) \cong \begin{cases} 1 & \text{if } \mathbb{R} \otimes_{vK} A \cong M_r(\mathbb{R}) \text{ for some } v\colon K \hookrightarrow \mathbb{R} \text{ and some } r \\ \mu_K & \text{otherwise.} \end{cases}$$

By Theorems 1.10(ii) and 4.13, Conjecture 4.14 holds at least whenever A is a simple summand of a group ring $L[G]$, for any finite G and any number field L. If Suslin's reduced norm homomorphism, when applied to a simple $\hat{\mathbb{Q}}_p$-algebra, could be shown always to factor through $K_2^c(-)$, then the proof of Theorem 4.13 above could easily be modified to prove the conjecture.

We now consider some easy consequences of Theorem 4.13. The next two theorems depend, in fact, not on the full description of $C(A) = \mathrm{Coker}[K_2(A) \longrightarrow \oplus_p K_2^c(\hat{A}_p)]$, but only on the property that each factor $K_2^c(\hat{A}_p)$ surjects onto $C(A)$. The first explains why we focus so much attention on $\mathbb{Z}$-orders: if any primes are inverted in a global order $\mathfrak{A}$, then $Cl_1(\mathfrak{A}) = 1$.

Theorem 4.15 *Let $\Lambda \subseteq \mathbb{Q}$ be any subring with $\Lambda \supsetneq \mathbb{Z}$. Then, if $\mathfrak{A}$ is any Λ-order in a semisimple $\mathbb{Q}$-algebra A, $Cl_1(\mathfrak{A}) = 1$. More* *precisely, if* $\mathcal{P}$ *denotes the set of primes not invertible in* Λ, *then*

$$SK_1(\mathfrak{A}) \cong \bigoplus_{p\in\mathcal{P}} SK_1(\hat{\mathfrak{A}}_p).$$

Proof Let $\mathfrak{M} \supseteq \mathfrak{A}$ be a maximal Λ-order in A, and set $n = [\mathfrak{M}:\mathfrak{A}]$. The same construction as was used in the proof of Theorem 3.9 yields an exact sequence

$$\varprojlim_I SK_1(\mathfrak{A},I) \longrightarrow SK_1(\mathfrak{A}) \xrightarrow{\ \ell\ } \bigoplus_{p\in\mathcal{P}} SK_1(\hat{\mathfrak{A}}_p) \longrightarrow 1; \tag{1}$$

where the limit is taken over all ideals $I \subseteq \mathfrak{A}$ of finite index, and where $\varprojlim SK_1(\mathfrak{A},I) \cong \varprojlim Cl_1(\mathfrak{M},I)$ for any maximal Λ-order $\mathfrak{M} \supseteq \mathfrak{A}$. Furthermore, the same construction as that used in Theorem 3.12 (based on Quillen's localization sequence for a maximal order) shows that

$$\varprojlim_I Cl_1(\mathfrak{M},I) \cong \mathrm{Coker}\Big[f_{\mathcal{P}}\colon K_2(A) \longrightarrow \bigoplus_{p\in\mathcal{P}} K_2^c(\hat{A}_p)\Big]. \tag{2}$$

By Theorem 4.13, under the isomorphism

$$C(A) \cong \mathrm{Coker}\Big[K_2(A) \longrightarrow \bigoplus_p K_2^c(\hat{A}_p)\Big],$$

each factor $K_2^c(\hat{A}_p)$ surjects onto $C(A)$. Hence, since $\mathcal{P}$ does not include all primes, the map $f_{\mathcal{P}}$ in (2) is onto. It follows that $\varprojlim SK_1(\mathfrak{A},I) = 1$ in (1), and hence that ℓ is an isomorphism. □

The next theorem allows us, among other things, to extend Kuku's description of $SK_1(\mathfrak{M})$ for a maximal $\hat{\mathbb{Z}}_p$-order $\mathfrak{M}$ (Theorem 2.11) to maximal $\mathbb{Z}$-orders.

Theorem 4.16 (Bass et al [1]; Keating [3]) *If* $\mathfrak{A}$ *is any* $\mathbb{Z}$*-order in a semisimple* $\mathbb{Q}$*-algebra* A, *then* $Cl_1(\mathfrak{A})$ *has* p*-torsion only at primes* p *for which* $\hat{\mathfrak{A}}_p$ *is not a maximal order. In particular:*

(i) $Cl_1(\mathfrak{A}) = 1$, *and* $SK_1(\mathfrak{A}) \cong \bigoplus_p SK_1(\hat{\mathfrak{A}}_p)$, *if* $\mathfrak{A}$ *is maximal;*

(ii) $SK_1(R) = 1$ *if* R *is the ring of integers in any number field;* *and*

(iii) $SK_1(R[G])$ *has p-torsion only for primes* $p \mid |G|$, *if* G *is a finite group and* R *is the ring of integers in any number field.*

Proof Let $\mathfrak{M} \supseteq \mathfrak{A}$ be any maximal order in A; and consider the localization exact sequence

$$\bigoplus_p K_2^c(\hat{\mathfrak{M}}_p) \xrightarrow{\varphi} C(A) \longrightarrow SK_1(\mathfrak{M}) \longrightarrow \bigoplus_p SK_1(\hat{\mathfrak{M}}_p) \longrightarrow 1$$

of Theorem 3.9. For each p, $\varphi|K_2^c(\hat{\mathfrak{M}}_p)$ is the composite

$$K_2^c(\hat{\mathfrak{M}}_p) \xrightarrow{\text{incl}} K_2^c(\hat{A}_p) \subseteq \bigoplus_p K_2^c(\hat{A}_p) \xrightarrow{\text{proj}} \operatorname{Coker}\Big[K_2(A) \longrightarrow \bigoplus_p K_2^c(\hat{A}_p)\Big] \cong C(A);$$

and $K_2^c(\hat{A}_p)$ surjects onto $C(A)$ by Theorem 4.13. Also, $K_2^c(\hat{\mathfrak{M}}_p) = K_2^c(\hat{A}_p)_{(p)}$ by Theorem 4.11, and so $\varphi(K_2^c(\hat{\mathfrak{M}}_p)) = C_p(A)$ (the p-power torsion in $C(A)$). Hence φ is onto, and $Cl_1(\mathfrak{M}) = 1$. Corollary 3.10 now applies to show that $Cl_1(\mathfrak{A}) = \operatorname{Ker}[Cl_1(\mathfrak{A}) \longrightarrow Cl_1(\mathfrak{M})]$ has p-torsion only for primes $p \mid [\mathfrak{M}:\mathfrak{A}]$.

It remains only to prove point (iii). For any group ring $R[G]$ as above, $\hat{R}_p[G]$ is a maximal order for all $p \nmid |G|$ by Theorem 1.4(v). In particular, $p \nmid |Cl_1(R[G])|$ for such p, and $p \nmid |SK_1(\hat{R}_p[G])|$ by Theorem 1.17(i). On the other hand, for each p, $SK_1(\hat{R}_p[G])$ is a p-group by Wall's theorem (Theorem 3.14). So $Cl_1(R[G])$ and $\bigoplus_p SK_1(\hat{R}_p[G])$ both have torsion only at primes dividing $|G|$. □

Point (iii) above will be strengthened in Corollary 5.7 in the next chapter: $SK_1(R[G])$ has p-torsion only for primes p such that the p-Sylow subgroup $S_p(G)$ is noncyclic.

We end the section with a somewhat more technical application of Theorem 4.13; one which often will be useful when working with group rings. For example, it allows us to compare $C(\mathbb{Q}[G])$, for a finite group G, with $C(K[G])$ when K is a splitting field.

Lemma 4.17 *Let K be any number field, and let A be a semisimple K-algebra. Then for any finite extension $L \supseteq K$, the transfer map*

$$\mathrm{trf}_K^L : C(L \otimes_K A) \twoheadrightarrow C(A)$$

is surjective. If L/K is a Galois extension, then the induced epimorphism

$$\mathrm{trf}_0 : H_0(\mathrm{Gal}(L/K); C(L \otimes_K A)) \twoheadrightarrow C(A)$$

is an isomorphism in odd torsion; and is an isomorphism in 2-power torsion if either (i) K has no real embedding and Conjecture 4.14 holds for each simple summand of A, or (ii) A is simple and $2 \mid |C(A)|$.

Proof Note first that trf_K^L is a sum of transfer maps, one for each simple summand of $L \otimes_K A$. When proving the surjectivity of trf_K^L, it thus suffices to consider the case where A is simple and $K = Z(A)$. By the description of $C(A)$ in Theorem 3.12, it then suffices to show that

$$\mathrm{trf} : K_2^c(\hat{L}_q \otimes_{\hat{K}_p} \hat{A}_p) \longrightarrow K_2^c(\hat{A}_p)$$

is surjective for any prime p in K, and any $q|p$ in L. And this follows since the following square commutes by Proposition 4.8(ii):

$$\begin{array}{ccc} K_2^c(\hat{L}_q) & \xrightarrow{\ \mathrm{trf}\ } & K_2^c(\hat{K}_p) \\ \downarrow{\scriptstyle \psi^c} & & \downarrow{\scriptstyle \psi^c} \\ K_2^c(\hat{L}_q \otimes_{\hat{K}_p} \hat{A}_p) & \xrightarrow{\ \mathrm{trf}\ } & K_2^c(\hat{A}_p); \end{array}$$

where the transfer for $\hat{L}_q \supseteq \hat{K}_p$ is onto by Theorem 4.6, and the two maps

ψ^c are onto by Theorem 4.11.

Now assume that L/K is Galois, and set $G = Gal(L/K)$ for short. It will suffice to show that if A is simple, then trf_0 is an isomorphism in odd torsion, and an isomorphism if $2 \,|\, |C(A)|$ (so $C(A) \cong \mu_{Z(A)}$). Write $L \otimes_K Z(A) = \prod_{i=1}^m L_i$, where each L_i is a finite Galois extension of $Z(A)$; then $L \otimes_K A \cong \prod_{i=1}^m L_i \otimes_{Z(A)} A$ and G permutes the factors transitively. Hence, if $G_1 \subsetneq G$ is the subgroup of elements which leave L_1 invariant (so $G_1 \cong Gal(L_1/Z(A))$), then

$$H_0(G;\, C(L \otimes_K A)) \cong H_0(G_1;\, C(L_1 \otimes_{Z(A)} A)).$$

In other words, we are reduced to the case where $K = Z(A)$ (and $G = G_1$, $L = L_1$).

In particular, $L \otimes_K A$ is now a simple algebra with center L. By Theorem 4.13, there are isomorphisms

$$\sigma_{L\otimes A} : C(L \otimes_K A) \xrightarrow{\cong} \mu_L/T_1 \qquad \text{and} \qquad \sigma_A : C(A) \xrightarrow{\cong} \mu_K/T_0$$

where $T_i \subsetneq \{\pm 1\}$. Furthermore, as abstract groups,

$$\mu_K = (\mu_L)^G \cong H_0(G;\, \mu_L);$$

since for any group action on a finite cyclic group, the group of coinvariants is isomorphic to the group of invariants. The domain and range of trf_0 are thus isomorphic (in odd torsion if $C(A) \not\cong \mu_K$). Since trf_0 is onto, it must be an isomorphism. □

Chapter 5 FIRST APPLICATIONS OF THE CONGRUENCE SUBGROUP PROBLEM

The results in this chapter are a rather miscellaneous mixture. Their main common feature is that they all are simple applications of the congruence subgroup problem (Theorem 4.13) to study Cl_1 of group rings; applications which do not require any of the tools of the later chapters.

In Section 5a, the group $G = C_4 \times C_2 \times C_2$ is used to illustrate the computation of $SK_1(\mathbb{Z}[G])$ ($\cong \mathbb{Z}/2$); as well as the procedures for constructing and detecting explicit matrices representing elements of $SK_1(\mathbb{Z}[G])$. Several vanishing results are then proven in Section 5b: for example, that $Cl_1(R[G]) = 1$ whenever G is cyclic and R is the ring of integers in an algebraic number field (Theorem 5.6), that $Cl_1(\mathbb{Z}[G]) = 1$ if G is any dihedral, quaternion, or symmetric group (Example 5.8 and Theorem 5.4), and that $Cl_1(R[G])$ is generated by induction from elementary subgroups of G (Theorem 5.3). These are all based on certain natural epimorphisms $\mathcal{J}_{RG}\colon R_{\mathbb{C}}(G) \twoheadrightarrow Cl_1(R[G])$; epimorphisms which are constructed in Proposition 5.2. In Section 5c, the "standard involution" on Whitehead groups is defined; and is shown, for example, to be the identity on $C(\mathbb{Q}[G])$ and $Cl_1(\mathbb{Z}[G])$ for any finite group G.

5a. Constructing and detecting elements in $SK_1(\mathbb{Z}[G])$: an example

We first focus attention on one particular group abelian G; and sketch the procedures for computing $SK_1(\mathbb{Z}[G])$ $(= Cl_1(\mathbb{Z}[G]))$, for constructing an explicit matrix to represent its nontrivial element, and for detecting whether a given matrix does or does not vanish in $SK_1(\mathbb{Z}[G])$.

Example 5.1 Set $G = C_4 \times C_2 \times C_2$. *Let* $g, h_1, h_2 \in G$ *be generators, where* $|g| = 4$ *and* $|h_1| = |h_2| = 2$. *Then* $SK_1(\mathbb{Z}[G]) \cong \mathbb{Z}/2$, *and is generated by the element*

$$\begin{bmatrix} 1 + 8(1-g^2)(1+h_1)(1+h_2)(1-g) & -(1-g^2)(1+h_1)(1+h_2)(3+g) \\ -13(1-g^2)(1+h_1)(1+h_2)(3-g) & 1 + 8(1-g^2)(1+h_1)(1+h_2)(1+g) \end{bmatrix} \in SK_1(\mathbb{Z}[G]).$$

Proof This will be shown in three steps. The actual computation of $SK_1(\mathbb{Z}[G])$ will be carried out in Step 1. In Step 2, the procedure for constructing an explicit nontrivial element in $SK_1(\mathbb{Z}[G])$ is described. Then, in Step 3, the matrix just constructed is used to illustrate the procedure for lifting it back to $C(\mathbb{Q}[G])$ and determining whether or not it vanishes in $SK_1(\mathbb{Z}[G])$. This is, of course, redundant in the present situation, but since the construction and detection procedures are very different, it seems important to give an example of each.

Step 1 An easy check shows that $\mathbb{Q}[G]$ splits as a product

$$\mathbb{Q}[G] \cong \mathbb{Q}^8 \times \mathbb{Q}(i)^4.$$

By Theorem 4.13, $C(\mathbb{Q}) \cong 1$ and $C(\mathbb{Q}(i)) \cong \langle i \rangle \cong \mathbb{Z}/4$. We must first determine

$$\mathrm{Im}\Big[\varphi_G : K_2^c(\hat{\mathbb{Z}}_2[G]) \longrightarrow C(\mathbb{Q}[G]) \cong (\langle i \rangle)^4\Big].$$

For each $r,s \in \{0,1\}$, let $\chi_{rs}: G \longrightarrow \langle i \rangle$ denote the character: $\chi_{rs}(g) = i$, $\chi_{rs}(h_1) = (-1)^r$, $\chi_{rs}(h_2) = (-1)^s$. Each of these four characters identifies one of the $\mathbb{Q}(i)$-summands of $\mathbb{Q}[G]$ with $\mathbb{Q}(i) \subseteq \mathbb{C}$. Let A_{rs} denote the summand of $\mathbb{Q}[G]$ mapped isomorphically under χ_{rs}, so that

$$\mathbb{Q}[G] = \mathbb{Q}[G/\langle g^2 \rangle] \times A_{00} \times A_{01} \times A_{10} \times A_{11}.$$

Recall that the isomorphism $\sigma: K_2^c(\hat{\mathbb{Q}}_2(i)) \cong C(\mathbb{Q}(i)) \xrightarrow{\cong} \langle i \rangle$ is induced by the norm residue symbol. For the purposes here, the formula

$$\sigma(\{i,u\}) = i^{(N(u)-1)/4} \qquad (N(a+bi) = a^2 + b^2) \tag{1}$$

of Theorem 4.7(ii) will be the most useful. Consider the following table, which (using (1)) lists values for $\varphi_G(x)$ at the $\mathbb{Q}(i)$-summands, for some chosen symbols $x \in K_2^c(\hat{\mathbb{Z}}_2[G])$.

x	$\sigma(\chi_{00}(x))$	$\sigma(\chi_{01}(x))$	$\sigma(\chi_{10}(x))$	$\sigma(\chi_{11}(x))$
$\{g,1+(1+h_1)g\}$	i	i	1	1
$\{g,1+(1+h_2)g\}$	i	1	i	1
$\{g,1+(1+h_1h_2)g\}$	i	1	1	i
$\{-h_2,1+(1+h_1)g\}$	-1	1	1	1

A quick inspection shows that $\mathrm{Im}(\varphi_G)$ has index at most 2.

To see that φ_G is not onto, we define a homomorphism

$$\alpha : C(\mathbb{Q}[G]) \twoheadrightarrow \{\pm1\}; \qquad \alpha(x) = \prod_{r,s} \sigma(\chi_{rs}(x))^2.$$

In other words, α sends each $C(\mathbb{Q}(i)) \cong \langle i\rangle$ onto $\{\pm1\}$; and $\mathrm{Im}(\varphi_G) \supseteq \mathrm{Ker}(\alpha)$ by the above table. To see that $\mathrm{Ker}(\alpha) = \mathrm{Im}(\varphi_G)$, recall first that by Corollary 3.4,

$$K_2^c(\hat{\mathbb{Z}}_2[G]) = \langle \{-1,u\},\ \{g,u\},\ \{h_1,u\},\ \{h_2,u\} \ :\ u \in (\hat{\mathbb{Z}}_2[G])^*\rangle.$$

The symbols $\varphi_G(\{h_i,u\})$ and $\varphi_G(\{-1,u\})$ have order at most 2, are thus squares in $C(\mathbb{Q}[G])$, and lie in $\mathrm{Ker}(\alpha)$. Also, for any $u \in (\hat{\mathbb{Z}}_2[G])^*$,

$$\alpha(\{g,u\}) = \sigma(\{i, \prod_{r,s} \chi_{rs}(u)\})^2 \in \{\pm1\}.$$

Let $\beta\colon \hat{\mathbb{Z}}_2[G] \longrightarrow \hat{\mathbb{Z}}_2[i][C_2 \times C_2]$ be induced by $\beta(g) = i$, and write

$$\beta(u) = a + bh_1 + ch_2 + dh_1h_2 \qquad (a,b,c,d \in \hat{\mathbb{Z}}_2[i]).$$

A direct calculation now gives

$$\prod_{r,s} \chi_{rs}(u) = (a+b+c+d)(a+b-c-d)(a-b+c-d)(a-b-c+d)$$

$$= (a^2+b^2-c^2-d^2)^2 - (2ab-2cd)^2 \equiv 1 \pmod{4\hat{\mathbb{Z}}_2[i]}.$$

Formula (1) now applies to show that $\alpha(\{g,u\}) = 1$.

This finishes the proof that $\mathrm{Im}(\varphi_G) = \mathrm{Ker}(\alpha)$. So by Theorem 3.15,

$$SK_1(\mathbb{Z}[G]) \cong \mathrm{Coker}(\varphi_G) \cong \mathbb{Z}/2.$$

Step 2 Let $\mathfrak{M} \subseteq \mathbb{Q}[G]$ be the maximal order. Then $\mathfrak{M} \supseteq \mathbb{Z}[G]$, and $\mathfrak{M} \cong (\mathbb{Z})^8 \times (\mathbb{Z}[i])^4$. Under this identification, the $\mathfrak{M}$-ideal

$$I = (16\mathbb{Z})^8 \times (8\mathbb{Z}[i])^4 = \langle 16\cdot\tfrac{1+g^2}{2},\ 8\cdot\tfrac{1-g^2}{2}\rangle_{\mathfrak{M}} \subseteq \mathfrak{M}$$

is in fact contained in $\mathbb{Z}[G]$: to see this, just note that $\mathfrak{M}$ is generated (over $\mathbb{Z}[G]$) by the twelve idempotents

$$\tfrac{1}{16}\cdot(1+g^2)(1\pm g)(1\pm h_1)(1\pm h_2) \quad \text{and} \quad \tfrac{1}{8}\cdot(1-g^2)(1\pm h_1)(1\pm h_2).$$

Consider the following homomorphisms:

$$\begin{array}{ccc} SK_1(\mathbb{Z}[G],I) & \xrightarrow{\ \partial\ } & SK_1(\mathbb{Z}[G]) \\ f\downarrow\cong & & \\ SK_1(\mathfrak{M},I) \cong SK_1(\mathbb{Z},16)^8 \times SK_1(\mathbb{Z}[i],8)^4. & & \end{array}$$

Here, f is an isomorphism by Alperin et al [2, Theorem 1.3]. By Step 1, $SK_1(\mathbb{Z}[G])$ is generated by $\partial\circ f^{-1}(x)$, for any $x \in SK_1(\mathfrak{M},I)$ which generates one of the $SK_1(\mathbb{Z}[i],8)$ factors and vanishes in the others. So an explicit generator of $SK_1(\mathbb{Z}[G])$ can be found by first constructing a matrix $A \in GL(\mathbb{Z}[i],8)$ such that $[A]$ generates $SK_1(\mathbb{Z}[i],8)$, and then regarding $GL(\mathbb{Z}[i],8)$ as a summand of $GL(\mathfrak{M},I) = GL(\mathbb{Z}[G],I) \subseteq GL(\mathbb{Z}[G])$.

To find a generator of $SK_1(\mathbb{Z}[i],8)$, consider the epimorphisms

$$K_2^c(\hat{\mathbb{Z}}_2[i]) \twoheadrightarrow K_2(\mathbb{Z}[i]/8) \xrightarrow{\partial} SK_1(\mathbb{Z}[i],8)$$

(recall that $SK_1(\mathbb{Z}[i]) = 1$ by Theorem 4.16(ii)). By (1) above, $K_2^c(\hat{\mathbb{Z}}_2[i])$, and hence also $K_2(\mathbb{Z}[i]/8)$, are generated by the symbol

$$\{i,1+2i\} = \left[\phi^{-1}(\mathrm{diag}(i,1,i^{-1}))\ ,\ \phi^{-1}(\mathrm{diag}(1+2i,(1+2i)^{-1},1))\right];$$

where $\phi\colon St(\mathbb{Z}[i]/8) \twoheadrightarrow E(\mathbb{Z}[i]/8)$ is the canonical surjection. Hence $SK_1(\mathbb{Z}[i],8)$ is generated by the commutator

$$\partial(\{i,1+2i\}) = [\mathrm{diag}(i,1,i^{-1})\ ,\ \mathrm{diag}(M,1,M)] = \left[\begin{pmatrix} i & 0 \\ 0 & 1 \end{pmatrix}, M\right] \in GL(\mathbb{Z}[i],8);$$

when $M \in GL_2(\mathbb{Z}[i])$ is any mod 8 approximation to $\mathrm{diag}(1+2i,(1+2i)^{-1})$. (Recall that $\mathrm{diag}(M,1,M^{-1}) \in E(\mathbb{Z}[i])$ by Theorem 1.13.)

To find M, we could take the usual decomposition

$$\mathrm{diag}(u,u^{-1}) = e_{12}^{u}\cdot e_{21}^{-u^{-1}}\cdot e_{12}^{u}\cdot e_{12}^{-1}\cdot e_{21}^{1}\cdot e_{12}^{-1} \in E_2(R),$$

then replace u by $1+2i$ and u^{-1} by any mod 8 approximation to $(1+2i)^{-1}$, and multiply it out. However, the ring $\mathbb{Z}[i]$ is small enough that it is easier to use trial and error. For example,

$$M = \begin{pmatrix} 1+2i & 8 \\ 8 & 13(1-2i) \end{pmatrix}$$

can be used; and shows that $SK_1(\mathbb{Z}[i],8)$ is generated by the matrix

$$A = \begin{pmatrix} i & 0 \\ 0 & 1 \end{pmatrix}\begin{pmatrix} 1+2i & 8 \\ 8 & 13(1-2i) \end{pmatrix}\begin{pmatrix} -i & 0 \\ 0 & 1 \end{pmatrix}\begin{pmatrix} 13(1-2i) & -8 \\ -8 & 1+2i \end{pmatrix}$$

$$= \begin{pmatrix} 65-64i & -8(3+i) \\ -104(3-i) & 65+64i \end{pmatrix} \in SL_2(\mathbb{Z}[i],8). \qquad (2)$$

Under the inclusion of $\mathbb{Q}(i)$ as the simple summand A_{00} of $\mathbb{Q}[G]$, A now lifts to the generator

$$\begin{bmatrix} 1 + 8(1-g^2)(1+h_1)(1+h_2)(1-g) & -(1-g^2)(1+h_1)(1+h_2)(3+g) \\ -13(1-g^2)(1+h_1)(1+h_2)(3-g) & 1 + 8(1-g^2)(1+h_1)(1+h_2)(1+g) \end{bmatrix} \in SK_1(\mathbb{Z}[G]).$$

<u>Step 3</u> We now reverse the process, and demonstrate how to detect whether or not a given matrix vanishes in $SK_1(\mathbb{Z}[G])$. We have seen in Step 1 that the two epimorphisms

$$Cl_1(\mathbb{Z}[G]) \overset{\partial}{\twoheadleftarrow} C(\mathbb{Q}[G]) \overset{\alpha}{\twoheadrightarrow} \{\pm 1\}$$

have the same kernel. So the idea is to first lift the matrix to an element $X \in C(\mathbb{Q}[G])$ using Proposition 3.13, and then compute $\alpha(X)$ using the formula for the tame symbol in Theorem 4.7(i).

Consider the matrix $A = \begin{pmatrix} a & b \\ c & d \end{pmatrix} \in SL_2(\mathbb{Z}[G])$ constructed in Step 2 above. Write $\mathbb{Q}[G] = A_{oo} \times B$, where $A_{oo} \cong \mathbb{Q}(i)$ is as in Step 1. Set $n = 130$: the product of the primes at which $\chi_{oo}(c) = -104(3-i)$ is not invertible. Write $\mathbb{Z}[\frac{1}{n}][G] = \mathfrak{A}_{oo} \times \mathcal{B}$ where $\mathfrak{A}_{oo} \subseteq A_{oo}$ and $\mathcal{B} \subseteq B$. Then $a \in \mathcal{B}^*$ $(a = 1$ in $\mathcal{B})$, $c \in (\mathfrak{A}_{oo})^* \cong (\mathbb{Z}[\frac{1}{n}][i])^*$, and $a \in (\hat{\mathbb{Z}}_p[G])^*$ for $p|n$. By Proposition 3.13, $[A] = \partial(X)$, where

$$X = \{\chi_{oo}(a), \chi_{oo}(c)\} = \{65-64i,\ -104(3-i)\}$$

$$\in \mathrm{Im}\Big[\bigoplus_{p\nmid n} K_2^c((A_{oo})\hat{}_p) \subseteq \bigoplus_{p} K_2^c((A_{oo})\hat{}_p) \xrightarrow{\mathrm{proj}} C(A_{oo}) \subseteq C(\mathbb{Q}[G])\Big].$$

It remains to show that $\alpha(X) = -1$. We are interested in 2-power torsion only, and at odd primes $p \nmid 130$. Hence, we can use the formula

$$(u,v)_{\mathfrak{p}} = \left((-1)^{\mathfrak{p}(u)\mathfrak{p}(v)} \cdot u^{\mathfrak{p}(v)}/v^{\mathfrak{p}(u)}\right)^{(N(\mathfrak{p})-1)/4} \in \langle i \rangle \subseteq (\mathbb{Z}[i]/\mathfrak{p})^* \qquad (3)$$

(Theorem 4.7(i)) for each prime ideal $\mathfrak{p}|p\nmid n$ in $\mathbb{Z}[i]$: where $N(\mathfrak{p}) = |\mathbb{Z}[i]/\mathfrak{p}|$ and $\mathfrak{p}(-)$ denotes the $\mathfrak{p}$-adic valuation. In particular, $(u,v)_{\mathfrak{p}} = 1$ if u and v are both units mod $\mathfrak{p}$. Since

$$N(65-64i) = 8321 = 53 \cdot 157,$$

we are left with only these two primes to consider. Both split in $\mathbb{Z}[i]$;

and a direct computation shows that $65-64i$ is divisible only by the prime ideals

$$\mathfrak{p}_1 = (7-2i): \qquad i = 30, \quad c = -104(3-i) = -1 \quad \text{in} \quad \mathbb{Z}[i]/\mathfrak{p}_1 \cong \mathbb{F}_{53}$$

$$\mathfrak{p}_2 = (11-6i): \qquad i = 28, \quad c = -104(3-i) = 88 \quad \text{in} \quad \mathbb{Z}[i]/\mathfrak{p}_2 \cong \mathbb{F}_{157}.$$

Formula (3), and the definition of α in Step 1, are now used to compute

$$\alpha(X) = \left[(65-64i,-104(3-i))_{\mathfrak{p}_1} \cdot (65-64i,-104(3-i))_{\mathfrak{p}_2}\right]^2$$

$$= \left(\frac{-1}{53}\right)\cdot\left(\frac{88}{157}\right) = (+1)\cdot(-1) = -1. \qquad \square$$

The above method for computing $\operatorname{Im}(\varphi_G) \subseteq C(\mathbb{Q}[G])$ is not very practical for large groups; and much of the rest of the book (Chapters 9 and 13, in particular) is devoted to finding more effective ways of doing this. Once $\operatorname{Im}(\varphi_G)$ is known, however, the construction and detection procedures in Steps 2 and 3 above can be directly applied to $SK_1(\mathbb{Z}[G])$ for an arbitrary finite abelian group G. Note in particular that any $M \in GL(\mathbb{Z}[G])$ can be reduced using elementary operations to a 2×2 matrix $\begin{pmatrix} a & b \\ c & d \end{pmatrix}$ with $ad-bc=1$ (see Bass [1, Proposition 11.2]). Also, when constructing matrices, it is most convenient to take as ideal $I \subseteq \mathfrak{M}$ (in Step 2) the *conductor*

$$I = \{x \in \mathfrak{M} : x\mathfrak{M} \subseteq \mathbb{Z}[G]\}$$

(i. e., the largest $\mathfrak{M}$-ideal contained in $\mathbb{Z}[G]$). Then Alperin et al [2, Theorem 1.3] applies to show that $SK_1(\mathbb{Z}[G],I) \cong SK_1(\mathfrak{M},I)$. Also, I and $\mathfrak{M}$ both factor as products, one for each simple component of $\mathbb{Q}[G]$, and the rest of the procedures are carried out exactly as above.

When G is nonabelian, the procedure for constructing explicit elements is similar. The main difference is that $SK_1(\mathbb{Z}[G],I)$ need not be isomorphic to $SK_1(\mathfrak{M},I)$; so it might be necessary to replace I by I^2 (see Lemma 2.4); or to use the description in Bass et al [1, Theorem 4.1 and Corollary 4.3] to determine whether $SK_1(\mathfrak{M},I)$ is large enough.

Theorem 4.11 can then be used to represent elements of $K_2(\mathfrak{M}/I)$ by symbols, which are lifted to $SK_1(\mathfrak{M},I)$ exactly as above.

The procedure for detecting a given $[A] \in Cl_1(\mathbb{Z}[G])$ is much harder in general in the nonabelian case. The main problem is that one must know, not only that $[A]$ lies in $Cl_1(\mathbb{Z}[G])$, but also why it lies there. One way of doing this (sometimes) is to first replace A by some $A' \equiv A \pmod{E(\mathbb{Z}[G])}$ such that $A' \equiv 1 \pmod{I^2}$; where $I \subseteq \mathbb{Z}[G]$ again denotes the conductor from the maximal order $\mathfrak{M}$. This is probably the hardest part of the procedure — the descriptions of $SK_1(\hat{\mathbb{Z}}_p[G])$ in Chapters 8 and 12 are unfortunately too indirect to be of much use for this — but $\mathbb{Z}[G]/I^2$ is after all a finite ring. Then A' can be split up and analyzed in the individual components, and in most cases reduced to elements in $SL_2(R,I)$ for some ring of integers R.

For some nonabelian groups, there are alternate ways of detecting elements in $Cl_1(\mathbb{Z}[G])$. Examples of such techniques can be extracted from the proofs of Propositions 16, 17, and 18 in Oliver [1].

5b. $Cl_1(R[G])$ and the complex representation ring

By Theorem 4.13, for any number field K and any finite group G, $C(K[G])$ is isomorphic to a product of roots of unity in certain field components of the center $Z(\mathbb{Q}[G])$. However, it is not always clear from this description how $C(K[G])$ acts with respect to, for example, group homomorphisms and transfer maps. One way of doing this is to use the complex representation ring $R_{\mathbb{C}}(G)$ for "bookkeeping" in $C(K[G])$.

Throughout this section, all number fields will be assumed to be subfields of $\mathbb{C}$. In particular, for any number field K, $R_K(G)$ can be identified as a subgroup of $R_{\mathbb{C}}(G)$; and $R_K(G) = R_{\mathbb{C}}(G)$ whenever K is a splitting field for G (i. e., whenever $K[G]$ is a product of matrix rings over K). For any number field K with no real embeddings, the norm residue symbol defines an isomorphism $\sigma_K\colon C(K) \xrightarrow{\cong} \mu_K \subseteq \mathbb{C}^*$ (Theorem 4.13). We fix a generator $c_K \in C(K)$ by setting $c_K = \sigma_K^{-1}(\exp(2\pi i/n))$ if $|\mu_K| = n$ and K has no real embeddings, and $c_K = 1$ otherwise. By

Theorem 4.6, for any pair L/K of number fields,

$$\mathrm{trf}_K^L(c_L) = c_K \in C(K).$$

If G is any finite group, then we regard $C(K)$ as a subgroup of $C(K[G])$ — the subgroup corresponding to the summand K of $K[G]$ with trivial action — and in this way regard c_K as an element of $C(K[G])$.

For fixed K and G, consider $C(K[G])$ as an $R_K(G)$-module in the usual way. In particular, multiplication by $[V]$, for any finite dimensional $K[G]$-module V, is the endomorphism induced by the functor

$$V\otimes_K : K[G]\text{-}\underline{\underline{\mathrm{mod}}} \longrightarrow K[G]\text{-}\underline{\underline{\mathrm{mod}}}.$$

Alternatively, in terms of Proposition 1.18, multiplication by $[V]$ is induced by the $(K[G],K[G])$-bimodule $V\otimes_K K[G]$, where the bimodule structure is induced by setting $g\cdot(v\otimes h)\cdot k = gv\otimes ghk$ for $g,h,k\in G$ and $v\in V$.

Similarly, if $R \subseteq K$ is the ring of integers, then tensor product over R by $R[G]$-modules makes $Cl_1(R[G])$ into a $G_0(R[G])$-module; where $G_0(R[G])$ is the Grothendieck group on all finitely generated (but not necessarily projective) $R[G]$-modules. There are surjections

$$G_0(R[G]) \xrightarrow{(K\otimes_R)_*} K_0(K[G]) = R_K(G) \qquad \text{and} \qquad C(K[G]) \xrightarrow{\partial} Cl_1(R[G]);$$

and ∂ is $G_0(R[G])$-linear by the description of ∂ in Theorem 3.12. In this way, $Cl_1(R[G])$ can, in fact, be regarded as an $R_K(G)$-module.

Now, if K is a splitting field for G, we define a homomorphism

$$\tilde{\mathcal{J}}_{KG} : R_{\mathbb{C}}(G) = R_K(G) \longrightarrow C(K[G])$$

by setting $\tilde{\mathcal{J}}_{KG}(v) = v\cdot c_K$ for $v \in R_K(G)$. If K and G are arbitrary, and if $L \supseteq K$ is a splitting field for G, we let $\tilde{\mathcal{J}}_{KG}$ be the composite

$$\tilde{\mathcal{J}}_{KG} = \mathrm{trf}\circ\tilde{\mathcal{J}}_{LG} : R_{\mathbb{C}}(G) \longrightarrow C(L[G]) \xrightarrow{\mathrm{trf}} C(K[G]).$$

Finally, if $R \subseteq K$ is the ring of integers, we write

$$\mathcal{J}_{RG} = \partial_{RG} \circ \tilde{\mathcal{J}}_{KG} : R_{\mathbb{C}}(G) \longrightarrow Cl_1(R[G]).$$

A more explicit formula for $\tilde{\mathcal{J}}_{RG}$ will be given in Lemma 5.9(ii).

Proposition 5.2 *Fix a number field* K *and a finite group* G, *and let* $R \subseteq K$ *be the ring of integers. Then* $\tilde{\mathcal{J}}_{KG}$ *and* $\mathcal{J}_{RG}$ *are well defined, independently of the choice of splitting field. Furthermore:*

(i) $\tilde{\mathcal{J}}_{KG}$ *and* $\mathcal{J}_{RG}$ *are both surjective.*

(ii) for any number field $L \supseteq K$ *with ring of integers* $S \subseteq L$, *the following two triangles commute:*

$$\begin{array}{ccc} & & C(L[G]) \\ & \overset{\tilde{\mathcal{J}}}{\nearrow} & \\ R_{\mathbb{C}}(G) & (1) & \big\downarrow \text{trf} \\ & \underset{\tilde{\mathcal{J}}}{\searrow} & \\ & & C(K[G]) \end{array} \qquad \text{and} \qquad \begin{array}{ccc} & & Cl_1(S[G]) \\ & \overset{\mathcal{J}}{\nearrow} & \\ R_{\mathbb{C}}(G) & (1a) & \big\downarrow \text{trf} \\ & \underset{\mathcal{J}}{\searrow} & \\ & & Cl_1(R[G]). \end{array}$$

(iii) For any $H \subseteq G$ *and any group homomorphism* $f: G' \longrightarrow G$, *the following diagrams commute:*

$$\begin{array}{ccc} R_{\mathbb{C}}(G') & \xrightarrow{\tilde{\mathcal{J}}} & C(K[G']) \\ \big\downarrow f_* & (2) & \big\downarrow C(f) \\ R_{\mathbb{C}}(G) & \xrightarrow{\tilde{\mathcal{J}}} & C(K[G]) \\ \big\downarrow \text{Res} & (3) & \big\downarrow \text{trf} \\ R_{\mathbb{C}}(H) & \xrightarrow{\tilde{\mathcal{J}}} & C(K[H]) \end{array} \qquad \text{and} \qquad \begin{array}{ccc} R_{\mathbb{C}}(G') & \xrightarrow{\mathcal{J}} & Cl_1(R[G']) \\ \big\downarrow f_* & (2a) & \big\downarrow Cl_1(f) \\ R_{\mathbb{C}}(G) & \xrightarrow{\mathcal{J}} & Cl_1(R[G]) \\ \big\downarrow \text{Res} & (3a) & \big\downarrow \text{trf} \\ R_{\mathbb{C}}(H) & \xrightarrow{\mathcal{J}} & Cl_1(R[H]). \end{array}$$

(iv) $\tilde{\mathcal{J}}_{KG} : R_{\mathbb{C}}(G) \longrightarrow C(K[G])$ *and* $\mathcal{J}_{RG} : R_{\mathbb{C}}(G) \longrightarrow Cl_1(R[G])$

are both $R_K(G)$*-linear.*

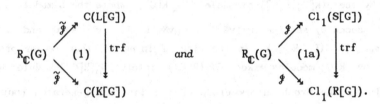

(v) *If* $K \subseteq \mathbb{R}$, *then* $R_{\mathbb{R}}(G) \subseteq \operatorname{Ker}(\tilde{\mathcal{I}}_{KG})$.

<u>Proof</u> By Theorem 3.15, the boundary maps ∂_{RG} are surjective, and are natural with respect to all of the induced maps used above. So it suffices to prove the claims for $\tilde{\mathcal{I}}$.

Note first that for any $L \supseteq K$, the transfer homomorphism

$$\operatorname{trf}_{KG}^{LG} : C(L[G]) \longrightarrow C(K[G])$$

is $R_K(G)$-linear $(R_K(G) \subseteq R_L(G))$. In other words,

$$\operatorname{trf}_{KG}^{LG}(v\cdot x) = v\cdot \operatorname{trf}_{KG}^{LG}(x) \qquad (v \in R_K(G),\ \ x \in C(L[G])). \tag{4}$$

This amounts to showing, for any $K[G]$-module V, the commutativity of the following square

$$\begin{array}{ccc} C(L[G]) & \xrightarrow{[L\otimes_K V]\cdot} & C(L[G]) \\ \Big\downarrow \operatorname{trf}_{KG}^{LG} & & \Big\downarrow \operatorname{trf}_{KG}^{LG} \\ C(K[G]) & \xrightarrow{[V]\cdot} & C(K[G]). \end{array}$$

This in turn follows from Proposition 1.18, since each side is induced by the $(K[G],L[G])$-bimodule $V\otimes_K L[G]$ (where $K[G]$ acts by left multiplication on both factors, and $L[G]$ by right multiplication on the second factor).

If $L \supseteq K$ are both splitting fields for G, then $R_{\mathbb{C}}(G) = R_K(G) = R_L(G)$. So using (4), for any $v \in R_{\mathbb{C}}(G)$,

$$\operatorname{trf}_{KG}^{LG}(\tilde{\mathcal{I}}_{LG}(v)) = \operatorname{trf}_{KG}^{LG}(v\cdot c_L) = v\cdot \operatorname{trf}_{KG}^{LG}(c_L) = v\cdot c_K = \tilde{\mathcal{I}}_{KG}(v).$$

In other words, triangle (1) commutes in this case. But by definition, $\tilde{\mathcal{I}}_{KG} = \operatorname{trf}_{KG}^{\tilde{K}G}\circ\tilde{\mathcal{I}}_{\tilde{K}G}$ for any splitting field $\tilde{K} \supseteq K$, and so (1) commutes for arbitrary $L \supseteq K$. This proves (ii), and also shows that $\tilde{\mathcal{I}}_{KG}$ is well defined, independently of the choice of splitting field.

To prove the surjectivity of $\tilde{\mathcal{J}}_{KG}$, again let $L \supseteq K$ be a splitting field for G. Then for each simple summand A of $L[G]$ with irreducible module V, tensoring by V is a Morita equivalence from L-mod to A-mod; and hence an isomorphism from $C(L)$ to $C(A)$. In other words, the $R_L(G)$-module structure on $C(L[G])$ restricts to an isomorphism

$$R_{\mathbb{C}}(G) \otimes C(L) = R_L(G) \otimes C(L) \subseteq R_L(G) \otimes C(L[G]) \xrightarrow{\;\cdot\;} C(L[G]);$$

and so $\tilde{\mathcal{J}}_{LG}$ is onto. Also, trf_{KG}^{LG} is onto by Lemma 4.17, and so $\tilde{\mathcal{J}}_{KG} = \mathrm{trf}_{KG}^{LG} \circ \tilde{\mathcal{J}}_{LG}$ is onto.

We next check point (iii). Using the commutativity of (1), it suffices to show that squares (2) and (3) commute when K is a splitting field for G', G, and H. This amounts to showing that the following diagram commutes:

$$\begin{array}{ccccc} R_K(G') & \xrightarrow{R_K(f)} & R_K(G) & \xrightarrow{\mathrm{Res}_H^G} & R_K(H) \\ \downarrow{\cdot c_K} & (5a) & \downarrow{\cdot c_K} & (5b) & \downarrow{\cdot c_K} \\ C(K[G']) & \xrightarrow{C(f)} & C(K[G]) & \xrightarrow{\mathrm{trf}} & C(K[H]). \end{array} \qquad (5)$$

For any $K[G']$-representation V, Proposition 1.18 again applies to show that

$$R_K(f)([V])\cdot c_K = [K[G] \otimes_{K[G']} V]\cdot c_K = \big[K[G] \otimes_{K[G']} V \otimes_K\big]_*(c_K)$$

$$= [K[G] \otimes_{K[G']}]_* \circ [V \otimes_K]_*(c_K) = C(f)([V]\cdot c_K).$$

So (5a) commutes, and the proof for (5b) is similar.

To prove (iv), let $L \supseteq K$ be a splitting field for G. Fix $w \in R_K(G)$ and $v \in R_L(G) = R_{\mathbb{C}}(G)$. Then by definition of $\tilde{\mathcal{J}}$ (and (4)),

$$\tilde{\mathcal{J}}_{KG}(w\cdot v) = \mathrm{trf}_{KG}^{LG}(w\cdot v\cdot c_L) = w\cdot \mathrm{trf}_{KG}^{LG}(v\cdot c_L) = w\cdot \tilde{\mathcal{J}}_{LG}(v);$$

and so $\tilde{\mathcal{J}}_{KG}$ is $R_K(G)$-linear.

Finally, assume $K \subseteq \mathbb{R}$ is such that $R_K(G) = R_{\mathbb{R}}(G)$, and let $L \supseteq K$ be a splitting field for K. Then $C(K) = 1$ (Theorem 4.13), so $c_K = 1$; and using (4):

$$\tilde{\mathcal{J}}_{KG}(R_{\mathbb{R}}(G)) = \mathrm{trf}_{KG}^{LG}(R_K(G)\cdot c_L) = R_K(G)\cdot \mathrm{trf}_{KG}^{LG}(c_L) = R_K(G)\cdot c_K = 1.$$

Thus, $R_{\mathbb{R}}(G) \subseteq \mathrm{Ker}(\tilde{\mathcal{J}}_{KG})$ in this case; and the commutativity of (1) allows us to extend this to arbitrary $K \subseteq \mathbb{R}$. □

These strong naturality properties of the $\tilde{\mathcal{J}}_{\mathbb{Z}G}\colon R_{\mathbb{C}}(G) \twoheadrightarrow C(\mathbb{Q}[G])$ make $\tilde{\mathcal{J}}$ into an excellent bookkeeping device for comparing, for example, $C(\mathbb{Q}[G])$ or $Cl_1(\mathbb{Z}[G])$ with $C(\mathbb{Q}[H])$ or $Cl_1(\mathbb{Z}[H])$ for subgroups $H \subseteq G$. The next few results present some applications of this, and more will be seen in later chapters.

For any prime p, a *p-elementary group* is a finite group of the form $C_n \times \pi$, where π is a p-group. According to Brauer's induction theorem (see Serre [2, §10, Theorems 18 and 19], or Theorem 11.2 below), for any finite group G, $R_{\mathbb{C}}(G)$ is generated by elements which are induced up from elementary subgroups of G — i. e., subgroups which are p-elementary for some prime p — and for each prime p, $R_{\mathbb{C}}(G)_{(p)}$ is generated by induction from p-elementary subgroups. So Proposition 5.2 has as an immediate corollary:

<u>Theorem 5.3</u> *Let R be the ring of integers in any number field K. Then for any finite group G, $C(K[G])$ and $Cl_1(R[G])$ are generated by induction from elementary subgroups of G. For each prime p, $C_p(K[G])$ and $Cl_1(R[G])_{(p)}$ are generated by induction from p-elementary subgroups of G.* □

The naturality properties of $\mathcal{J}_{RG}$ in Proposition 5.2 can also be used to show that $SK_1(\mathbb{Z}[G])$ vanishes in many concrete cases. We start with a very simple result, one which also could be shown directly using Theorem 4.13(i).

Theorem 5.4 *Let* G *be a finite group such that* $\mathbb{R}[G]$ *is a product of matrix rings over* $\mathbb{R}$. *Then* $Cl_1(\mathbb{Z}[G]) = 1$.

Proof By hypothesis, $R_{\mathbb{C}}(G) = R_{\mathbb{R}}(G)$. By Proposition 5.2(v),

$$R_{\mathbb{R}}(G) \subseteq \operatorname{Ker}\left[R_{\mathbb{C}}(G) \twoheadrightarrow C(\mathbb{Q}[G])\right];$$

and so $C(\mathbb{Q}[G]) \cong Cl_1(\mathbb{Z}[G]) = 1$. □

Note in particular that Theorem 5.4 applies to elementary abelian 2-groups, to all dihedral groups, and to any symmetric group S_n ($\mathbb{Q}[S_n]$ is a product of matrix algebras over $\mathbb{Q}$: see James & Kerber [1, Theorem 2.1.12]). This result will be sharpened in Theorem 14.1, with the help of later results about $SK_1(\hat{\mathbb{Z}}_p[G])$ and $Wh'(G)$.

We next consider cyclic groups, and show that $SK_1(R[C_n]) = 1$ when R is the ring of integers in any number field. Clearly, to do this, some information about $K_2^c(\hat{R}_p[C_n])$ is needed, and this is provided by the following technical lemma.

Lemma 5.5 *Fix a prime* p *and a finite extension* F *of* $\hat{\mathbb{Q}}_p$, *and let* $R \subseteq F$ *be the ring of integers. Then for any cyclic* p*-group* G, *the transfer homomorphism*

$$\mathrm{trf}_R^{RG} : K_2^c(R[G]) \longrightarrow K_2^c(R)$$

is surjective.

Proof Let $E \supseteq F$ be any finite extension, and let $S \subseteq E$ be the ring of integers. Then $\mathrm{trf}_R^S \circ \mathrm{trf}_S^{SG} = \mathrm{trf}_R^{RG} \circ \mathrm{trf}_{RG}^{SG}$, and trf_R^S is onto by Theorem 4.6. This shows that trf_R^{RG} is onto if trf_S^{SG} is. In particular, if $p^k = |G|$, it will suffice to prove the lemma under the assumption that $\zeta_{p^{k+1}}, p^{1/p^{k-1}} \in F$.

Step 1 Let $\mathfrak{p} \subseteq R$ be the maximal ideal, and let $v: F^* \longrightarrow \mathbb{Z}$ be the valuation. Fix a primitive p-th root of unity ζ. Let $e = v(p)$ be the ramification index; i. e., $pR = \mathfrak{p}^e$. By assumption,

$$e \geq e(\hat{\mathbb{Q}}_p(\zeta_{p^{k+1}})) = p^k(p-1). \tag{1}$$

Choose any x such that

$$v(x) = e/(p^{k-1}(p-1)) - 1 > 0, \tag{2}$$

and set $u = 1 - x^{p^k} \in R^*$ $(x \in \mathfrak{p})$. From (1) and (2) we get inequalities

$$v(x^{2p^k}) \geq v(px^{p^k}) \geq v(px^{2p^{k-1}}) \geq pe/(p-1) = v(p(1-\zeta));$$

and so x^{2p^k}, px^{p^k}, $px^{2p^{k-1}} \in p(1-\zeta)R$. It follows that

$$(1 + x^{p^{k-1}})^p \cdot (1 - x^{p^k}) \equiv 1 + p \cdot x^{p^{k-1}} \equiv 1 - (x_{k-1})^{p^{k-1}} \qquad (\mathrm{mod}\ \ p(1-\zeta)R)$$

where $x_{k-1} = (-p)^{1/p^{k-1}} \cdot x$ and $v(x_{k-1}) = e/p^{k-1} + v(x) = e/p^{k-2}(p-1) - 1$.

Upon repeating this procedure, we get sequences

$$x = x_k, x_{k-1}, \ldots, x_0 \in R \qquad \text{and} \qquad u = u_k, u_{k-1}, \ldots, u_0 \in R^*;$$

where for each $0 \leq i \leq k-1$,

$$x_i = (-p)^{1/p^i} \cdot x_{i+1}, \qquad v(x_i) = e/(p^{i-1}(p-1)) - 1; \quad \text{and} \tag{3}$$

$$u_i = \left(1 + (x_{i+1})^{p^i}\right)^p \cdot u_{i+1} \equiv 1 - (x_i)^{p^i} \qquad (\mathrm{mod}\ \ p(1-\zeta)R).$$

In particular, $u_0 \equiv 1 - x_0$ $(\mathrm{mod}\ \ p(1-\zeta)R)$, and $p \nmid v(x_0) < v(p(1-\zeta))$. If u_0 is a p-th power, then there exists $y \in \mathfrak{p}$ such that

$$u_0 \equiv 1 - x_0 \equiv (1 + y)^p = 1 + py + \ldots + y^p \qquad (\mathrm{mod}\ \ p(1-\zeta)R = \mathfrak{p}^{pe/(p-1)}).$$

Then $v(y) < e/(p-1)$, so $v(py) > p \cdot v(y) = v(y^p)$. It follows that $v(x_0) = v(y^p)$. But $p \nmid v(x_0)$ by (3), and this is a contradiction.

In other words, u_0 is not a p-th power in F. The same argument shows that u_0 is not a p-th power in any unramified extension of F, since the valuations remain unchanged. So $F(u_0^{1/p}) = F(u^{1/p})$ is ramified over F.

<u>Step 2</u> Now set $E = F(u^{1/p})$, and let $S \subseteq E$ be the ring of integers. Since E/F is ramified,

$$R^*/N_{E/F}(S^*) \cong F^*/N_{E/F}(E^*) \cong \mathbb{Z}/p$$

by Proposition 1.8(ii). For any $v \in R^* \setminus N_{E/F}(S^*)$, if $(,)_p$ denotes the norm residue symbol with values in $\langle \zeta_p \rangle$, then $(v,u)_p \neq 1$ by definition. Hence, by Moore's theorem (Theorem 4.4), $\{v,u\}$ generates $K_2^c(R)$. Furthermore, if $g \in G$ is any generator, then

$$\{v,u\} = \{v, 1-x^{p^k}\} = \left\{v, \prod_{i=1}^{p^k} (1-\xi^i x)\right\} \qquad (\xi = \zeta_{p^k})$$

$$= \left\{v, \mathrm{trf}_R^{RG}(1-gx)\right\} = \mathrm{trf}_R^{RG}(\{v,1-gx\}); \qquad \text{(Theorem 3.1(v))}$$

and so $\mathrm{trf}_R^{RG}: K_2^c(R[G]) \longrightarrow K_2^c(R)$ is onto. □

For any cyclic p-group G, and any $R \subseteq K$ such that K splits G, Proposition 5.2(iv) can be used to make $Cl_1(R[G]) = SK_1(R[G])$ into a quotient ring of the local ring $R_{\mathbb{C}}(G)_{(p)}$. So to show that $SK_1(R[G]) = 1$, it suffices to find any element $x \in \mathrm{Ker}\left[\wp_{RG}: R_{\mathbb{C}}(G)_{(p)} \twoheadrightarrow SK_1(R[G])\right]$ which is not contained in the unique maximal ideal of $R_{\mathbb{C}}(G)_{(p)}$. This is the idea behind the proof of the following theorem.

<u>Theorem 5.6</u> *Let* R *be the ring of integers in any number field* K. *Then, for any finite cyclic group* C_n, $SK_1(R[C_n]) = 1$.

<u>Proof</u> Set $G = C_n$, for short. If S is the ring of integers in any finite extension $L \supseteq K$, then the transfer map from $SK_1(S[G])$ to $SK_1(R[G])$ is surjective by Proposition 5.2(i,ii). It therefore suffices

to prove that $SK_1(R[G]) = 1$ when R contains the n-th roots of unity.

Assume first that G is a p-group for some prime p. Set $p^\ell = |(\mu_K)_p|$. Consider the following commutative diagram:

$$\begin{array}{ccccc} K_2^c(\hat{R}_p[G]) & \xrightarrow{\varphi} & C_p(K[G]) & \underset{\cong}{\xleftarrow{\tilde{\mathcal{J}}_{KG}}} & R_{\mathbb{C}}(G)/p^\ell \cong \mathbb{Z}/p^\ell[G^*] \\ \downarrow{\scriptstyle trf_1} & & \downarrow{\scriptstyle trf_2} & & \downarrow{\scriptstyle Res} \\ K_2^c(\hat{R}_p) & \xrightarrow{\varphi_0} & C_p(K) & \underset{\cong}{\xleftarrow{\tilde{\mathcal{J}}_{K}}} & R_{\mathbb{C}}(1)/p^\ell \cong \mathbb{Z}/p^\ell . \end{array} \qquad (G^* = \operatorname{Hom}(G,\mathbb{C}^*))$$

Here, trf_1 is onto by Lemma 5.5, $\operatorname{Coker}(\varphi_0) \cong Cl_1(R) = 1$ by Theorem 4.16(ii), $\tilde{\mathcal{J}}_{KG}$ and $\tilde{\mathcal{J}}_K$ induce isomorphisms on $\mathbb{Z}/p^\ell[G^*]$ and $\mathbb{Z}/p^\ell$ by Theorem 4.13 (K is a splitting field by assumption); and the right-hand square commutes by Proposition 5.2(iii). If we identify $C_p(K[G])$ with the ring $\mathbb{Z}/p^\ell[G^*]$, then $\operatorname{Ker}(trf_2)$ is contained in the unique maximal ideal by Example 1.12. Also, since $\mathcal{J}_{RG}: R_{\mathbb{C}}(G) \twoheadrightarrow SK_1(R[G])$ is $R_{\mathbb{C}}(G)$-linear by Proposition 5.2(iv) $(R_{\mathbb{C}}(G) = R_K(G))$,

$$\operatorname{Ker}\Big[\partial: C_p(K[G]) \twoheadrightarrow SK_1(R[G])_{(p)}\Big] = \operatorname{Im}\Big[\varphi: K_2^c(\hat{R}_p[G]) \longrightarrow C_p(K[G])\Big]$$

is an ideal in $C_p(K[G]) \cong \mathbb{Z}/p^\ell[G^*]$. But $C_p(K[G]) = \operatorname{Im}(\varphi) + \operatorname{Ker}(trf_2)$, since $\varphi_0 \circ trf_1$ is surjective, and so $SK_1(R[G])_{(p)} \cong \operatorname{Coker}(\varphi) = 1$.

Now assume that $n = |G|$ is arbitrary. Fix a prime $p|n$; we will show that $SK_1(R[G])$ is p-torsion free. Write $n = p^k \cdot m$ where $p\nmid m$. Then $K[C_m] \cong K^m$ ($\zeta_n \in K$ by assumption); and by Theorem 1.4(v) there is an inclusion $R[C_m] \subseteq R^m$ of index prime to p. So by Corollary 3.10,

$$SK_1(R[G])_{(p)} = SK_1(R[C_m \times C_{p^k}]) \cong \overset{m}{\bigoplus} SK_1(R[C_{p^k}]) = 1. \qquad \square$$

If G is a finite group, and if $S_p(G)$ (the p-Sylow subgroup) is cyclic, then any p-elementary subgroup of G is cyclic. So Theorem 5.3 can be combined with Theorem 5.6 to give:

__Corollary 5.7__ *Let* R *be the ring of integers in any number field. Then for any finite group* G, *and any prime* p *such that* $S_p(G)$ *is cyclic,* $Cl_1(R[G])_{(p)} = 1$. □

By Theorem 5.6, together with the naturality properties of $\mathcal{J}_{RG}: R_{\mathbb{C}}(G) \twoheadrightarrow Cl_1(R[G])$ in Proposition 5.2, $\mathcal{J}_{RG}$ factors through the *complex Artin cokernel*

$$A_{\mathbb{C}}(G) = \mathrm{Coker}\Big[\bigoplus_{\substack{H\subseteq G\\ \text{cyclic}}} R_{\mathbb{C}}(H) \xrightarrow{\mathrm{Ind}} R_{\mathbb{C}}(G)\Big] = R_{\mathbb{C}}(G)\Big/\Big(\sum_{\substack{H\subseteq G\\ \text{cyclic}}} \mathrm{Ind}_H^G(R_{\mathbb{C}}(H))\Big).$$

In fact, for any fixed G, $\mathcal{J}'_{RG}: A_{\mathbb{C}}(G) \twoheadrightarrow Cl_1(R[G])$ is an isomorphism for R large enough (see Oliver [7, Theorem 5.4]); so that $A_{\mathbb{C}}(G)$ represents the "upper bound" on the size of $Cl_1(R[G])$ as R varies.

The next example deals with some other familiar classes of finite groups, and illustrates the use of the Artin cokernel to get upper bounds on the order of $Cl_1(R[G])$.

__Example 5.8__ *Let* G *be any finite dihedral, quaternionic, or semidihedral group (not necessarily of 2-power order), and let* R *be the ring of integers in any number field* K. *Then* $|Cl_1(R[G])| \le 2$, *and* $Cl_1(R[G]) = 1$ *if either* R *has a real embedding, or if* G^{ab} *is cyclic.*

__Proof__ As remarked above, by Proposition 5.2 and Theorem 5.6, there is a surjection

$$\mathcal{J}'_{RG} : A_{\mathbb{C}}(G) = R_{\mathbb{C}}(G)\Big/\Big(\sum_{\substack{H\subseteq G\\ \text{cyclic}}} \mathrm{Ind}_H^G(R_{\mathbb{C}}(H))\Big) \longrightarrow\!\!\!\!\!\rightarrow Cl_1(R[G]).$$

By assumption, G contains a normal subgroup $H \triangleleft G$ of index 2. All nonabelian irreducible $\mathbb{C}[G]$-representations — i. e., those which do not factor through G^{ab} — are 2-dimensional, and are induced up from representations of H. In particular, this shows that

$$A_{\mathbb{C}}(G) \cong A_{\mathbb{C}}(G^{ab}) \cong \begin{cases} 1 & \text{if } G^{ab} \text{ is cyclic} \\ \mathbb{Z}/2 & \text{if } G^{ab} \cong C_2 \times C_2. \end{cases}$$

Thus, $|Cl_1(R[G])| \le 2$, and $Cl_1(R[G]) = 1$ if G^{ab} is cyclic.

If K has a real embedding and $G^{ab} \cong C_2 \times C_2$, then the composite $R_{\mathbb{R}}(G) \subseteq R_{\mathbb{C}}(G) \twoheadrightarrow A_{\mathbb{C}}(G)$ is onto $(R_{\mathbb{R}}(G^{ab}) = R_{\mathbb{C}}(G^{ab}))$. But by Proposition 5.2(v), $R_{\mathbb{R}}(G) \subseteq \operatorname{Ker}(\mathcal{J}_{RG})$, and so $Cl_1(R[G]) = 1$ in this case. □

In fact, in Lemma 14.3, we will see that $Cl_1(R[G]) \cong \mathbb{Z}/2$ in the above situation, whenever $G^{ab} \cong C_2 \times C_2$ and K has no real embedding.

The results in this section have been obtained mostly without using the precise computation of $C(A)$ in Theorem 4.13. But it is sometimes useful to have $C(\mathbb{Q}[G])$ presented as an explicit quotient group of $R_{\mathbb{C}}(G)$. Recall that for any group G and any $\mathbb{Z}[G]$-module M, M_G denotes the group of G-*coinvariants*; i. e.,

$$M_G = M/\langle gm - m : g \in G, \ m \in M\rangle \cong H_0(G;M).$$

By Brauer's theorem (Theorem 1.5(ii) above), for any finite G, if $n = \exp(G)$, then $\mathbb{Q}\zeta_n$ is a splitting field for G. In particular, $(\mathbb{Z}/n)^* = \operatorname{Gal}(\mathbb{Q}\zeta_n/\mathbb{Q})$ acts on $R_{\mathbb{C}}(G) = R_{\mathbb{Q}\zeta_n}(G)$ in this case.

<u>Lemma 5.9</u> *Fix a finite group* G.

(*i*) *Write* $R_{\mathbb{C}/\mathbb{R}}(G) = R_{\mathbb{C}}(G)/R_{\mathbb{R}}(G)$ *for short; and fix any even* n *such that* $\exp(G)|n$. *Then* $(\mathbb{Z}/n)^* \cong \operatorname{Gal}(\mathbb{Q}\zeta_n/\mathbb{Q})$ *acts on* $R_{\mathbb{C}}(G) = R_{\mathbb{Q}\zeta_n}(G)$ *by Galois conjugation and on* $\mathbb{Z}/n$ *by multiplication, and* $\tilde{\mathcal{J}}_{\mathbb{Q}G}$ *factors through an isomorphism*

$$\mathcal{J}_G^* : \left[R_{\mathbb{C}/\mathbb{R}}(G) \otimes \mathbb{Z}/n\right]_{(\mathbb{Z}/n)^*} \xrightarrow{\cong} C(\mathbb{Q}[G]).$$

(*ii*) *For any irreducible* $\mathbb{C}[G]$*-representation* V, *there is a unique simple summand* A *of* $\mathbb{Q}[G]$ *and a unique embedding* $\alpha\colon Z(A) \hookrightarrow \mathbb{C}$, *such* *that* V *is a* $\mathbb{C}\otimes_{\alpha Z(A)} A$-*module. Then* $\tilde{\mathfrak{z}}_{QG}([V]) \in C(A) \subseteq C(\mathbb{Q}[G])$. *If* $C(A) \neq 1$, *so that* $\sigma_A\colon C(A) \xrightarrow{\cong} \mu_{Z(A)}$ *is an isomorphism, and if* $n = |\mu_{Z(A)}|$, *then*

$$\tilde{\mathfrak{z}}_{QG}([V]) = \sigma_A^{-1}\circ\alpha^{-1}(\exp(2\pi i/n)) \in C(A).$$

<u>Proof</u> Set $K = \mathbb{Q}\zeta_n$ $(\subseteq \mathbb{C})$ for short.

(i) By construction, $\tilde{\mathfrak{z}}_{QG}$ factors through the composite

$$\mathfrak{z}' : R_{\mathbb{C}}(G)\otimes \mathbb{Z}/n \cong R_K(G)\otimes C(K) \xrightarrow{\cong} C(K[G]) \xrightarrow{trf} C(\mathbb{Q}[G]).$$

Any element of $\mathrm{Gal}(K/\mathbb{Q}) = (\mathbb{Z}/n)^*$ acts on $R_{\mathbb{C}}(G)\otimes\mathbb{Z}/n$ via the diagonal action, and acts trivially on $C(\mathbb{Q}[G])$. Also, $R_{\mathbb{R}}(G)\otimes\mathbb{Z}/n \subseteq \mathrm{Ker}(\mathfrak{z}')$ by Proposition 5.2(v); and so $\mathfrak{z}'$ factors through an epimorphism

$$\mathfrak{z}_G^* : \left[R_{\mathbb{C}/\mathbb{R}}(G)\otimes \mathbb{Z}/n\right]_{(\mathbb{Z}/n)^*} \twoheadrightarrow C(\mathbb{Q}[G]).$$

To see that $\mathfrak{z}_G^*$ is an isomorphism, it remains only to compare

$$\left[\frac{K_0(\mathbb{C}\otimes A)}{K_0(\mathbb{R}\otimes A)}\otimes\mathbb{Z}/n\right]_{(\mathbb{Z}/n)^*} \quad\text{and}\quad C(A)$$

(as abstract groups), separately for each simple summand A of $\mathbb{Q}[G]$. If $2\,\big|\,|C(A)|$, then

$$\left[K_0(\mathbb{C}\otimes_{\mathbb{Q}} A)\otimes\mathbb{Z}/n\right]_{(\mathbb{Z}/n)^*} \cong H_0(\mathrm{Gal}(K/\mathbb{Q}); C(K\otimes_{\mathbb{Q}} A)) \cong C(A)$$

by Lemma 4.17. If $2\nmid |C(A)|$, then there is an embedding $Z(A) \hookrightarrow \mathbb{R}$ such that $\mathbb{R}\otimes_{Z(A)} A$ is a matrix algebra over $\mathbb{R}$ (Theorems 4.13(ii) and 1.10(ii)); and so $H_0\Big((\mathbb{Z}/n)^*; \frac{K_0(\mathbb{C}\otimes A)}{K_0(\mathbb{R}\otimes A)}\otimes\mathbb{Z}/n\Big)$ and $C(A)$ both vanish.

(ii) Write $\mathbb{Q}[G] = \prod_{i=1}^k A_i$, where each A_i is simple with center

K_i. For each i, let

$$\alpha_{i1},\ldots,\alpha_{im_i} : K_i \hookrightarrow K = \mathbb{Q}\zeta_n \subseteq \mathbb{C}$$

be the distinct embeddings, and set $B_{ij} = K \otimes_{\alpha_{ij}K_i} A_i$ for each j. Then

$$K[G] \cong \prod_{i=1}^{k} K \otimes_{\mathbb{Q}} A_i \cong \prod_{i=1}^{k}\Bigl(\prod_{j=1}^{m_i} B_{ij}\Bigr);$$

where each B_{ij} is simple by Theorem 1.1(ii). In particular, for each irreducible $K[G]$-representation V, V is the irreducible B_{ij}-module for some unique i and j; and $\tilde{\mathfrak{J}}_{\mathbb{Q}G}([V]) \in C(A_i)$.

If $C(A_i) \neq 1$, then consider the following diagram:

$$\begin{array}{ccccc} c_K \in C(K) & \xrightarrow[\cong]{[V\otimes_K]_*} & C(B_{ij}) & \xrightarrow{\text{trf}} & C(A_i) \\ & \searrow^{\cong\,\sigma_K} & \downarrow{\scriptstyle \cong\,\sigma_{B_{ij}}} & & \downarrow{\scriptstyle \cong\,\sigma_{A_i}} \\ & & \mu_K & \xrightarrow{\tau} & \mu_{K_i}\,. \end{array}$$

Here, $\tau(\zeta) = \alpha_{ij}^{-1}(\zeta^r)$ for any $\zeta \in \mu_K$, where $r = [\mu_K:\mu_{K_i}] = n/|\mu_{K_i}|$. Since B_{ij} is a matrix algebra over K, by assumption, $[V\otimes_K]_*$ is a Morita equivalence. The triangle commutes by definition of σ (and Proposition 4.8(iv)), and the square by Theorem 4.6 and Proposition 4.8(ii). So by definition of $\tilde{\mathfrak{J}}$,

$$\tilde{\mathfrak{J}}_{\mathbb{Q}G}([V]) = \text{trf}\circ[V\otimes_K]_*(c_K) = \sigma_{A_i}^{-1}(\tau\circ\sigma_K(c_K)) = \sigma_{A_i}^{-1}\circ\alpha_{ij}^{-1}(\exp(2\pi i/|\mu_{K_i}|)). \qquad \square$$

Both parts of Lemma 5.9 can easily be generalized to apply to $C(K[G])$, for any finite G and any number field K.

5c. The "standard involution" on Wh(G)

As was seen in the introduction, the algorithms for describing the odd torsion in $SK_1(\mathbb{Z}[G])$, for a finite group G, are much more complete than those (discovered so far) for describing the 2-power torsion. The reason in almost every case is that the standard involution on $SK_1(\mathbb{Z}[G])$, $Cl_1(\mathbb{Z}[G])$, $C(\mathbb{Q}[G])$, etc., can be used to split the terms in the localization sequence of Theorem 3.15 — in odd torsion — into their ± 1 eigenspaces. It is of particular importance that $C(\mathbb{Q}[G])$ and $Cl_1(\mathbb{Z}[G])$, as well as $Wh'(G)$, all are fixed by the involution.

Throughout this chapter, an *involution* on a ring R will mean an antiautomorphism $r \mapsto \bar{r}$ of order 2 (i. e., $\bar{\bar{r}} = r$ and $(rs)^- = \bar{s}\cdot\bar{r}$). If R is any ring with involution $r \mapsto \bar{r}$, there are induced involutions on $GL(R)$ and $St(R)$ defined by setting

$$\bar{M} = (\bar{r}_{ji}) \quad \text{if} \quad M = (r_{ij}) \in GL_n(R)$$

(i. e., conjugate transpose), and

$$\overline{x_{ij}^{r}} = x_{ji}^{\bar{r}} \in St(R) \quad \text{if} \quad i,j \geq 1, \quad i \neq j, \quad r \in R.$$

Then $\phi\colon St(R) \longrightarrow GL(R)$ commutes with the involutions, and so this defines induced involutions on $K_1(R) = \mathrm{Coker}(\phi)$ and $K_2(R) = \mathrm{Ker}(\phi)$.

We first note the following general properties of involutions:

Lemma 5.10 *(i) If R is a ring with involution $r \mapsto \bar{r}$, and if $a,b \in R^*$ are commuting units, then*

$$\overline{\{a,b\}} = \{\bar{b},\bar{a}\} = \{\bar{a},\bar{b}\}^{-1} \in K_2(R).$$

(ii) If A is a central simple F-algebra with involution, then the reduced norm map $nr_{A/F}\colon A^* \longrightarrow F^*$ commutes with the involutions on A and on $F = Z(A)$.

<u>Proof</u> (i) Fix $x,y \in St(R)$ such that

$$\phi(x) = \mathrm{diag}(a,a^{-1},1) \quad \text{and} \quad \phi(y) = \mathrm{diag}(b,1,b^{-1}).$$

Then $\{a,b\} = [x,y] \in K_2(R)$ (see Section 3a). Clearly,

$$\phi(\bar{x}) = \mathrm{diag}(\bar{a},\bar{a}^{-1},1) \quad \text{and} \quad \phi(\bar{y}) = \mathrm{diag}(\bar{b},1,\bar{b}^{-1});$$

and so (using Theorem 3.1(ii,iv))

$$\overline{\{a,b\}} = \overline{[x,y]} = (\bar{y})^{-1}(\bar{x})^{-1}\cdot\bar{y}\cdot\bar{x} = [\bar{y}^{-1},\bar{x}^{-1}] = \{\bar{b}^{-1},\bar{a}^{-1}\} = \{\bar{b},\bar{a}\}.$$

(ii) Let $\tau_A\colon A \longrightarrow A$ denote the involution, set $\tau = \tau_A|F$, and let $F^\tau \subseteq F$ be the fixed field of τ. Let $E \supseteq F$ be a splitting field for A such that E/F^τ is Galois. Fix an isomorphism $f\colon E\otimes_F A \xrightarrow{\cong} M_n(E)$, let $\sigma \in \mathrm{Gal}(E/F^\tau)$ be any extension of τ, and set

$$\alpha = f\circ(\sigma\otimes\tau_A)\circ f^{-1} : M_n(E) \xrightarrow{\cong} M_n(E).$$

Then α and $(M \longmapsto \sigma(M)^t)$ are two antiautomorphisms of $M_n(E)$ with the same action on the center. By the Skolem-Noether theorem (Theorem 1.1(iv)), they differ by an inner automorphism.

Fix $a \in A^*$, and set $M = f(1\otimes a) \in M_n(E)$. By definition of $nr_{A/F}$,

$$nr_{A/F}(\tau_A(a)) = \det_E(f(1\otimes\tau_A(a))) = \det_E(\alpha(M)) = \det_E(\sigma(M)^t)$$

$$= \sigma(\det_E(M)) = \sigma(nr_{A/F}(a)) = \tau(nr_{A/F}(a));$$

and so $nr_{A/F}$ commutes with the involutions. □

When G is a group and R is any commutative ring, the "standard involution" on $R[G]$ is the involution $\sum a_ig_i \longmapsto \sum a_ig_i^{-1}$. When G is finite and R is the ring of integers in any algebraic number field K, then this induces involutions on $SK_1(R[G])$, $Cl_1(R[G])$, and $C(K[G])$;

as well as on $SK_1(\hat{R}_p[G])$ and $K_2^c(\hat{R}_p[G])$ for all primes p. In other words, all terms in the localization sequences of Theorem 3.15 carry the involution.

Proposition 5.11 *The following hold for any finite group* G.

(i) If R *is the ring of integers in some algebraic number field* K, *then all homomorphisms in the localization sequences for* $SK_1(R[G])$ *of Theorem 3.15 commute with the involutions.*

(ii) Write $\mathbb{Q}[G] = \prod_{i=1}^k A_i$, *where each* A_i *is simple with center* F_i. *Then the involution on* $\mathbb{Q}[G]$ *leaves each* A_i *invariant, and acts via complex conjugation on each* F_i.

Proof (i) This is clear, except for the boundary homomorphism

$$\partial : C(K[G]) \cong \mathrm{Coker}\Big[K_2(K[G]) \longrightarrow \bigoplus_p K_2(\hat{K}_p[G])\Big] \twoheadrightarrow Cl_1(R[G]).$$

For any $[M] \in Cl_1(R[G])$, the formula in Theorem 3.12 says that $\partial^{-1}([M]) = x^{-1}y \in \bigoplus_p K_2(\hat{K}_p[G])$, where $x \in St(K[G])$ and $y = (y_p) \in \prod_p St(\hat{R}_p[G])$ are liftings of $M \in GL(R[G])$, such that $x = y_p$ in $St(\hat{K}_p[G])$ for almost all p. Then

$$\overline{\partial^{-1}([M])} = \bar{y}\cdot\bar{x}^{-1} = \bar{x}^{-1}\cdot\bar{y} = \partial^{-1}([\bar{M}])$$

($\bar{y}\cdot\bar{x}^{-1} = \bar{x}^{-1}\cdot\bar{y}$ since $K_2(\hat{K}_p[G])$ is central). So ∂ commutes with the involution.

(ii) It suffices to study the action of the involution on the center $Z(\mathbb{Q}[G])$ (see Theorem 1.1). In other words, it suffices to show that the composite

$$Z(\mathbb{Q}[G]) \xrightarrow{pr_i} F_i \xhookrightarrow{v} \mathbb{C}$$

(for each i and v) commutes with the involutions on $\mathbb{C}$ and $\mathbb{Q}[G]$.

Fix such i and v. Then $\mathbb{C}\otimes_{vF_i} A_i$ is simple with center $\mathbb{C}$ (Theorem 1.1(ii)), hence is a matrix algebra over $\mathbb{C}$, and so its irreducible representation V is an irreducible $\mathbb{C}[G]$-representation. If $d = \dim_{\mathbb{C}}(V)$, and $\chi_V\colon G \longrightarrow \mathbb{C}$ is the character, then for any $x = \sum a_i g_i \in Z(\mathbb{Q}[G])$,

$$v \circ pr_i(x) = \frac{1}{d}\cdot \mathrm{Tr}_{\mathbb{C}\otimes_{vF_i} A}(1\otimes x) = \frac{1}{d}\cdot \sum a_i \cdot \chi_V(g_i) \in \mathbb{C}.$$

But for any $g \in G$, $\chi_V(g^{-1}) = \overline{\chi_V(g)}$; and so $v \circ pr_i(\bar{x}) = \overline{v \circ pr_i(x)}$. □

The above results will now be applied to describe the involution on $C(\mathbb{Q}[G])$ and $Cl_1(\mathbb{Z}[G])$. This will be important when describing the odd torsion in $Cl_1(\mathbb{Z}[G])$ in Chapters 9 and 13.

Theorem 5.12 (Bak [1]) *For any finite group G, the standard involution acts on $C(\mathbb{Q}[G])$ and $Cl_1(\mathbb{Z}[G])$ via the identity. More generally, if K is an algebraic number field such that p is unramified in K for all primes $p \mid |G|$, and if $R \subseteq K$ is the ring of integers, then the standard involution acts via the identity on $Cl_1(R[G])$, and on $C_p(K[G])$ for $p \mid |G|$.*

Proof For convenience, let $\mathcal{P}$ be the set of all primes if $K = \mathbb{Q}$, and the set of primes $p \mid |G|$ otherwise. By Theorem 4.16(iii), $Cl_1(R[G])$ has p-torsion only for $p \in \mathcal{P}$. So by Theorem 3.15 and Proposition 5.11(i), it suffices to show that the involution on $C_p(K[G])$ is trivial for all $p \in \mathcal{P}$.

Let A be a simple summand of $\mathbb{Q}[G]$, and set $F = Z(A)$. Then by Brauer's splitting theorem (Theorem 1.5(ii)), $F \subseteq \mathbb{Q}(\zeta_n)$, where $n = \exp(G)$ and $\zeta_n = \exp(2\pi i/n)$. So the assumption on K implies that $F \cap K = \mathbb{Q}$ under any embeddings into $\mathbb{C}$. Hence, $F' = K\otimes_{\mathbb{Q}} F$ is a field and $A' = K\otimes_{\mathbb{Q}} A$ is simple (Theorem 1.1). Furthermore, for any $p \in \mathcal{P}$, F' has the same p-th power roots of unity as F.

Fix $p \in \mathcal{P}$, and let $(\mu_F)_p$ be the groups of p-th power roots of unity in F (and F'). Assume that $C(A') \neq 1$ (otherwise there is nothing to prove). Let

$$\sigma : C_p(A') = \operatorname{Coker}\Big[K_2(A') \longrightarrow \bigoplus_q K_2^c(\hat{A}'_q)\Big]_{(p)} \xrightarrow{\ \cong\ } (\mu_F)_p$$

be the isomorphism (p-locally) of Theorem 4.13. Set $\mu = (\mu_F)_p = (\mu_{F'})_p$. Then for each (finite) prime q in F', and for any $a \in (\hat{F}'_q)^*$ and any $u \in (\hat{A}'_q)^*$ $(\hat{A}'_q = \prod \hat{A}'_q)$,

$$\sigma(\{a,u\}) = (a, nr_{A'/F'}(u))_{\mu,\hat{F}'_q}.$$

By Proposition 5.11(ii), the involution leaves F (and hence F') invariant, and acts on $\mu = (\mu_F)_p$ via $(\xi \mapsto \xi^{-1})$. So by Lemma 5.10,

$$\sigma(\overline{\{a,u\}}) = \sigma(\{\bar{a},\bar{u}\}^{-1}) = (\bar{a}, nr_{A'/F'}(\bar{u}))_\mu = (\bar{a}, \overline{nr_{A'/F'}(u)})_\mu$$

$$= (a,\ nr_{A'/F'}(u))_\mu \qquad \text{(by naturality: } \bar{\xi} = \xi^{-1} \text{ for } \xi \in \mu)$$

$$= \sigma(\{a,u\}).$$

Thus, $\overline{\{a,u\}} = \{a,u\}$ in $C_p(A)$. Since $C_p(A)$ is generated by such symbols, the involution on $C_p(A)$ is trivial. □

Note that Theorem 5.12 does not hold for arbitrary R[G]. Without the above restrictions, it is easy to construct examples where the standard involution on K[G] does not even leave all simple components invariant. It is not hard to show that $\mathcal{J}_{KG}: R_{\mathbb{C}}(G) \twoheadrightarrow C(K[G])$ is always negative equivariant with respect to the involution (note that the involution on C(K) is (-1), by Lemma 5.10(i)).

In Chapter 7 (Corollary 7.5), we will see that the involution also acts via the identity on Wh'(G) (= $Wh(G)/SK_1(\mathbb{Z}[G])$) for any finite G.

Chapter 6 THE INTEGRAL P-ADIC LOGARITHM

In Chapter 2, p-adic logarithms were used to get information about the structure of $K_1(\mathfrak{A})$ for a $\hat{\mathbb{Z}}_p$-order $\mathfrak{A}$. When $\mathfrak{A}$ is a group ring, there is also an "integral p-adic logarithm": defined by composing the usual p-adic logarithm with a linear endomorphism to make it integral valued. This yields a simple additive description of $K_1'(\hat{\mathbb{Z}}_p[G])$ for any p-group G (Theorems 6.6 and 6.7); and in later chapters will play a key role in studying $Wh'(G)$ and $SK_1(\mathbb{Z}[G])$, as well as $K_1'(\hat{\mathbb{Z}}_p[G])$ itself.

Integral logarithms have also been important when studying class groups $D(\mathbb{Z}[G]) \subseteq \tilde{K}_0(\mathbb{Z}[G])$ for finite G. One example of this is Martin Taylor's proof in [2] of the Fröhlich conjecture, which identifies the class $[R] \in \tilde{K}_0(\mathbb{Z}[G])$, when R is the ring of integers in a number field L, L/K is a tamely ramified Galois extension, and $G = Gal(L/K)$. Another application is the logarithmic description of $D(\mathbb{Z}[G])$, when G is a p-group and p a regular prime, in Oliver & Taylor [1].

Throughout this chapter, p will denote a fixed prime. We will be working with group rings of the form $R[G]$, where G is a finite group and R is the ring of integers in a finite extension F of $\hat{\mathbb{Q}}_p$. Recall from Example 1.12 that if G is a p-group, and if $\mathfrak{p} \subseteq R$ is the maximal ideal, then

$$J(R[G]) = \langle \mathfrak{p}, r(1-g) : g \in G,\ r \in R \rangle = \{\textstyle\sum r_i g_i \in R[G] : \sum r_i \in \mathfrak{p}\}.$$

6a. The integral logarithm for p-adic group rings

In Theorem 2.8, a logarithm homomorphism

$$\log_I : K_1(\mathfrak{A}, I) \longrightarrow \mathbb{Q} \otimes_{\mathbb{Z}} (I/[\mathfrak{A}, I])$$

was constructed, for any ideal I in a $\hat{\mathbb{Z}}_p$-order $\mathfrak{A}$. When applying this to a p-adic group ring $R[G]$, where R is the ring of integers in any extension F of $\hat{\mathbb{Q}}_p$, it is convenient to identify

$$F[G]/[F[G],F[G]] \cong H_0(G;F[G]) \quad \text{and} \quad I/[R[G],I] \cong H_0(G;I)$$

(for any ideal $I \subseteq R[G]$), where the homology in both cases is taken with respect to the conjugation action of G on $F[G]$. In particular, $H_0(G;F[G])$ and $H_0(G;R[G])$ can be regarded as the free F- and R-modules with basis the set of conjugacy classes of elements of G.

Mostly, we will be working with $R \subseteq F$ for which F is unramified over $\hat{\mathbb{Q}}_p$; so that R/pR is a field and $Gal(F/\hat{\mathbb{Q}}_p) \cong Gal((R/pR)/\mathbb{F}_p)$. Hence, in this case, there is a unique generator $\varphi \in Gal(F/\hat{\mathbb{Q}}_p)$ — the *Frobenius automorphism* — such that $\varphi(r) \equiv r^p \pmod{pR}$ for any $r \in R$.

<u>Definition 6.1</u> (Compare M. Taylor [1, Section 1] and Oliver [2, Section 2]). *Let* R *be the ring of integers in any finite unramified extension* F *of* $\hat{\mathbb{Q}}_p$. *Define* $\Phi\colon H_0(G;F[G]) \longrightarrow H_0(G;F[G])$, *for any finite group* G, *by setting*

$$\Phi\Big(\sum_{i=1}^k a_i g_i\Big) = \sum_{i=1}^k \varphi(a_i) g_i^p. \qquad (a_i \in F,\ \ g_i \in G)$$

Define

$$\Gamma = \Gamma_{RG} : K_1(R[G]) \longrightarrow H_0(G;F[G])$$

by setting $\Gamma(u) = \log(u) - \frac{1}{p}\cdot\Phi(\log(u))$ *for* $u \in K_1(R[G])$.

To help motivate this construction, consider the case where G is abelian. Then Φ is a ring endomorphism, and so

$$\Gamma([u]) = \log(u) - \frac{1}{p}\cdot\log(\Phi(u)) = \frac{1}{p}\cdot\log(u^p/\Phi(u))$$

for $u \in 1+J(R[G])$ (J = Jacobson radical). But $u^p \equiv \Phi(u) \pmod{pR[G]}$,

$$\log(u^p/\Phi(u)) \in \log(1+pR[G]) \subseteq pR[G],$$

and so $\Gamma([u]) \in R[G]$.

The proof that Γ_{RG} also is integral valued in the nonabelian case is more complicated.

Theorem 6.2 *Let R be the ring of integers in some finite unramified extension F of $\hat{\mathbb{Q}}_p$. Then for any finite group G,*

$$\Gamma_{RG}(K_1(R[G])) \subseteq H_0(G;R[G]).$$

The map Γ is natural with respect to maps induced by group homomorphisms and Galois automorphisms of F. For any G and any extension K/F (both finite and unramified over $\hat{\mathbb{Q}}_p$), if $S \subseteq K$ and $R \subseteq F$ denote the rings of integers, then the following squares commute:

$$\begin{array}{ccc} K_1(R[G]) & \xrightarrow{\Gamma} & H_0(G;R[G]) \\ \downarrow{\scriptstyle \text{incl}} & & \downarrow{\scriptstyle \text{incl}} \\ K_1(S[G]) & \xrightarrow{\Gamma} & H_0(G;S[G]) \end{array} \qquad \begin{array}{ccc} K_1(S[G]) & \xrightarrow{\Gamma} & H_0(G;S[G]) \\ \downarrow{\scriptstyle \text{trf}} & & \downarrow{\scriptstyle \text{Tr}} \\ K_1(R[G]) & \xrightarrow{\Gamma} & H_0(G;R[G]) \end{array}$$

Proof Let $J = J(R[G])$ denote the Jacobson radical. For any $x \in J$,

$$\Gamma(1-x) = -\left[x + \frac{x^2}{2} + \frac{x^3}{3} + \dots\right] + \left[\frac{\Phi(x)}{p} + \frac{\Phi(x^2)}{2p} + \frac{\Phi(x^3)}{3p} + \dots\right]$$

$$\equiv -\sum_{k=1}^{\infty} \frac{1}{pk}\cdot[x^{pk} - \Phi(x^k)] \qquad (\text{mod } H_0(G;R[G])).$$

So it suffices to show that $pk|[x^{pk} - \Phi(x^k)]$ for all k; or (since all primes other than p are invertible in R) that

$$p^n|[x^{p^n} - \Phi(x^{p^{n-1}})]$$

(in $H_0(G;R[G])$) for all $n \geq 1$ and all $x \in R[G]$.

Write $x = \sum r_i g_i$, set $q = p^n$, and consider a typical term in x^q:

$$r_{i_1}\cdots r_{i_q}\cdot g_{i_1}\cdots g_{i_q}.$$

Let $\mathbb{Z}/p^n$ act by cyclically permuting the g_i's, so that we get a total of p^{n-t} conjugate terms, where p^t is the number of cyclic permutations leaving each term invariant. Then $g_{i_1}\cdots g_{i_q}$ is a p^t-th power, and the sum of the conjugate terms has the form

$$p^{n-t}\cdot\hat{r}^{p^t}\cdot\hat{g}^{p^t} \in H_0(G;R[G]) \qquad \Big(\hat{r} = \prod_{j=1}^{p^{n-t}} r_{i_j}\ ,\quad \hat{g} = \prod_{j=1}^{p^{n-t}} g_{i_j}\Big).$$

If $t = 0$, then this is a multiple of p^n. If $t > 0$, then there is a corresponding term $p^{n-t}\cdot\hat{r}^{p^{t-1}}\cdot\hat{g}^{p^{t-1}}$ in the expansion of $x^{p^{n-1}}$. It remains only to show that

$$p^{n-t}\cdot\hat{r}^{p^t}\cdot\hat{g}^{p^t} \equiv p^{n-t}\cdot\Phi(\hat{r}^{p^{t-1}}\cdot\hat{g}^{p^{t-1}}) = p^{n-t}\varphi(\hat{r}^{p^{t-1}})\cdot\hat{g}^{p^t} \qquad (\text{mod } p^n).$$

But $p^t\,|\,[\hat{r}^{p^t}-\varphi(\hat{r}^{p^{t-1}})]$, since $p\,|\,[\hat{r}^p-\varphi(\hat{r})]$.

Naturality with respect to group homomorphisms is immediate from the definitions, and naturality with respect to Galois automorphisms holds since they all commute with the Frobenius automorphism φ (note that $\text{Gal}(F/\hat{\mathbb{Q}}_p)$ is cyclic since F is unramified). If $S \supseteq R$, then Γ commutes with the inclusion maps since $\varphi_S|R = \varphi_R$.

To see naturality with respect to the trace and transfer maps, first note that Φ commutes with the trace (since it commutes with Galois automorphisms). It suffices therefore to show that $\log\circ\text{trf} = \text{Tr}\circ\log$. For $s \in S$, $\text{Tr}(s) = \text{Tr}_{S/R}(s)$ is the trace of the matrix for multiplication by s as an R-linear map (see Reiner [1, Section 1a]). Hence, for any $x \in S[G]$ and any $n > 0$,

$$\begin{aligned}\log(\text{trf}(1+p^n x)) &\equiv \log(1+p^n\cdot\text{Tr}(x))\\ &\equiv p^n\cdot\text{Tr}(x) \equiv \text{Tr}(\log(1+p^n x)) \qquad (\text{mod } p^{2n-1}).\end{aligned}$$

For any $u \in 1+J(R[G])$, $u^{p^k} \in 1+pR[G]$ for some k (u has p-power order in $(R/p[G])^*$). Then $u^{p^{k+n}} \in 1+p^{n+1}R[G]$ for all $n>0$, and so if $n \geq k$:

$$\log(\mathrm{trf}(u)) = p^{-n-k}\cdot\log(\mathrm{trf}(u^{p^{k+n}}))$$

$$\equiv p^{-k-n}\cdot\mathrm{Tr}(\log(u^{p^{k+n}})) = \mathrm{Tr}(\log(u)) \qquad (\mathrm{mod}\ \ p^{2n+1}/p^{n+k} = p^{n-k+1}).$$

Since this holds for any $n \geq k$, the congruence is an equality. □

In fact, Γ_{RG} is also natural with respect to transfer homomorphisms for inclusions of groups, although in this case the corresponding restriction map on $H_0(G;R[G])$ is much less obvious. This will be shown, for p-groups at least, in Theorem 6.8 below.

The next lemma collects some miscellaneous relations which will be needed.

<u>Lemma 6.3</u> *(i) For any group* G *and any any element* $g \in G$,

$$(1-g)^p \equiv (1-g^p) - p(1-g) \qquad (\mathrm{mod}\ \ p(1-g)^2\mathbb{Z}[G]).$$

(ii) Let K *be any finite field of p-power order. Then the sequence*

$$0 \longrightarrow \mathbb{F}_p \xrightarrow{\text{incl}} K \xrightarrow{1-\varphi} K \xrightarrow{\text{Tr}} \mathbb{F}_p \longrightarrow 0 \qquad (1)$$

is exact, where Tr *denotes the trace map.*

<u>Proof</u> (i) Just note that

$$g^p = [1 - (1-g)]^p \equiv 1 - p(1-g) + (-1)^p(1-g)^p$$

$$\equiv 1 - p(1-g) - (1-g)^p \qquad (\mathrm{mod}\ \ p(1-g)^2\mathbb{Z}[G]).$$

(ii) The trace map is onto by Proposition 1.8(iii), $\mathrm{Tr}\circ(1-\varphi) = 0$ by definition of the trace, and $\mathrm{Ker}(1-\varphi) = \mathbb{F}_p$ since φ generates $\mathrm{Gal}(K/\mathbb{F}_p)$. A counting argument then shows that (1) is exact. □

Attention will now be restricted to p-groups. Both here, when identifying the image of Γ_G (or of its restrictions to certain subgroups), and later when studying $SK_1(R[G])$, one of the main techniques is to work inductively by comparing $K_1(R[G])$ with $K_1(R[G/z])$ for $z \in Z(G)$ central of order p. In particular, the case where z is a *commutator* — i. e., $z=[g,h]$ for some $g,h \in G$ (as opposed to a product of such elements) plays a key role when doing this. The reason for this is (in part) seen in the next proposition.

<u>Proposition 6.4</u> *Let* R *be the ring of integers in any finite extension* F *of* $\hat{\mathbb{Q}}_p$, *let* $\mathfrak{p} \subseteq R$ *be the maximal ideal, and let* τ *denote the composite*

$$\tau : R \twoheadrightarrow R/\mathfrak{p} \xrightarrow{\mathrm{Tr}} \mathbb{F}_p.$$

Then for any p*-group* G *and any central element* $z \in G$ *of order* p, *there is an exact sequence*

$$1 \to \langle z\rangle \longrightarrow K_1(R[G],(1-z)R[G]) \xrightarrow{\log} H_0(G;(1-z)R[G]) \xrightarrow{\omega} \mathbb{F}_p \to 0; \quad (1)$$

where $\omega((1-z)\sum r_ig_i) = \tau(\sum r_i)$ *for any* $r_i \in R$ *and* $g_i \in G$. *If* $F/\hat{\mathbb{Q}}_p$ *is unramified, and if we set*

$$\overline{H}_0(G;(1-z)R[G]) = \mathrm{Im}\Big[H_0(G;(1-z)R[G]) \longrightarrow H_0(G;R[G])\Big]$$

$$= \mathrm{Ker}\Big[H_0(G;R[G]) \longrightarrow H_0(G/z;R[G/z])\Big];$$

then $\Gamma_{RG}(1+(1-z)\xi) = \log(1+(1-z)\xi)$ *in* $\overline{H}_0(G;(1-z)R[G])$ *for all* $\xi \in R[G]$ *and*

$$\Big[\overline{H}_0(G;(1-z)R[G]) : \Gamma_G(1+(1-z)R[G])\Big] = \begin{cases} 1 & \text{if } z \text{ is a commutator} \\ p & \text{otherwise.} \end{cases}$$

<u>Proof</u> Set $I = (1-z)R[G]$, for short, and let $J = J(R[G])$ denote the Jacobson radical. Note that $(1-z)^p \in p(1-z)R[G]$ by Lemma 6.3(i). So Theorem 2.8 applies to show that the p-adic logarithm induces a homomorphism $\log^I$ and an isomorphism $\log^{IJ}$, which sit in the following

commutative diagram with exact rows:

$$\begin{array}{ccccccccc}
 & K_1(R[G],(1-z)J) & \longrightarrow & K_1(R[G],(1-z)R[G]) & \longrightarrow & K_1\left(\frac{R[G]}{(1-z)J}, \frac{(1-z)R[G]}{(1-z)J}\right) & \longrightarrow & 1 & \\
 & \cong\Big\downarrow \log^{IJ} & & \Big\downarrow \log^{I} & & \Big\downarrow \log_0 & & & (2)\\
0 \longrightarrow & H_0(G;(1-z)J) & \longrightarrow & H_0(G;(1-z)R[G]) & \longrightarrow & H_0\left(G, \frac{(1-z)R[G]}{(1-z)J}\right) & \longrightarrow & 0. &
\end{array}$$

Also, by Theorem 1.15 and Example 1.12, there are isomorphisms

$$K_1\left(\frac{R[G]}{(1-z)J}, \frac{(1-z)R[G]}{(1-z)J}\right) \xrightarrow[\cong]{\alpha} R[G]/J \cong R/p \cong H_0\left(G, \frac{(1-z)R[G]}{(1-z)J}\right);$$

where $\alpha(1+(1-z)\xi) = \xi$ for $\xi \in R[G]/J$.

Now consider the following diagram

$$\begin{array}{ccccccccc}
 & & & K_1(R[G],(1-z)R[G]) & \xrightarrow{\log^I} & H_0(G;(1-z)R[G]) & & & \\
 & & & \Big\downarrow \alpha' & (3) & \Big\downarrow \alpha'' & \searrow \omega & & \\
0 \longrightarrow & \mathbb{F}_p & \hookrightarrow & R/p & \xrightarrow{1-\varphi} & R/p & \xrightarrow{\mathrm{Tr}} & \mathbb{F}_p & \longrightarrow 0
\end{array}$$

where $\alpha'(1+(1-z)\sum r_i g_i) = \sum \bar{r}_i$ and $\alpha''((1-z)\sum r_i g_i) = \sum \bar{r}_i$. Here, $\bar{r} \in R/p$ denotes the reduction of $r \in R$. The bottom row is exact by Lemma 6.3(ii); and square (3) commutes since for $r \in R$ and $g \in G$,

$$\alpha''(\mathrm{Log}(1 + (1-z)rg)) = \alpha''((1-z)(rg - r^p g^p)) = r - \varphi(r) \in R/p.$$

Then (3) is a pullback square by diagram (2), and so $\log^I$ and $1-\varphi$ have isomorphic kernel and cokernel. The exactness of (1) now follows since $\omega = \mathrm{Tr} \circ \alpha''$, and since α' maps $\langle z \rangle$ isomorphically to $\mathbb{F}_p = \mathrm{Ker}(1-\varphi)$.

If $F/\hat{\mathbb{Q}}_p$ is unramified (so Γ_G is defined), then for any $\xi \in R[G]$, $\log(1+(1-z)\xi) = (1-z)\eta$ for some η, and $\Phi((1-z)\eta) = (1-z^p)\Phi(\eta) = 0$. So $\Gamma_G(1+(1-z)\xi) = \log(1+(1-z)\xi)$ in this case. By the exactness of (1),

$$\left[\bar{H}_0(G;(1-z)R[G]) : \Gamma_G(1+(1-z)R[G])\right] = \begin{cases} 1 & \text{if } (1-z)g = 0 \in H_0(G;R[G]) \text{ some } g \in G \\ p & \text{otherwise.} \end{cases}$$

In other words, the index is 1 if and only if g is conjugate to zg for some g, if and only if $z = [h,g]$ for some $h,g \in G$. □

The next lemma, on the existence of central commutators, will be needed to apply Proposition 6.4.

Lemma 6.5 *Let G be a p-group, and let $H \triangleleft G$ be a nontrivial normal subgroup generated by commutators in G. Then H contains a commutator $z \in Z(G)$ of order p. In particular, any nonabelian p-group G contains a central commutator of order p.*

Proof Fix any commutator $x_0 \in H \smallsetminus 1$. If x_0 is not central, then choose any $g_0 \in G$ not commuting with x_0, and set $x_1 = [x_0, g_0] \in H \smallsetminus 1$. Since G is nilpotent, this procedure can be continued, setting $x_i = [x_{i-1}, g_{i-1}] \in H \smallsetminus 1$, until $x_k \in H \smallsetminus 1$ is central for some $k \geq 0$. Then, if $x_k = [g,h]$ and has order p^n for some $n \geq 1$, $x_k^{p^{n-1}} = [g, h^{p^{n-1}}]$ and has order p. □

The main result of this chapter can now be shown. It gives a very simple description of the image of the integral logarithm on $K_1(R[G])$.

Theorem 6.6 *Fix a p-group G, and a finite unramified extension F of $\hat{\mathbb{Q}}_p$ with ring of integers $R \subseteq F$. Set $\epsilon = (-1)^{p-1}$, and define*

$$\omega = \omega_{RG} : H_0(G;R[G]) \longrightarrow \langle\epsilon\rangle \times G^{ab} \quad \text{by} \quad \omega(\textstyle\sum a_i g_i) = \prod(\epsilon g_i)^{\mathrm{Tr}(a_i)}.$$

Then the sequence

$$1 \longrightarrow K_1(R[G])/\text{torsion} \xrightarrow{\ \Gamma\ } H_0(G;R[G]) \xrightarrow{\ \omega\ } \langle\epsilon\rangle \times G^{ab} \longrightarrow 1 \qquad (1)$$

is exact.

Proof Assume first that $G = 1$, the trivial group. By Theorem 2.8, $\mathrm{Log}(1+pR) = pR$ if p is odd, and $\mathrm{Log}(1+4R) = 4R$ if $p = 2$. Also, if $p=2$, then $\mathrm{Log}(1+2r) \equiv 2(r-r^2) \equiv 2(r-\varphi(r)) \pmod{4R}$ for any $r \in R$. It follows that

$$\text{Log}(R^*) = \text{Log}(1+pR) = \left\{ \begin{array}{l} pR \\ \\ 4R + 2(1-\varphi)R \end{array} \right\} = p \cdot \text{Ker}(\omega) \qquad \begin{array}{l} \text{if} \quad p \quad \text{is odd} \\ \\ \text{if} \quad p = 2. \end{array}$$

Furthermore, since $\text{Log}(1+pR)$ is φ-invariant,

$$\Gamma(R^*) = (1 - \tfrac{1}{p} \cdot \Phi)(\text{Log}(R^*)) = \tfrac{1}{p} \cdot \text{Log}(R^*) \qquad \left((1 - \tfrac{1}{p} \cdot \Phi)^{-1}(r) = - \sum_{i=1}^{\infty} p^i \varphi^{-i}(r) \right).$$

So $\text{Im}(\Gamma) = \text{Ker}(\omega)$, and (1) is exact in this case.

Now assume that G is a nontrivial p-group. We first show that $\omega_G \circ \Gamma_G = 1$; it suffices by naturality to do this when G is abelian and $R = \hat{\mathbb{Z}}_p$. Let $I = \{ \sum r_i g_i \in R[G] : \sum r_i = 0 \}$ denote the augmentation ideal. For any $u = 1 + \sum r_i(1-a_i)g_i \in 1 + I$, where $r_i \in \hat{\mathbb{Z}}_p$,

$$u^p \equiv 1 + p\sum r_i(1-a_i)g_i + \sum r_i^p(1-a_i)^p g_i^p \qquad (\text{mod} \quad pI^2)$$

$$\equiv 1 + p\sum r_i(1-a_i)g_i + \sum r_i[(1-a_i^p) - p(1-a_i)]g_i^p \qquad (\text{Lemma 6.3(i)})$$

$$\equiv \Phi(u) + p\sum r_i(1-a_i)(g_i - g_i^p) \equiv \Phi(u) \qquad (\varphi(r_i) = r_i)$$

This shows that $u^p/\Phi(u) \in 1 + pI^2$, and hence that

$$\Gamma(u) = \log(u) - \tfrac{1}{p} \cdot \Phi(\log(u)) = \tfrac{1}{p} \cdot \log(u^p/\Phi(u)) \in I^2.$$

On the other hand, for any $r \in \hat{\mathbb{Z}}_p$ and any $a,b,g \in G$,

$$\omega(r(1-a)(1-b)g) = (\epsilon g)^r (\epsilon a g)^{-r} (\epsilon b g)^{-r} (\epsilon a b g)^r = 1 \in \langle \epsilon \rangle \times G^{ab}.$$

Thus, $\Gamma(1+I) \subseteq I^2 \subseteq \text{Ker}(\omega)$, and so

$$\Gamma(K_1(R[G])) = \Gamma(R^* \times (1+I)) = \langle \Gamma(R^*), \Gamma(1+I) \rangle \subseteq \text{Ker}(\omega). \qquad (3)$$

Now fix some central element $z \in Z(G)$ of order p, such that z is a commutator if G is nonabelian (Lemma 6.5). Set $\hat{G} = G/z$, assume inductively that the theorem holds for $\hat{G}$, and consider the following diagram (where $\alpha: G \twoheadrightarrow \hat{G}$ denotes the projection):

$$
\begin{array}{ccccccccc}
 & & 1 & & 1 & & 1 & & \\
 & & \downarrow & & \downarrow & & \downarrow & & \\
1 & \longrightarrow & K_1(R[G],(1-z)R[G])/\mathrm{tors} & \xrightarrow{\Gamma_o} & \overline{H}_0(G;(1-z)R[G]) & \xrightarrow{\omega_o} & \mathrm{Ker}(\alpha^{ab}) & \longrightarrow & 1 \\
 & & \downarrow & & \downarrow & & \downarrow & & \\
1 & \longrightarrow & K_1(R[G])/\mathrm{tors} & \xrightarrow{\Gamma_G} & H_0(G;R[G]) & \xrightarrow{\omega_G} & \langle\epsilon\rangle\times G^{ab} & \longrightarrow & 1 \\
 & & \downarrow{\scriptstyle K(\alpha)} & & \downarrow{\scriptstyle H(\alpha)} & & \downarrow{\scriptstyle \alpha^{ab}} & & \\
1 & \longrightarrow & K_1(R[G])/\mathrm{tors} & \xrightarrow{\Gamma_{\hat{G}}} & H_0(\hat{G};R[\hat{G}]) & \xrightarrow{\omega_{\hat{G}}} & \langle\epsilon\rangle\times \hat{G}^{ab} & \longrightarrow & 1 \\
 & & \downarrow & & \downarrow & & \downarrow & & \\
 & & 1 & & 1 & & 1 & &
\end{array}
$$

Since $K(\alpha)$ is onto (Theorem 1.14(iii)), the columns are all exact. The bottom row is exact by the induction hypothesis. Also, the top row is exact: ω_o is clearly onto, Γ_o is injective by Proposition 6.4, $\mathrm{Im}(\Gamma_o) \subseteq \mathrm{Ker}(\omega_o)$ by (3); and using Proposition 6.4 again:

$$
|\mathrm{Ker}(\alpha^{ab})| = \left\{ \begin{array}{ll} 1 & \text{if } z \text{ is a commutator} \\ p & \text{otherwise (i. e., if } G \text{ is abelian)} \end{array} \right\} = |\mathrm{Coker}(\Gamma_o)|.
$$

Since $\omega_G \circ \Gamma_G = 1$ by (3), the middle row is exact by the 3×3 lemma. □

One simple application of Theorem 6.6 is to the following question of Wall, which arises when computing surgery groups. Let G be an arbitrary 2-group, and set $Wh'(\hat{\mathbb{Z}}_2[G]) = K_1(\hat{\mathbb{Z}}_2[G])/\langle\{\pm1\}\times G^{ab}\times SK_1(\hat{\mathbb{Z}}_2[G])\rangle$. The problem is to describe the cohomology group $H^1(\mathbb{Z}/2;Wh'(\hat{\mathbb{Z}}_2[G]))$, where $\mathbb{Z}/2$ acts via the standard involution $(g \mapsto g^{-1})$ (see Section 5c). Assuming Theorem 7.3 below, $Wh'(\hat{\mathbb{Z}}_2[G])$ is torsion free, and so the exact sequence of Theorem 6.6 takes the form

$$
1 \longrightarrow Wh'(\hat{\mathbb{Z}}_2[G]) \longrightarrow H_0(G;\hat{\mathbb{Z}}_2[G]) \longrightarrow \{\pm1\}\times G^{ab} \longrightarrow 1;
$$

with the obvious involution on each term. Also, $H^1(\mathbb{Z}/2;H_0(G;\hat{\mathbb{Z}}_2[G])) = 1$, since the involution permutes a $\hat{\mathbb{Z}}_2$-basis of $H_0(G;\hat{\mathbb{Z}}_2[G])$. We thus get an exact sequence

$$H^0(\mathbb{Z}/2;H_0(G;\hat{\mathbb{Z}}_2[G])) \longrightarrow H^0(\mathbb{Z}/2;\{\pm1\}\times G^{ab}) \longrightarrow H^1(\mathbb{Z}/2;Wh'(\hat{\mathbb{Z}}_2[G])) \longrightarrow 1;$$

and this yields the simple formula

$$H^1(\mathbb{Z}/2;Wh'(\hat{\mathbb{Z}}_2[G])) \cong \frac{\{[g]\in G^{ab} : [g^2]=1\}}{\langle [g]: g \text{ conjugate } g^{-1}\rangle}.$$

Theorem 6.6 gives a very simple description of $K_1(\hat{\mathbb{Z}}_p[G])$/torsion, and the torsion subgroup of $K_1(\hat{\mathbb{Z}}_p[G])$ will be identified in the next chapter (Theorem 7.3). This suffices for many applications; for example, to prove the results on $Cl_1(\mathbb{Z}[G])$ and $SK_1(\mathbb{Z}[G])$ in Chapters 8 and 9 below. But sometimes, a description of $K_1'(R[G])$ $(= K_1(R[G])/SK_1(R[G]))$ up to extension only is not sufficient. The following version of the logarithmic exact sequence helps take care of this problem.

<u>Theorem 6.7</u> *Let* R *be the ring of integers in any finite unramified extension* F *of* $\hat{\mathbb{Q}}_p$. *For any p-group* G, *define*

$$(\nu,\hat{\theta}):\ K_1'(R[G]) \longrightarrow (G^{ab}\otimes R)\oplus(R/2) \qquad \textit{and}$$

$$(\omega,\theta):\ H_0(G;R[G]) \longrightarrow (G^{ab}\otimes R)\oplus(R/2)$$

by setting, for $g_i \in G$ *and* $a, a_i \in R$ *(with reductions* $\bar{a}, \bar{a}_i \in R/2$*),*

$$(\nu,\hat{\theta})((1+pa)(1+\textstyle\sum a_i(g_i-1))) = (\textstyle\sum g_i\otimes a_i,\ \bar{a}), \qquad \textit{and}$$

$$(\omega,\theta)(\textstyle\sum a_ig_i) = (\textstyle\sum g_i\otimes a_i,\ \textstyle\sum\bar{a}_i).$$

Then ν, $\hat{\theta}$, *and* Γ *are all well defined on* $K_1'(R[G])$, *and the sequence*

$$1 \longrightarrow K_1'(R[G])_{(p)} \xrightarrow{(\Gamma,\upsilon,\varphi\hat{\theta})} H_0(G;R[G]) \oplus (G^{ab} \otimes R) \oplus (R/2)$$

$$\xrightarrow{\begin{pmatrix}\omega & 1\otimes(\varphi-1) & 0\\ \theta & 0 & 1\otimes(\varphi-1)\end{pmatrix}} (G^{ab} \otimes R) \oplus (R/2) \longrightarrow 0$$

is exact.

<u>Proof</u> The main step is to show that the composite of the above two homomorphisms is zero, and this is a direct calculation. The injectivity of $(\Gamma,\omega,\varphi\hat{\theta})$ is a consequence of Theorem 7.3 below, which says that

$$\mathrm{Ker}(\Gamma_{RG}) = \mathrm{tors}(K_1'(R[G])) = \mathrm{tors}(R^*) \times G^{ab}.$$

The exactness of the whole sequence then follows easily from Theorem 6.6. See Oliver [8, Theorem 1.2] for more details. □

When G is an arbitrary finite group, then Γ_{RG} sits in exact sequences analogous to, but more complicated than, those in Theorems 6.6 and 6.7. Since their construction depends on induction theory, we wait until Chapter 12 (Theorem 12.9) to state them.

Several naturality properties for Γ were shown in Theorem 6.2. One more property, describing its behavior with respect to transfer maps for inclusions of groups, is also often useful. To state this, we define, for any prime p and any pair $H \subseteq G$ of p-groups, a homomorphism

$$\mathrm{Res}_H^G : H_0(G;\hat{\mathbb{Z}}_p[G]) \longrightarrow H_0(H;\hat{\mathbb{Z}}_p[H])$$

as follows. Fix $g \in G$, let $x_1,\ldots,x_k$ be double coset representatives for $H\backslash G/\langle g\rangle$, and set $n_i = \min\{n>0 : g^n \in x_i^{-1}Hx_i\}$ for $1 \le i \le k$. Then define

$$\mathrm{Res}_H^G(g) = \sum_{i=1}^{k} x_i g^{n_i} x_i^{-1} \in H_0(H;R[H]).$$

For example, if G and H are p-groups and $[G:H] = p$, then

$$\mathrm{Res}_H^G(g) = \begin{cases} \sum_{i=0}^{p-1} x^i g x^{-i} \quad (\text{any } x \in G\smallsetminus H) & \text{if } g \in H \\ g^p & \text{if } g \notin H. \end{cases}$$

Theorem 6.8 *For any pair* $H \subseteq G$ *of* p*-groups, the diagram*

$$\begin{array}{ccccccc} K_1'(\hat{\mathbb{Z}}_p[G]) & \xrightarrow{\Gamma_G} & H_0(G;\hat{\mathbb{Z}}_p[G]) & \xrightarrow{\omega_G} & \langle\epsilon\rangle \times G^{ab} & \longrightarrow & 1 \\ \Big\downarrow \mathrm{trf} & (1) & \Big\downarrow \mathrm{Res}_H^G & & \Big\downarrow R_H^G & & \\ K_1'(\hat{\mathbb{Z}}_p[H]) & \xrightarrow{\Gamma_H} & H_0(H;\hat{\mathbb{Z}}_p[H]) & \xrightarrow{\omega_G} & \langle\epsilon\rangle \times H^{ab} & \longrightarrow & 1 \end{array}$$

commutes. Here, $\epsilon = (-1)^{p-1}$, $\langle\epsilon\rangle \times G^{ab}$ *and* $\langle\epsilon\rangle \times H^{ab}$ *are identified as subgroups of* $K_1'(R[G])$ *and* $K_1'(R[H])$, *and* R_H^G *is the restriction of the transfer map.*

Proof The easiest way to prove the commutativity of (1) is to split it up into two squares:

$$\begin{array}{ccccc} K_1'(\hat{\mathbb{Z}}_p[G]) & \xrightarrow{\log} & H_0(G;\hat{\mathbb{Q}}_p[G]) & \xrightarrow{1-\frac{1}{p}\cdot\Phi} & H_0(G;\hat{\mathbb{Q}}_p[G]) \\ \Big\downarrow \mathrm{trf} & (1a) & \Big\downarrow \mathrm{res} & (1b) & \Big\downarrow \mathrm{Res}_H^G \\ K_1'(\hat{\mathbb{Z}}_p[H]) & \xrightarrow{\log} & H_0(H;\hat{\mathbb{Q}}_p[H]) & \xrightarrow{1-\frac{1}{p}\cdot\Phi} & H_0(H;\hat{\mathbb{Q}}_p[H]). \end{array}$$

Here, if $a_1,\ldots,a_m$ denote right coset representatives for $H \subseteq G$, then

$$\mathrm{res}(g) = \sum\{a_i g a_i^{-1} : 1 \le i \le m,\ a_i g a_i^{-1} \in H\} \in H_0(H;\hat{\mathbb{Q}}_p[H])$$

for any $g \in G$. The commutativity of (1b) is straightforward, and the commutativity of (1a) follows from the relations

$$\log(u) = \lim_{n\to\infty} \frac{1}{p^n}\cdot(u^{p^n}-1); \qquad \mathrm{trf}(u) = \lim_{n\to\infty} \left(1+\mathrm{res}(u^{p^n}-1)\right)^{1/p^n}.$$

See Oliver & Taylor [1, Theorem 1.4] for details. □

In Oliver & Taylor [1, Theorem 1.4], Res_H^G is in fact defined for an arbitrary pair $H \subseteq G$ of finite groups (not only for p-groups). The above formulas can also be extended to include the case of $R[G]$, for any R unramified over $\hat{\mathbb{Z}}_p$; in this case the Frobenius automorphism for R appears in the formula for Res_H^G.

6b. Variants of the integral logarithm

We list here, mostly without proof, some useful variants of the integral logarithm, and of the exact sequence describing its image. The first theorem is a generalization of Proposition 6.4 and Theorem 6.6. It is used in Oliver [4] to detect elements in $K_2^c(R[G])$.

Theorem 6.9 *Let R be the ring of integers in any finite unramified extension F of $\hat{\mathbb{Q}}_p$, let $\alpha: \tilde{G} \longrightarrow G$ be any surjection of p-groups, and set*

$$I_\alpha = \mathrm{Ker}\Big[R[\tilde{G}] \longrightarrow R[G]\Big].$$

Set $K = \mathrm{Ker}(\alpha) \subseteq \tilde{G}$. Then there is an exact sequence

$$K_1(R[\tilde{G}], I_\alpha) \xrightarrow{\ \Gamma_\alpha\ } H_0(G; I_\alpha) \xrightarrow{\ \omega_\alpha\ } K/[\tilde{G},K] \longrightarrow 1.$$

Here, for any $r \in R$, $g \in \tilde{G}$, and $w \in K$, $\omega_\alpha(r(1-w)g) = w^{\mathrm{Tr}(r)}$.

Proof This is an easy consequence of results in Oliver [4]; but since it was not stated explicitly there we sketch the proof here. Define a p-group $\hat{G}$ and a $\hat{\mathbb{Z}}_p$-order $\mathfrak{A}$ to be the pullbacks

$$\begin{array}{ccc} \hat{G} & \xrightarrow{\ \beta_1\ } & \tilde{G} \\ \downarrow{\scriptstyle \beta_2} & & \downarrow{\scriptstyle \alpha} \\ \tilde{G} & \xrightarrow{\ \alpha\ } & G \end{array} \qquad\qquad \begin{array}{ccc} \mathfrak{A} & \xrightarrow{\ b_1\ } & R[\tilde{G}] \\ \downarrow{\scriptstyle b_2} & & \downarrow{\scriptstyle R\alpha} \\ R[\tilde{G}] & \xrightarrow{\ R\alpha\ } & R[G]. \end{array}$$

Set $I_i = \mathrm{Ker}\Big[R\beta_i : R[\hat{G}] \longrightarrow R[\tilde{G}]\Big]$ $(i = 1,2)$, and let $\psi: R[\hat{G}] \longrightarrow \mathfrak{A}$ be the obvious projection. Then

$$\mathrm{Ker}(\psi) = I_1 \cap I_2 = I_1 I_2$$

(see Oliver [4, Lemma 2.4]); so $\mathfrak{A} \cong R[\hat{G}]/I_1I_2$ and

$$K_1(\mathfrak{A}) \cong K_1(R[\hat{G}])/(1 + I_1I_2). \tag{1}$$

Also, by Oliver [4, Theorem 1.1] (and this is the difficult point):

$$\Gamma_{R\hat{G}}(1 + I_1I_2) = \mathrm{Im}\Big[I_1I_2 \longrightarrow H_0(\hat{G};R[\hat{G}])\Big]. \tag{2}$$

Formulas (1) and (2), together with the exact sequence of Theorem 6.6 (applied to $R[\hat{G}]$), now combine to give an exact sequence

$$K_1(\mathfrak{A}) \xrightarrow{\Gamma_{\mathfrak{A}}} H_0(\hat{G};\mathfrak{A}) \xrightarrow{\omega_{\mathfrak{A}}} \langle\epsilon\rangle \times \hat{G}^{ab} \longrightarrow 1$$

(where $\epsilon = (-1)^{p-1}$, as usual). We thus get a commutative diagram

$$\begin{array}{ccccccc}
K_1(\mathfrak{A}) & \xrightarrow{\Gamma_{\mathfrak{A}}} & H_0(\hat{G};\mathfrak{A}) & \xrightarrow{\omega_{\mathfrak{A}}} & \langle\epsilon\rangle \times \hat{G}^{ab} & \longrightarrow & 1 \\
\downarrow{\scriptstyle K_1(b_2)} & & \downarrow{\scriptstyle H(b_2)} & & \downarrow{\scriptstyle \beta_2^{ab}} & & \\
K_1(R[\tilde{G}]) & \xrightarrow{\Gamma_{R\tilde{G}}} & H_0(\tilde{G};R[\tilde{G}]) & \xrightarrow{\omega_{R\tilde{G}}} & \langle\epsilon\rangle \times \tilde{G}^{ab} & \longrightarrow & 1,
\end{array}$$

where the vertical maps are split surjective (split by the diagonal map $\tilde{G} \longrightarrow \hat{G} \subseteq \tilde{G}\times\tilde{G}$). Then

$$\mathrm{Ker}(K_1(b_2)) \cong K_1(R[\tilde{G}],I_\alpha), \quad \mathrm{Ker}(H(b_2)) \cong H_0(\tilde{G};I_\alpha), \quad \mathrm{Ker}(\beta_2^{ab}) \cong K/[\tilde{G},K]$$

and this proves the theorem. □

Logarithms can be used to study, not only the abelianization of $(R[G])^*$, but also its center. The following theorem is in a sense dual to Theorem 6.6.

Theorem 6.10 *Let R be the ring of integers in any finite unramified extension F of $\hat{\mathbb{Q}}_p$. Then for any p-group G, there is an exact sequence*

$$1 \longrightarrow \langle\epsilon\rangle \times Z(G) \xrightarrow{\text{incl}} 1 + J(Z(R[G])) \xrightarrow{\Gamma} Z(R[G]) \xrightarrow{\omega} \langle\epsilon\rangle \times Z(G) \longrightarrow 1.$$

Proof See Oliver [9]. □

The last result described here involves polynomial extensions of the base ring.

Theorem 6.11 *Let $\mathbb{Z}[s]\hat{}_p$ denote the p-adic completion of the polynomial algebra $\mathbb{Z}[s]$. For any p-group G, let $I \subseteq \mathbb{Z}[s]\hat{}_p[G]$ denote the augmentation ideal, and define*

$$K_1'(\mathbb{Z}[s]\hat{}_p[G],I) = \operatorname{Im}\Big[K_1(\mathbb{Z}[s]\hat{}_p[G],I) \longrightarrow K_1(\mathbb{Z}[s]\hat{}_p[\tfrac{1}{p}][G])\Big].$$

Then there is a short exact sequence

$$1 \longrightarrow K_1'(\mathbb{Z}[s]\hat{}_p[G],I) \xrightarrow{(\Gamma,\omega)} H_0(G;I) \oplus (\mathbb{Z}[s]\hat{}_p \otimes G^{ab})$$
$$\xrightarrow{(\upsilon,\Phi-1)} (\mathbb{Z}[s]\hat{}_p \otimes G^{ab}) \longrightarrow 1.$$

Proof See Milgram & Oliver [1]. The homomorphisms are analogous to those in Theorem 6.7 above. □

6c. Logarithms defined on $K_2^c(\hat{\mathbb{Z}}_p[G])$

The "logarithm" homomorphisms discussed here are not needed for describing the odd torsion in $Cl_1(\mathbb{Z}[G])$, but they could be important in describing the 2-power torsion, and do help to motivate the conjectures in Chapter 9. In any case, Theorem 6.12 below does give a complete description of $K_2^c(\hat{\mathbb{Z}}_p[G])$ when p is any prime and G is an abelian p-group; and Conjecture 6.13 would give an analogous description (though only up to extension, in general) for arbitrary p-groups.

The natural target group for K_2 integral logarithms turns out to be Connes' cyclic homology group $HC_1(-)$. If R is any commutative ring and G is any finite group, then

$$HC_1(R[G]) \cong H_1(G;R[G])/\langle g\otimes rg\colon g\in G,\ r\in R\rangle \quad\text{and}\quad HC_0(R[G]) \cong H_0(G;R[G]);$$

where $H_n(G;R[G])$ is as usual defined with respect to the conjugation action of G on R[G]. We identify $H_1(G;R[G])$ with $G\otimes R[G]$ whenever G is abelian; and for arbitrary G this allows us to define elements $g\otimes rh \in H_1(G;R[G])$ for any $r \in R$ and any commuting pair $g,h \in G$.

Theorem 6.12 *Fix an unramified extension* F *of* $\mathbb{Q}$, *and let* $R \subseteq F$ *be the ring of integers. Let* $\mathrm{Tr}\colon R \longrightarrow \hat{\mathbb{Z}}_p$ *denote the trace map. Then, for any abelian p-group* G, *there is a short exact sequence*

$$1 \longrightarrow K_2^c(R[G])/\{\pm G,\pm G\} \xrightarrow{\Gamma_2} HC_1(R[G]) \xrightarrow{\omega_2} \frac{G\otimes G}{\langle g\otimes h+h\otimes g\rangle} \longrightarrow 0 \qquad (1)$$

which is natural in G, *and such that* Γ_2 *and* ω_2 *satisfy the following two formulas:*

(*i*) *For any* $g \in G$ *and any* $u \in (R[G])^*$, $\Gamma_2(\{g,u\}) = g\otimes\Gamma(u)$.

(*ii*) *For any* $g,h \in G$ *and any* $r \in R$, $\omega_2(g\otimes ah) = \mathrm{Tr}(r)\cdot g\otimes g^{-1}h$.

Proof See Oliver [6, Theorems 3.7 and 3.9]. □

Other explicit formulas for Γ_2 are also given in Oliver [6]: for example, formulas for $\Gamma_2(\{a,u\})$ when $a \in R^*$ and $u \in (R[G])^*$ (Oliver [6, Theorem 4.3]). Also, Γ_2 has been shown (Oliver [6, Theorem 4.8]) to be natural with respect to transfer maps.

The obvious hope now is that similar natural exact sequences exist for nonabelian p-groups. Some more definitions are needed before a precise conjecture can be stated.

For an arbitrary group G, Dennis has defined an abelian group

$\tilde{H}_2(G)$, which sits in a short exact sequence

$$0 \longrightarrow \mathbb{Z}/2 \otimes G^{ab} \longrightarrow \tilde{H}_2(G) \longrightarrow H_2(G) \longrightarrow 0,$$

and such that $\tilde{H}_2(G) \cong (G \otimes G)/\langle g\otimes h + h\otimes g\rangle$ whenever G is abelian. The easiest way to define $\tilde{H}_2(G)$ for nonabelian G is as the pullback

$$\begin{array}{ccc} \tilde{H}_2(G) & \longrightarrow & (G^{ab}\otimes G^{ab})/\langle g\otimes h+h\otimes g\rangle \\ \downarrow & & \downarrow \\ H_2(G) \longrightarrow H_2(G^{ab}) & \cong & (G^{ab}\otimes G^{ab})/\langle g\otimes g\rangle. \end{array}$$

For any commuting pair $g,h \in G$, $g\wedge h \in H_2(G)$ and $g\tilde{\wedge}h \in \tilde{H}_2(G)$ will denote the images of $g\otimes h \in H_2(\langle g,h\rangle)$ and $g\otimes h \in \tilde{H}_2(\langle g,h\rangle)$, respectively ($\langle g,h\rangle$ being an abelian group).

For any G, Loday [1] has defined a natural homomorphism

$$\lambda_G : H_2(G) \longrightarrow K_2(\mathbb{Z}[G])/\{-1,G\},$$

which will be considered in more detail in Section 13b. For now, we just note that $\lambda_G(g\wedge h) = \{g,h\}$ for any commuting pair $g,h \in G$ (recall that $\{g,g\} = \{-1,g\}$). If R is the ring of integers in any finite extension of $\hat{\mathbb{Q}}_p$, then for the purposes here we set

$$Wh_2^c(R[G]) = \mathrm{Coker}\Big[H_2(G) \xrightarrow{\lambda} K_2(\mathbb{Z}[G])/\{-1,\pm G\} \xrightarrow{\mathbb{Z}\hookrightarrow R} K_2(R[G])/\{-1,\pm G\}\Big].$$

Note that when G is abelian, $Wh_2^c(R[G]) = K_2^c(R[G])/\{\pm G,\pm G\}$.

The obvious conjecture is now:

<u>Conjecture 6.13</u> *For any p-group* G, *and any unramified extension* F *of* $\hat{\mathbb{Q}}_p$ *with ring of integers* R, *there is an exact sequence*

$$HC_2(R[G]) \xrightarrow{\omega_3} H_3(G) \xrightarrow{\eta} Wh_2^c(R[G]) \xrightarrow{\Gamma_2} HC_1(R[G]) \xrightarrow{\omega_2} \tilde{H}_2(G)$$

$$\longrightarrow Wh(R[G]) \xrightarrow{\Gamma} HC_0(R[G]) \xrightarrow{\omega} H_1(G) \longrightarrow 0,$$

natural in G, *and satisfying the formulas:*

(*i*) $\Gamma_2(\{g,u\}) = g\otimes\Gamma_{RH}(u)$ *if* $g\in G$, $H=C_G(g)$, *and* $u\in R[H]^*$

(*ii*) $\omega_2(g\otimes rh) = \mathrm{Tr}(r)\cdot g\tilde{\wedge}g^{-1}h$ *for commuting* $g,h\in G$.

Note that the existence and exactness of the last half of the above sequence follows as a consequence of Theorems 6.6, 7.3, and 8.6 (and is included here only to show how it connects with the first half). For example, $\mathrm{Coker}(\omega_2) \cong H_2(G)/H_2^{ab}(G) \cong SK_1(R[G])$ by Theorem 8.6. See Oliver [6, Conjectures 0.1 and 5.1] for some more detailed conjectures.

The results in Oliver [4] also help to motivate Conjecture 6.13. In particular, by Oliver [4, Theorem 3.6], there is an exact sequence

$$H_3(G) \longrightarrow Wh_2^*(R[G]) \xrightarrow{\Gamma_2^*} \mathcal{U}(R[G]) \longrightarrow H_2(G);$$

where $Wh_2^*(R[G])$ is a certain quotient of $Wh_2^c(R[G])$, and where

$$\mathcal{U}(R[G]) = H_1(G;R[G])/\langle g\otimes rg^n:\ g\in G,\ r\in R,\ n\geq 1\rangle$$

(recall that $HC_1(R[G]) \cong H_1(G;R[G])/\langle g\otimes rg\rangle$). This helps to motivate the conjectured contribution of $H_3(G)$ to $\mathrm{Ker}(\Gamma_2)$, and shows that Γ_2 is at least defined to this quotient group $\mathcal{U}(R[G])$ of $HC_1(R[G])$. This sequence can also be combined with Theorem 6.12 to prove Conjecture 6.13 for some nonabelian groups, including some cases — such as $G\cong Q(8)$ — where $\eta\neq 1$. But presumably completely different methods will be needed to do this in general.

PART II: GROUP RINGS OF P-GROUPS

We are now ready to study the more detailed structure of $K_1(\mathbb{Z}[G])$ for finite G. For various reasons, both the results themselves (e. g., the formulas for $Cl_1(\mathbb{Z}[G])$ and $SK_1(\hat{\mathbb{Z}}_p[G])$), as well as the methods used to obtain them, are simplest when G is a p-group. For example, in some of the induction proofs, it is important that G is nilpotent and $\hat{\mathbb{Z}}_p[G]$ is a local ring. Also, the image of the integral logarithm Γ_G (Theorem 6.6), and the structure of $\mathbb{Q}[G]$ (Theorem 9.1), are simpler when G is a p-group.

The central chapters, Chapters 8 and 9, deal with the computations of $SK_1(\hat{\mathbb{Z}}_p[G])$ and $Cl_1(\mathbb{Z}[G])$, respectively. The most important results are Theorem 8.6, where $SK_1(\hat{\mathbb{Z}}_p[G])$ is described in terms of $H_2(G)$; and Theorems 9.5 and 9.6, where formulas for $Cl_1(\mathbb{Z}[G])$ are derived. Some examples are also worked out at the end of each of these chapters.

Chapter 7 is centered around Wall's theorem (Theorem 7.4) that $SK_1(\mathbb{Z}[G])$ is the full torsion subgroup of $Wh(G)$ for any finite group G. In contrast, the torsion free part of $Wh(G)$ is studied in Chapter 10, mostly using logarithmic methods. Also, the problem of representing arbitrary elements of $Wh'(G)$ $(= Wh(G)/SK_1(\mathbb{Z}[G]))$ by units in $\mathbb{Z}[G]$ is discussed at the end of Chapter 10. Note that Chapters 7 and 10, while dealing predominantly with p-groups, are not completely limited to this case.

These four chapters are mostly independent of each other. The main exception is Theorem 7.1 (and Corollary 7.2), which give upper bounds on the torsion in $Wh(\hat{\mathbb{Z}}_p[G])$ for a p-group G. These are used, both later in Chapter 7 when showing that $Wh'(\hat{\mathbb{Z}}_p[G])$ is torsion free, and in Section 8b when establishing upper bounds on the size of $SK_1(\hat{\mathbb{Z}}_p[G])$.

Chapter 7 THE TORSION SUBGROUP OF WHITEHEAD GROUPS

If G is any finite group, and if R is the ring of integers in any finite extension F of $\mathbb{Q}$ or $\hat{\mathbb{Q}}_p$, then obvious torsion elements in $K_1(R[G])$ include roots of unity in F, elements of G, and elements in $SK_1(R[G])$. These elements generate a subgroup of the form

$$\mu_F \times G^{ab} \times SK_1(R[G]) \subseteq K_1(R[G]). \tag{1}$$

To see that this is, in fact, a subgroup, note that $\mu_F \times G^{ab}$ injects into $K_1(F[G])$ — since it is a subgroup of $(F[G^{ab}])^* \cong K_1(F[G^{ab}])$ — and hence that $\mu_F \times G^{ab} \subseteq K_1(R[G])$ and $(\mu_F \times G^{ab}) \cap SK_1(R[G]) = 1$.

In particular, if we define the Whitehead group $Wh(R[G])$ by setting

$$Wh(R[G]) = K_1(R[G])/(\mu_F \times G^{ab}),$$

then $SK_1(R[G])$ can also be regarded as a subgroup of $Wh(R[G])$, and

$$Wh'(R[G]) = Wh(R[G])/SK_1(R[G]) = K_1'(R[G])/(\mu_F \times G^{ab}).$$

Note that when $R \neq \mathbb{Z}$, this notation is far from standard (sometimes one divides out by all units in R).

When G is abelian, then $K_1'(\mathbb{Z}[G]) \cong (\mathbb{Z}[G])^*$; and Higman [1] showed that the only torsion in $(\mathbb{Z}[G])^*$ is given by the units $\pm g$ for $g \in G$. In particular, $Wh'(G)$ is torsion free in this case. This provided the motivation for Wall [1] to show that $Wh'(G)$ is torsion free for any finite group G; i. e., that the subgroup in (1) above is the full torsion subgroup of $K_1(\mathbb{Z}[G])$. More generally, Wall showed that $Wh'(R[G])$ is torsion free whenever F is a number field and G is finite (Theorem 7.4 below), or whenever F is a finite extension of $\hat{\mathbb{Q}}_p$

and G is a p-group (Theorem 7.3)

If one only is interested in the results on torsion in $Wh'(R[G])$, then Theorem 7.1 and Corollary 7.2 below can be skipped. These are directed towards showing that $Wh(R[G])$ is torsion free when R is the ring of integers in a finite extension of $\hat{\mathbb{Q}}_p$, and G is a p-group with a normal abelian subgroup of index p. Wall's proof in [1] that $Wh'(R[G])$ is torsion free in this situation is simpler than the proof given here in Corollary 7.2. But the additional information in Corollary 7.2 (and in Theorem 7.1 as well) about $Wh(R[G])$ itself will be needed in Chapter 8 to get upper bounds on the size of $SK_1(R[G])$.

Recall the exact sequence of Proposition 6.4: if R is the ring of integers in any finite extension of $\hat{\mathbb{Q}}_p$, if G is any p-group, and if $z \in G$ is central of order p, then there is an exact sequence

$$1 \longrightarrow \langle z\rangle \longrightarrow K_1(R[G],(1-z)R[G]) \xrightarrow{\log} H_0(G;(1-z)R[G]) \longrightarrow \mathbb{F}_p \longrightarrow 0.$$

In particular, this gives a precise description of the torsion subgroup of $K_1(R[G],(1-z)R[G])$. The next theorem gives an upper bound for the number of those torsion elements which survive in $K_1(R[G])$.

<u>Theorem 7.1</u> *Fix a prime* p *and a p-group* G, *and let* $z \in G$ *be central of order* p. *Set*

$$\Omega = \{g \in G : g \text{ conjugate } zg\} = \{g \in G : [g,h] = z, \text{ some } h \in G\},$$

and let $\sim$ *be the equivalence relation on* Ω *generated by:*

$$g \sim h \text{ if } \begin{cases} g \text{ is conjugate to } h, \text{ or} \\ [g,h] = z^i \text{ for any } i \text{ prime to } p. \end{cases}$$

Then, if R *is the ring of integers in any finite extension of* $\hat{\mathbb{Q}}_p$,

$$\text{Ker}\Big[\text{tors } Wh(R[G]) \longrightarrow \text{tors } Wh(R[G/z])\Big] \cong (\mathbb{Z}/p)^N,$$

where

$N = 0$ *if* $\Omega = \emptyset$

$N \le |\Omega/\!\sim\!| - 1$ *if* $\Omega \ne \emptyset$ *(i. e., if* z *is a commutator).*

More precisely, if $\Omega \ne \emptyset$, *let* $p \subseteq R$ *be the maximal ideal, let*

$$\tau : R \twoheadrightarrow R/p \xrightarrow{\text{Tr}} \mathbb{F}_p$$

be as in Proposition 6.4, and fix any $r \in R$ *with* $\tau(r) \ne 0$. *Then if* $\{g_0,\ldots,g_k\}$ *are* $\sim$*-equivalence class representatives in* Ω, *the elements*

$$\text{Exp}(r(1-z)(g_0-g_i)) \qquad (\textit{for}\ \ 1 \le i \le k)$$

generate $\text{Ker}\Big[\text{tors Wh}(R[G]) \longrightarrow \text{tors Wh}(R[G/z])\Big]$.

<u>Proof</u> By Proposition 6.4, the logarithm induces a homomorphism

$$\log : K_1(R[G],(1-z)R[G]) \longrightarrow H_0(G;(1-z)R[G]);$$

where $\text{Ker}(\log) = \langle z \rangle$ and

$$\text{Im}(\log) = \Big\{(1-z)\textstyle\sum r_i g_i \ : \ r_i \in R,\ \ g_i \in G,\ \ \sum r_i \in \text{Ker}(\tau)\Big\}.$$

By Theorem 2.9, for any $u \in 1+(1-z)R[G]$, $[u]$ is torsion in $\text{Wh}(R[G])$ if and only if $[u] \in \text{Ker}[\log_{RG}: \text{Wh}(R[G]) \longrightarrow H_0(G;R[G])]$, if and only if

$$\log(u) \in \text{Ker}\Big[H_0(G;(1-z)R[G]) \longrightarrow H_0(G;R[G])\Big]$$

$$= \Big\langle r(1-z)g \in H_0(G;(1-z)R[G]) \ : \ g \text{ conj. } gz,\ \ r \in R\Big\rangle = H_0(G;(1-z)R(\Omega)).$$

So if we set

$$D = \Big\{\xi \in R(\Omega) \ : \ (1-z)\xi \in \log(1+(1-z)R[G])\Big\} = \Big\{\textstyle\sum r_i g_i \in R(\Omega) \ : \ \sum r_i \in \text{Ker}(\tau)\Big\}$$

and

$$C = \Big\{\xi \in R(\Omega) \;:\; (1-z)\xi = \mathrm{Log}(u),\ \text{some } u \in \mathrm{Ker}\big[1+(1-z)R[G] \longrightarrow Wh(R[G])\big]\Big\};$$

then

$$\mathrm{Ker}\Big[\mathrm{tors}\ Wh(R[G]) \longrightarrow \mathrm{tors}\ Wh(R[G/z])\Big] \cong D/C.$$

The theorem will now follow if we can show that

$$\begin{aligned} C \supseteq\ &\mathrm{Ker}\Big[R(\Omega) \xrightarrow{\ \mathrm{proj}\ } R(\Omega/\!\sim) \xrightarrow{\ \tau\ } \mathbb{F}_p(\Omega/\!\sim)\Big] \\ &= \langle sg,\ r(g-h) \;:\; r,s \in R,\ g,h \in G,\ g\sim h,\ s \in \mathrm{Ker}(\tau)\rangle. \end{aligned} \tag{1}$$

Note that since $\mathrm{Ker}[\mathrm{Tr}\colon R/p \longrightarrow \mathbb{F}_p] = (1-\varphi)R/p$ by Lemma 6.3(ii),

$$\mathrm{Ker}[\tau\colon R \twoheadrightarrow R/p \xrightarrow{\ \mathrm{Tr}\ } \mathbb{F}_p] = \{r-r^p\colon r \in R\} + p = \langle r-r^p\colon r \in R\rangle. \tag{2}$$

For any $g \in \Omega$ and $r \in R$, and any $k \geq 2$,

$$r(1-z)^k g = r(1-z)^{k-1}(g-zg) = 0 \in H_0(G;(1-z)R[G]).$$

In particular, by Lemma 6.3(i),

$$pr(1-z)g = -r(1-z)^p g = 0 \in H_0(G;(1-z)R[G]);$$

and so

(a) $p\cdot R(\Omega) + (1-z)R(\Omega) \subseteq C$.

Also, by definition,

(b) $r(g-h) \in C$ if $r \in R$ and g is conjugate to h.

So using (2), (1) will follow once we show, for all $r \in R$ and all $g,h \in \Omega$, that

(c) $(r-r^p)g \in C$, and

(d) $r(g-h) \in C$ if $[g,h] = z^i$ and $p \nmid i$.

Fix $g,h \in \Omega$ with $[g,h] = z^i$ and $p \nmid i$. It suffices to prove (c) and (d) when $G = \langle g,h \rangle$; in particular, when G/z is abelian. To simplify calculations, set

$$C' = C \oplus R(G \setminus \Omega) \subseteq R[G].$$

Since G/z is abelian, all p-th powers in G lie in $G \setminus \Omega$; and so by (a), all p-th powers of elements in $R[G]$ lie in C'. Hence, for any $\xi \in R[G]$,

$$\log(1+(1-z)\xi) = (1-z)\xi - \frac{(1-z)^2}{2}\cdot\xi^2 + \ldots \equiv (1-z)\xi \pmod{(1-z)C'};$$

and so

$$[1+(1-z)\xi] = 1 \in Wh(R[G]) \quad \text{implies} \quad \xi \in C'. \tag{3}$$

We now consider some specific commutators. For any $k \geq 0$,

$$[g^{-1}h, 1-r(g-h)^k] = 1 - \left(r(z^{-i}g-z^{-i}h)^k - r(g-h)^k\right)\cdot\left(1-r(g-h)^k\right)^{-1}$$

$$= 1 + (1-z^{-ik})r(g-h)^k\cdot\left(1-r(g-h)^k\right)^{-1}$$

$$\equiv 1 - (1-z)ik\cdot\left(r(g-h)^k + r^2(g-h)^{2k} + r^3(g-h)^{3k} + \ldots\right) \pmod{(1-z)^2}.$$

Since $p \nmid i$, (3) shows that

$$kr(g-h)^k + kr^2(g-h)^{2k} + kr^3(g-h)^{3k} + \ldots \in C'$$

for any $k \geq 1$. For k large enough, $r(g-h)^k \in p\cdot R[G] \subseteq C'$ by (a). A downwards induction on k now shows that

$$r(g-h)^k \in C' \quad \text{for all } k \geq 0$$

(when $p|k$ this holds since C' contains all p-th powers). In particular, $r(g-h) \in C' \cap R(\Omega) = C$.

This proves (d). To prove (c), fix j such that $[h^j,g] = z$, and consider the commutator

$$[h^j,1-r(1-g)] = 1 - r\Big((1-zg) - (1-g)\Big)\cdot\Big(1-r(1-g)\Big)^{-1}$$

$$= 1 - (1-z)\Big(rg + r^2g(1-g) + r^3g(1-g)^2 + \ldots\Big).$$

By (3),

$$rg - r^2g(1-g) + r^3g(1-g)^2 - \ldots \in C'. \tag{4}$$

By (d), $r^kg^\ell \equiv r^kh \equiv r^kg \pmod{C}$ for any k and any ℓ prime to p; and so (4) reduces to give

$$0 \equiv rg + (-1)^{p-1}r^pg(1-g)^{p-1} \equiv (r-r^p)g + r^pg^p \equiv (r-r^p)g \pmod{C'}.$$

It follows that $(r-r^p)g \in C' \cap R(\Omega) = C$. □

Later, in Section 8b, Theorem 7.1 will play a key role when obtaining upper bounds for the size of $SK_1(R[G])$. But for now, its main interest lies in the following corollary.

Corollary 7.2 *Fix a prime p, and let R be the ring of integers in any finite extension of $\hat{\mathbb{Q}}_p$. Let G be any p-group which contains an abelian normal subgroup $H \lhd G$ such that G/H is cyclic. Then $Wh(R[G])$ is torsion free. In particular, $SK_1(R[G]) = 1$.*

Proof This is clear if $G = 1$. Otherwise, we may assume $H \neq 1$, choose $z \in H \cap Z(G)$ of order p, and assume inductively that $Wh(R[G/z])$ is torsion free. Define

$$\Omega = \{g \in G : [g,h] = z, \text{ some } h \in G\},$$

and let $\sim$ be the equivalence relation on Ω defined in Theorem 7.1. By Theorem 7.1, we will be done upon showing that $\sim$ is transitive on Ω.

If $\Omega \neq \emptyset$, then fix any $g \in \Omega$, and any $x \in G \smallsetminus H$ which generates G/H. Choose $h \in \Omega$ such that $[g,h] = z$. Either gh^i or g^ih lies in

H for some i (G/H being cyclic); we may assume by symmetry that $gh^i = a \in H$. If we write $h = bx^j$ for some $b \in H$, then

$$z = [g,h] = [gh^i,h] = [a,bx^j] = [a,x^j] = [ax,x^j]$$

$$= [ax,x^j(ax)^{-j}] = [x,x^j(ax)^{-j}];$$

the last step since $x^j(ax)^{-j} \in H$. It follows that

$$g \sim h \sim gh^i = a \sim x^j \sim ax \sim x^j(ax)^{-j} \sim x$$

in Ω; and hence that the relation is transitive. □

We are now ready to describe the torsion in $Wh(R[G])$ in the p-adic case.

<u>Theorem 7.3</u> (Wall [1]) *Fix a prime* p, *and let* R *be the ring of integers in any finite extension* F *of* $\hat{\mathbb{Q}}_p$. *Then for any p-group* G, $Wh'(R[G])$ *is torsion free. In other words,*

$$\operatorname{tors}(K_1(R[G])) = \mu_F \times G^{ab} \times SK_1(R[G]);$$

where $\mu_F \subseteq R^*$ *is the group of roots of unity in* F.

<u>Proof</u> If G is abelian, then the theorem holds by Corollary 7.2. So the result is equivalent to showing, for arbitrary G, that

$$pr_* : \operatorname{tors} K_1'(R[G]) \longrightarrow \operatorname{tors} K_1'(R[G^{ab}])$$

is injective on torsion.

Fix G, and assume inductively that the theorem holds for all of its proper subgroups and quotients. If G is cyclic, dihedral, quaternionic, or semidihedral, then the theorem holds by Corollary 7.2. Otherwise, all simple summands of $F[G]$ are detected by restriction to proper subgroups and projection to proper quotients (see Roquette [1], Oliver & Taylor [1, Proposition 2.5], or Theorem 9.1 below). In other words, the restriction maps and quotient maps define a monomorphism

$$\sum \mathrm{Res}_H^G \oplus \sum \mathrm{Proj}_{G/N}^G : K_1(F[G]) \rightarrowtail \bigoplus_{\substack{H \subseteq G \\ [G:H]=p}} K_1(F[H]) \oplus \bigoplus_{\substack{N \triangleleft G \\ |N|=p}} K_1(F[G/N]) \qquad (1)$$

So the corresponding homomorphism for $K_1'(R[G])$ is also injective.

For any $H \subsetneq G$ of index p, consider the following commutative diagram:

$$\begin{array}{ccccc} \mathrm{tors}\, K_1'(R[G]) & \xrightarrow{pr_1} & \mathrm{tors}\, K_1'(R[G/[H,H]]) & \overset{pr_3}{\rightarrowtail} & \mathrm{tors}\, K_1'(R[G^{ab}]) \\ \downarrow t_1 & & \downarrow t_2 & & \\ \mathrm{tors}\, K_1'(R[H]) & \overset{pr_2}{\rightarrowtail} & \mathrm{tors}\, K_1'(R[H^{ab}]). & & \end{array}$$

Here, the t_i are transfer maps and the pr_i are induced by projection; pr_2 is injective by the induction assumption, and pr_3 by Corollary 7.2 ($G/[H,H]$ contains an abelian subgroup of index p). Hence, for any $u \in \mathrm{Ker}(pr_3 \circ pr_1)$, $t_1(u) = 1 \in K_1'(R[H])$.

Thus, for any $u \in \mathrm{Ker}(pr_*)$, $\mathrm{Trf}_H^G(u) = 1$ for all $H \subsetneq G$ of index p. Also, $\mathrm{Proj}_{G/N}^G(u) = 1$ for all $N \triangleleft G$ of order p (by the induction hypothesis again); and so $u = 1$ by (1). □

Note that Theorem 7.3 only holds for p-groups. Formulas describing the torsion in $K_1'(R[G])$ in the non-p-group case are given in Theorems 12.5 and 12.9 below.

In order to prove the corresponding theorem for global group rings (in particular, for $Wh(G)$), some induction theory is needed. For this reason, the next theorem might technically fit better after Chapter 11, but organizationally it seems more appropriate to include it here.

<u>Theorem 7.4</u> (Wall [1]) *For any finite group* G, $Wh'(G)$ *is torsion free. More generally, if* R *is the ring of integers in any number field* K, *and if* $\mu_K \subseteq R^*$ *denotes the group of roots of unity in* K, *then*

$$\mathrm{tors}(K_1(R[G])) = \mu_K \times G^{ab} \times SK_1(R[G]). \qquad (1)$$

Proof Fix a prime p. The proof of (1) for p-power torsion will be carried out in three steps: when G is a p-group, when G is p-hyperelementary, and when G is arbitrary. Note that for any prime ideal $\mathfrak{p} \subseteq R$, the completion homomorphisms

$$K_1(F[G]) \rightarrowtail K_1(\hat{F}_{\mathfrak{p}}[G]), \qquad K_1'(R[G]) \rightarrowtail K_1'(\hat{R}_{\mathfrak{p}}[G])$$

are injective (the reduced norm maps are injective by Theorem 2.3).

Step 1 Assume G is a p-group. For any prime $\mathfrak{p}|p$ in R,

$$\mu_K \times G^{ab} \subseteq \text{tors } K_1'(R[G]) \rightarrowtail \text{tors } K_1'(\hat{R}_{\mathfrak{p}}[G]) = \mu(\hat{K}_{\mathfrak{p}}) \times G^{ab}.$$

Since the inclusion $K_1'(R[G]) \rightarrowtail K_1'(\hat{R}_{\mathfrak{p}}[G])$ contains $(R^* \hookrightarrow (\hat{R}_{\mathfrak{p}})^*)$ as a direct summand, this shows that $\text{tors } K_1'(R[G]) = \mu_K \times G^{ab}$.

Step 2 Assume G is p-hyperelementary — i. e., G contains a normal cyclic subgroup of p-power index — but not a p-group. Fix some prime $q \neq p$ dividing $|G|$, and let $H \triangleleft G$ be the q-Sylow subgroup. We may assume inductively that the theorem holds for G/H.

Let $\mathfrak{q} \subseteq R$ be any prime ideal dividing q, and set

$$I = \text{Ker}\left[\hat{R}_{\mathfrak{q}}[G] \longrightarrow \hat{R}_{\mathfrak{q}}[G/H]\right].$$

Then I is a radical ideal, since $\hat{R}_{\mathfrak{q}} \supseteq \hat{\mathbb{Z}}_q$ and H is a q-group (this follows from Example 1.12). Hence $K_1(\hat{R}_{\mathfrak{q}}[G],I)$ is a pro-q-group (Theorem 2.10(ii)), and so

$$\text{tors}_p K_1'(\hat{R}_{\mathfrak{q}}[G]) \cong \text{tors}_p K_1'(\hat{R}_{\mathfrak{q}}[G/H])$$

$(p \neq q)$. But $\text{tors}_p K_1'(R[G/H]) = (\mu_K \times (G/H)^{ab})_{(p)}$ by the induction hypothesis, and so $\text{tors}_p K_1'(R[G]) = (\mu_K \times G^{ab})_{(p)}$.

Step 3 By standard induction theory (see Lam [1, Chapter 4], Bass [2, Chapter XI], or Theorem 11.2 below), for any finite group G,

$\operatorname{tors}_p K_1'(R[G])$ is generated by induction from p-hyperelementary subgroups. So $\operatorname{tors}_p K_1'(R[G]) = (\mu_K \times G^{ab})_{(p)}$ by Step 2. □

As an easy consequence of Theorem 7.4, we now get:

<u>Corollary 7.5</u> (Wall) *For any finite group* G, *the standard involution acts on* Wh'(G) *by the identity.*

<u>Proof</u> Write $Z(\mathbb{Q}[G]) = \prod F_i$, where the F_i are fields; and let $R_i \subseteq F_i$ denote the rings of integers. The involution on $\mathbb{Q}[G]$ acts on each F_i via complex conjugation (Proposition 5.11(ii)), and the reduced norm homomorphism

$$nr_{\mathbb{Z}[G]} : K_1(\mathbb{Z}[G]) \longrightarrow \prod (R_i)^*$$

commutes with the involutions by Lemma 5.10(ii). Also, $\operatorname{Ker}(nr_{\mathbb{Z}[G]}) = SK_1(\mathbb{Z}[G])$ is finite; and for each i, $(R_i)^*$/torsion is fixed by complex conjugation. So $Wh'(G) = K_1(\mathbb{Z}[G])$/torsion is also fixed by the involution. □

Chapter 8 THE P-ADIC QUOTIENT OF $SK_1(\mathbb{Z}[G])$: P-GROUPS

The central result of this chapter is the construction of an isomorphism

$$\Theta_G : SK_1(\hat{\mathbb{Z}}_p[G]) \cong SK_1(\mathbb{Z}[G])/Cl_1(\mathbb{Z}[G]) \xrightarrow{\cong} H_2(G)/H_2^{ab}(G)$$

for any prime p and any p-group G. Here, $H_2^{ab}(G) \subseteq H_2(G)$ is the subgroup generated by elements induced up from abelian subgroups of G. In fact, in Theorem 8.7, we will see that $SK_1(R[G]) \cong H_2(G)/H_2^{ab}(G)$ whenever R is the ring of integers in any finite extension of $\hat{\mathbb{Q}}_p$.

In Section 8a, the homomorphisms Θ_{RG} are constructed, and shown to be surjective. The definition of Θ_{RG} involves lifting elements of $SK_1(R[G])$ to $K_1(R[\tilde{G}])$, for some appropriate $\tilde{G}$ surjecting onto G; and then taking their integral logarithms (see Proposition 8.4). Section 8b then deals mostly with the proof that Θ_{RG} is an isomorphism. In Section 8c, some examples are given, both of groups for which $SK_1(\hat{\mathbb{Z}}_p[G]) = 1$, and of groups for which it is nonvanishing. The last result, Theorem 8.13, gives one way of constructing explicit nonvanishing elements of $SK_1(\hat{\mathbb{Z}}_p[G])$ in certain cases.

Throughout this chapter, p will denote a fixed prime.

8a. Detection of elements

The following proposition is the basis for detecting all elements in $SK_1(\hat{\mathbb{Z}}_p[G])$.

Proposition 8.1 *Let R be the ring of integers in any unramified extension F of $\hat{\mathbb{Q}}_p$. Then, for any extension* $1 \longrightarrow K \longrightarrow \tilde{G} \xrightarrow{\alpha} G \longrightarrow 1$

of p-groups,

$$\operatorname{Coker}\Big[SK_1(R\alpha) : SK_1(R[\tilde{G}]) \longrightarrow SK_1(R[G])\Big] \cong \frac{K\cap[\tilde{G},\tilde{G}]}{\langle [g,h]\in K:\ g,h\in\tilde{G}\rangle}. \tag{1}$$

More precisely, for any $u \in SK_1(R[G])$, *and any lifting of* u *to* $\tilde{u} \in K_1(R[\tilde{G}])$, *then* $\Gamma_{R\tilde{G}}(\tilde{u}) = \sum r_i(z_i-1)g_i$ *for some* $g_i \in G$, $r_i \in R$, *and* $z_i \in K$; *and* u *corresponds under (1) to the element*

$$\prod z_i^{\operatorname{Tr}(r_i)} \in K\cap[\tilde{G},\tilde{G}].$$

Here, $\operatorname{Tr}: R \longrightarrow \hat{\mathbb{Z}}_p$ *is the trace map.*

Proof For convenience, set

$$K_0 = \langle [g,h] \in K:\ g,h \in \tilde{G}\rangle \qquad \text{and} \qquad I_\alpha = \operatorname{Ker}\Big[R[\tilde{G}] \longrightarrow R[G]\Big].$$

The snake lemma applied to the diagram

$$\begin{array}{ccccccccc}
1 \longrightarrow & SK_1(R[\tilde{G}]) \times \mu_F \times \tilde{G}^{ab} & \longrightarrow & K_1(R[\tilde{G}]) & \longrightarrow & Wh'(R[\tilde{G}]) & \longrightarrow 1 \\
& \Big\downarrow SK_1(R\alpha)\times\alpha^{ab} & & \Big\downarrow K(\alpha) & & \Big\downarrow Wh'(\alpha) & \\
1 \longrightarrow & SK_1(R[G]) \times \mu_F \times G^{ab} & \longrightarrow & K_1(R[G]) & \longrightarrow & Wh'(R[G]) & \longrightarrow 1
\end{array}$$

induces an exact sequence

$$K_1(R[\tilde{G}], I_\alpha) \longrightarrow \operatorname{Ker}(Wh'(\alpha)) \longrightarrow \operatorname{Coker}(SK_1(R\alpha)) \longrightarrow 1. \tag{2}$$

Also, the following diagram with exact rows

$$\begin{array}{ccccccccc}
1 \longrightarrow & Wh'(R[\tilde{G}]) & \xrightarrow{\Gamma_{\tilde{G}}} & H_0(\tilde{G};R[\tilde{G}]) & \xrightarrow{\omega_{\tilde{G}}} & \langle\epsilon\rangle \times \tilde{G}^{ab} & \longrightarrow 1 \\
& \Big\downarrow Wh'(\alpha) & & \Big\downarrow H(\alpha) & & \Big\downarrow \alpha^{ab} & \\
1 \longrightarrow & Wh'(R[G]) & \xrightarrow{\Gamma_G} & H_0(G;R[G]) & \xrightarrow{\omega_G} & \langle\epsilon\rangle \times G^{ab} & \longrightarrow 1
\end{array}$$

(see Theorems 6.6 and 7.3) induces a short exact sequence of kernels

$$1 \longrightarrow \text{Ker}(Wh'(\alpha)) \xrightarrow{\ \Gamma\ } \overline{H}_0(\tilde{G};I_\alpha) \xrightarrow{\ \omega\ } \frac{K}{K\cap[\tilde{G},\tilde{G}]} \longrightarrow 1 \tag{3}$$

where $\overline{H}_0(\tilde{G};I_\alpha) = \text{Ker}(H(\alpha)) = \text{Im}\left[H_0(\tilde{G};I_\alpha) \longrightarrow H_0(\tilde{G};R[\tilde{G}])\right]$.

It remains to describe $\Gamma_{\tilde{G}}(K_1(R[\tilde{G}],I_\alpha))$. This could be done using the exact sequence of Theorem 6.9, but we take an alternate approach here to emphasize that the difficult part of that theorem is not needed.

We first check that there is a well defined homomorphism

$$\tilde{\omega} : \overline{H}_0(\tilde{G};I_\alpha) \longrightarrow K/K_0$$

such that $\tilde{\omega}(\sum r_i(z_i-1)g_i) = \prod z_i^{\text{Tr}(r_i)}$ for $r_i \in R$, $z_i \in K$, and $g_i \in \tilde{G}$. It suffices to check this when $K_0 = 1$; i. e., when K is central and contains no commutators. In particular, $\overline{H}_0(G;I_\alpha) \cong H_0(G;I_\alpha)$ in this case, since two distinct elements of $\tilde{G}$ in the same coset of K cannot be conjugate. And $\tilde{\omega}$ is well defined on $H_0(G;I_\alpha)$, since it is well defined on I_α itself (and $H_0(G;K) = K/[G,K] = K$).

Now define $\hat{G} = \{(g,h) \in \tilde{G}\times\tilde{G}:\ \alpha(g) = \alpha(h)\}$, so that

$$\begin{array}{ccc} \hat{G} & \xrightarrow{\ \beta_2\ } & \tilde{G} \\ {\scriptstyle\beta_1}\big\downarrow & & \big\downarrow{\scriptstyle\alpha} \\ \tilde{G} & \xrightarrow{\ \alpha\ } & G \end{array}$$

is a pullback square. Set

$$K_i = \text{Ker}(\beta_i)\ (\cong K) \quad\text{and}\quad I_i = \text{Ker}\left[R\beta_i : R[\hat{G}] \twoheadrightarrow R[\tilde{G}]\right] \qquad (i = 1,2).$$

Then β_2 is split by the diagonal map from $\tilde{G}$ to $\hat{G}$. In particular,

$$R[\hat{G}] \cong R[\tilde{G}] \oplus I_2, \qquad K_1(R[\hat{G}]) \cong K_1(R[\tilde{G}]) \oplus K_1(R[\hat{G}],I_2),$$

and $\hat{G}^{ab} \cong \tilde{G}^{ab} \times (K_2/[\tilde{G},K_2])$.

Consider the following diagram:

$$\begin{array}{ccccccc}
K_1(R[\hat{G}],I_2) & \xrightarrow{\Gamma_{\hat{G}}|} & H_0(\hat{G};I_2) & \xrightarrow{\omega_{\hat{G}}|} & K_2/[\hat{G},K_2] & \longrightarrow & 1 \\
\downarrow f_1 & & \downarrow f_2 & & \downarrow f_3 & & \\
K_1(R[\tilde{G}],I_\alpha) & \xrightarrow{\Gamma_{\tilde{G}}|} & \bar{H}_0(\tilde{G};I_\alpha) & \xrightarrow{\tilde{\omega}} & K/K_0 & \longrightarrow & 1;
\end{array} \tag{4}$$

where the f_i are induced by $\beta_1\colon \hat{G} \twoheadrightarrow \tilde{G}$. The top row is a direct summand of the exact sequence of Theorem 6.6 applied to $K_1(R[\hat{G}])$, and hence is exact. Furthermore, $K_2 \cong K$, and so $\mathrm{Ker}(f_3)$ is generated by elements of the form

$$([g,h],1) = (ghg^{-1},h)\cdot(h,h)^{-1} = \omega_{\hat{G}}\Big(r\cdot((ghg^{-1},h)-(h,h))\Big) \in \omega_{\hat{G}}(\mathrm{Ker}(f_2))$$

for $g,h \in \tilde{G}$ such that $[g,h] \in K$ (and where $\mathrm{Tr}(r) = 1$). In other words, $\omega_{\hat{G}}(\mathrm{Ker}(f_2)) = \mathrm{Ker}(f_3)$, and so the bottom row in (4) is exact.

It now follows that

$$\mathrm{Coker}(SK_1(R\alpha)) \cong \mathrm{Coker}\Big[K_1(R[\tilde{G}],I_\alpha) \longrightarrow \mathrm{Ker}(Wh'(\alpha))\Big] \qquad \text{(by (2))}$$

$$\cong \Gamma_{\tilde{G}}(\mathrm{Ker}(Wh'(\alpha)))/\Gamma_{\tilde{G}}(K_1(R[\tilde{G}],I_\alpha)) \qquad \text{(by (3))}$$

$$\cong \tilde{\omega}\circ\Gamma_{\tilde{G}}(\mathrm{Ker}(Wh'(\alpha))) \qquad \text{(by (4))}$$

$$= (K\cap[\tilde{G},\tilde{G}])/K_0 = \frac{K\cap[\tilde{G},\tilde{G}]}{\langle [g,h]\in K:\ g,h\in\tilde{G}\rangle}. \qquad \text{(by (3))}$$

The description of the isomorphism follows from the definition of $\tilde{\omega}$. □

Proposition 8.1 shows that elements in $SK_1(R[G])$ are detected by the difference between commutators in K (when $G \cong \tilde{G}/K$), and products of commutators in K. The functor $H_2(G)$ will now be used to provide a "universal group" for $\mathrm{Coker}(SK_1(\alpha))$, for all surjections α of p-groups onto G.

If G is any group, and $G \cong F/R$ where F is free, then by a formula of Hopf (see, e. g., Hilton & Stammbach [1, Section VI.9]),

$$H_2(G) \cong (R\cap[F,F])/[R,F].$$

If g, h is any pair of commuting elements in G, we let

$$g\wedge h \in H_2(G)$$

denote the element corresponding to $[\tilde{g},\tilde{h}] \in R\cap[F,F]$ for any liftings of g and h to $\tilde{g},\tilde{h}\in F$. If G is abelian, then $H_2(G) \cong \Lambda_2(G)$ is generated by such elements. So for arbitrary G,

$$H_2^{ab}(G) = \operatorname{Im}\Big[\sum\{H_2(H):\ H\subseteq G,\ H \text{ abelian}\} \xrightarrow{\Sigma\mathrm{Ind}} H_2(G)\Big]$$

$$= \langle g\wedge h:\ g,h\in G,\ gh=hg\rangle \subseteq H_2(G).$$

<u>Theorem 8.2</u> *Let* $1 \longrightarrow K \longrightarrow \tilde{G} \xrightarrow{\alpha} G \longrightarrow 1$ *be any extension of groups. Then for any* $\mathbb{Z}[G]$*-module* M*, there is a "five term homology exact sequence"*

$$H_2(\tilde{G};M) \xrightarrow{\alpha_*} H_2(G;M) \xrightarrow{\delta^\alpha_M} K^{ab}\otimes_{\mathbb{Z}[G]} M \longrightarrow H_1(\tilde{G};M) \xrightarrow{\alpha_*} H_1(G;M) \longrightarrow 0.$$

In particular, when $M = \mathbb{Z}$*, this takes the form*

$$H_2(\tilde{G}) \xrightarrow{H_2(\alpha)} H_2(G) \xrightarrow{\delta^\alpha} K/[\tilde{G},K] \longrightarrow \tilde{G}^{ab} \xrightarrow{\alpha^{ab}} G^{ab} \longrightarrow 1;$$

where for any commuting pair $g,h \in G$ *and any liftings to* $\tilde{g},\tilde{h} \in \tilde{G}$,

$$\delta^\alpha(g\wedge h) = [\tilde{g},\tilde{h}] \pmod{[\tilde{G},K]}.$$

If K *is central, then this can be extended to a 6-term exact sequence*

$$K\otimes\tilde{G} \xrightarrow{\gamma} H_2(\tilde{G}) \xrightarrow{H_2(\alpha)} H_2(G) \xrightarrow{\delta^\alpha} K \longrightarrow \tilde{G}^{ab} \xrightarrow{\alpha^{ab}} G^{ab} \longrightarrow 1,$$

where $\gamma(h\otimes g) = h\wedge g \in H_2^{ab}(G)$ *for any* $h \in K$ *and* $g \in \tilde{G}$.

<u>Proof</u> The 5-term sequences are shown in Hilton & Stammbach [1, Theorem VI.8.1 and Corollary VI.8.2]; and the formula for δ^α follows from the definition of $g\wedge h$ and naturality. The 6-term sequence, and the formula for γ, are shown in Stammbach [1, V.2.2 and V.2.1]. □

When $1 \longrightarrow K \longrightarrow \tilde{G} \xrightarrow{\alpha} G \longrightarrow 1$ is a central extension, then δ^α can be regarded as the image of the extension $[\alpha] \in H^2(G;K)$ under the epimorphism

$$H^2(G;K) \twoheadrightarrow \mathrm{Hom}(H_2(G),K)$$

in the universal coefficient theorem. So it is not surprising that central extensions can be constructed to realize any given homomorphism $H_2(G) \longrightarrow K$.

<u>Lemma 8.3</u> *(i) For any finite group G and any subgroup $T \subseteq H_2(G)$, there is a central extension $1 \longrightarrow K \longrightarrow \tilde{G} \xrightarrow{\alpha} G \longrightarrow 1$ such that $\delta^\alpha\colon H_2(G) \twoheadrightarrow K$ is surjective with kernel T.*

(ii) For any pair $H \subseteq G$ of finite groups, there is an extension

$$1 \longrightarrow K \longrightarrow \tilde{G} \xrightarrow{\alpha} G \longrightarrow 1$$

of finite groups, such that if we set $\tilde{H} = \alpha^{-1}(H)$ and $\alpha_0 = \alpha|\tilde{H}\colon \tilde{H} \twoheadrightarrow H$, then $H_2(\alpha_0) = 0$. If $H \lhd G$, then $\tilde{G}$ can be chosen such that $K \subseteq Z(\tilde{H})$; and if G is a p-group then $\tilde{G}$ can also be taken to be a p-group.

(iii) For any finite group G, and any finitely generated $\mathbb{Z}[G]$- or $\hat{\mathbb{Z}}_p[G]$-module M, there is an extension $1 \longrightarrow K \longrightarrow \tilde{G} \xrightarrow{\alpha} G \longrightarrow 1$ of finite groups such that

$$H_2(\alpha;M) = 0 : H_2(\tilde{G};M) \longrightarrow H_2(G;M).$$

<u>Proof</u> (i) Write $G = F/R$, where F is free. By Theorem 8.2, there is an exact sequence

$$0 = H_2(F) \longrightarrow H_2(G) \xrightarrow{\delta^F} R/[F,R] \longrightarrow F^{ab} \longrightarrow G^{ab} \longrightarrow 1.$$

Furthermore, F^{ab} and all of its subgroups are free abelian groups, and so $R/[F,R]$ splits as a product

$$R/[F,R] \cong R_0/[F,R] \times \delta^F(H_2(G))$$

for some $R_0 \lhd F$ where $[F,R] \subseteq R_0 \subseteq R$. If we now set $\tilde{G} = F/R_0$, and let $\alpha\colon \tilde{G} \twoheadrightarrow G$ be the projection, then $\delta^\alpha\colon H_2(G) \xrightarrow{\cong} R/R_0$ is an isomorphism. So for any $T \subseteq H_2(G)$, $\alpha_T\colon \tilde{G}/\delta^\alpha(T) \twoheadrightarrow G$ has the property that δ^{α_T} is surjective with kernel T.

(iii) Again write $G = F/R$, where F is free and finitely generated. By Theorem 8.2, there is an exact sequence

$$0 = H_2(F;M) \longrightarrow H_2(G;M) \longrightarrow R^{ab} \otimes_{\mathbb{Z}[G]} M \longrightarrow H_1(F;M).$$

Here, $H_2(G;M)$ is a finite p-group and $R^{ab} \otimes_{\mathbb{Z}[G]} M$ is a finitely generated $\mathbb{Z}$- or $\hat{\mathbb{Z}}_p$-module. So there is a normal subgroup $T \lhd F$ of finite index such that $[R,R] \subseteq T \subseteq R$, and such that $H_2(G;M)$ still injects into $(R/T) \otimes_{\mathbb{Z}[G]} M$. If we now set $\tilde{G} = F/T$, $K = R/T$, and let $\alpha\colon \tilde{G} \twoheadrightarrow G$ be the surjection, then δ^α is injective in the exact sequence

$$H_2(\tilde{G};M) \xrightarrow{H_2(\alpha;M)} H_2(G;M) \xrightarrow{\delta^\alpha} K \otimes_{\mathbb{Z}[G]} M.$$

So $H_2(\alpha;M) = 0$.

(ii) Now fix $H \subseteq G$, and set $M = \mathbb{Z}(G/H)$. Then $H_2(G;M) \cong H_2(H)$; and $H_2(\tilde{G};M) \cong H_2(\alpha^{-1}H)$ for any $\alpha\colon \tilde{G} \twoheadrightarrow G$. So by (iii), there is an extension $1 \longrightarrow K \longrightarrow \tilde{G} \xrightarrow{\alpha} G \longrightarrow 1$, such that if we set $\tilde{H} = \alpha^{-1}(H)$ and $\alpha_0 = \alpha|\tilde{H}$, then $H_2(\alpha_0) = 0$ and δ^{α_0} is injective. If $H \lhd G$, then $[\tilde{H},K] \lhd \tilde{G}$, and we can replace $\tilde{G}$ by $\tilde{G}/[\tilde{H},K]$ (so $K \subseteq Z(\tilde{H})$)

without changing the injectivity of δ^{α_0}. If G is a p-group, then $\tilde{G}$ can be replaced by any p-Sylow subgroup. □

Now, for any extension $1 \longrightarrow K \longrightarrow \tilde{G} \xrightarrow{\alpha} G \longrightarrow 1$ of p-groups, the 5-term exact sequence of Theorem 8.2 induces an exact sequence

$$H_2(\tilde{G}) \longrightarrow H_2(G)/H_2^{ab}(G) \xrightarrow{\delta_*} \frac{K \cap [\tilde{G},\tilde{G}]}{\langle [g,h] \in K:\ g,h \in \tilde{G} \rangle} \longrightarrow 1.$$
$$\cong \mathrm{Coker}(SK_1(R\alpha))$$

By Lemma 8.3(i), for any G, there exists $\tilde{G} \overset{\alpha}{\twoheadrightarrow} G$ such that $H_2(\alpha) = 0$. So $H_2(G)/H_2^{ab}(G)$ represents the largest possible group $\mathrm{Coker}(SK_1(R\alpha))$, among all $\alpha: \tilde{G} \twoheadrightarrow G$. This is the basis of the following proposition:

Proposition 8.4 *Let R be the ring of integers in any finite unramified extension of $\hat{\mathbb{Q}}_p$. Then for any p-group G, there is a natural surjection*

$$\Theta_{RG} : SK_1(R[G]) \twoheadrightarrow H_2(G)/H_2^{ab}(G),$$

characterized by the following property. For any extension

$$1 \longrightarrow K \longrightarrow \tilde{G} \xrightarrow{\alpha} G \longrightarrow 1$$

of p-groups, for any $u \in SK_1(R[G])$, and for any lifting $\tilde{u} \in K_1(R[\tilde{G}])$ of u, if we write $\Gamma_{RG}(\tilde{u}) = \sum r_i(z_i-1)g_i$ (where $r_i \in R$, $z_i \in K$, and $g_i \in \tilde{G}$), then

$$\delta^{\alpha}(\Theta_{RG}(u)) = \prod z_i^{\mathrm{Tr}(r_i)} \in K/\langle [g,h] \in K:\ g,h \in \tilde{G} \rangle. \qquad (\delta^{\alpha}: H_2(G) \longrightarrow K/[\tilde{G},K])$$

Furthermore, Θ_{RG} is an isomorphism if $\Theta_{R\tilde{G}}$ is an isomorphism for any p-group $\tilde{G}$ surjecting onto G.

__Proof__ The only thing left to check is the last statement. By the above discussion, Θ_{RG} is an isomorphism if and only if $SK_1(R\alpha) = 1$ for some surjection $\tilde{G} \xrightarrow{\alpha} G$ of p-groups. Clearly, this property holds for G if it holds for any p-group surjecting onto G. □

8b. Establishing upper bounds

It remains to show that the epimorphism Θ_{RG} of Proposition 8.4 is an isomorphism. While lower bounds for $SK_1(R[G])$ were found by studying $\mathrm{Coker}(SK_1(R\alpha))$ for surjections $\alpha: \tilde{G} \twoheadrightarrow G$, the upper bounds will be established by studying $\mathrm{Coker}(SK_1(f))$ when $f: H \hookrightarrow G$ is an inclusion of a subgroup of index p. The following lemma provides the main induction step.

Lemma 8.5 *Let R be the ring of integers in any finite unramified extension of $\hat{\mathbb{Q}}_p$. Then for any pair $H \lhd G$ of p-groups with $[G:H] = p$, if $SK_1(R[H]) = 1$, then Θ_{RG} is an isomorphism.*

Proof For the purposes of induction, the following stronger statement will be shown: for any pair $G \supseteq H$ of p-groups with $[G:H] = p$, Θ_{RG} factors through an isomorphism

$$\Theta_0 : \mathrm{Coker}\Big[SK_1(R[H]) \to SK_1(R[G])\Big] \xrightarrow{\cong} \mathrm{Coker}\Big[H_2(H)/H_2^{ab}(H) \to H_2(G)/H_2^{ab}(G)\Big]. \qquad (1)$$

Note that Θ_0 is onto by Proposition 8.4. Let $f: H \longrightarrow G$ denote the inclusion, and let

$$SK_1(f): SK_1(R[H]) \longrightarrow SK_1(R[G]), \quad Wh(f): Wh(R[H]) \longrightarrow Wh(R[G]),$$

$$H_2(f): H_2(H) \longrightarrow H_2(G), \quad \text{and} \quad H_2/H_2^{ab}(f): H_2(H)/H_2^{ab}(H) \longrightarrow H_2(G)/H_2^{ab}(G)$$

denote the induced homomorphisms. Fix some $x \in G \smallsetminus H$; and fix $r \in R$ such that $\mathrm{Tr}(r) = 1 \in \hat{\mathbb{Z}}_p$ (Proposition 1.8(iii)).

Choose any $\xi \in SK_1(R[G])$ such that $\Theta_{RG}(\xi) \in \mathrm{Im}(H_2/H_2^{ab}(f))$. We must show that $\xi \in \mathrm{Im}(SK_1(f))$. This will be done in three steps. In Step 1, Theorem 7.1 will be used to show that $\xi = Wh(f)(\xi_0)$, for some $\xi_0 \in Wh(R[H])$ such that $\Gamma_{RH}(\xi_0) = \sum_{i=1}^{n} r(h_i - xh_ix^{-1})$, and where the $h_i \in H$ satisfy $[h_i,x] \in [H,H]$. In Step 2, we first identify $\mathrm{Coker}(H_2/H_2^{ab}(f))$ with a certain subquotient of H; and then show that $\Theta_0([\xi])$ corresponds under this identification to $h_1 \cdots h_n$. Then, in Step 3, this is used to show that $\xi \in \mathrm{Im}(SK_1(f))$.

<u>Step 1</u> We can assume that H is nonabelian: otherwise $SK_1(R[G]) = 1$ by Theorem 1.14(ii). Fix $z \in [H,H]$ which is a central commutator of order p in G (Lemma 6.5), set $\hat{H} = H/z$, $\hat{G} = G/z$, and let

$$\begin{array}{ccc} H & \xrightarrow{f} & G \\ \downarrow{\scriptstyle \alpha_H} & & \downarrow{\scriptstyle \alpha} \\ \hat{H} & \xrightarrow{\hat{f}} & \hat{G} \end{array}$$

be the induced maps. Consider the following commutative diagram:

$$\begin{array}{ccc} \mathrm{Coker}(SK_1(f)) & \xrightarrow{\Theta_0} & \mathrm{Coker}(H_2/H_2^{ab}(f)) \\ \downarrow{\scriptstyle S(\alpha)} & & \downarrow{\scriptstyle H(\alpha)} \\ \mathrm{Coker}(SK_1(\hat{f})) & \xrightarrow[\cong]{\hat{\Theta}_0} & \mathrm{Coker}(H_2/H_2^{ab}(\hat{f})). \end{array}$$

Here, $\hat{\Theta}_0$ is induced by $\Theta_{R\hat{G}}$ (and is assumed inductively to be an isomorphism); and $S(\alpha)$ and $H(\alpha)$ are induced by α. In particular, $[\xi] \in \mathrm{Ker}(\Theta_0) \subseteq \mathrm{Ker}(S(\alpha))$.

Consider the following homomorphisms:

$$\begin{array}{ccccccc} SK_1(R[H]) & \xrightarrow{SK_1(f)} & SK_1(R[G]) & \xrightarrow{pr} & \mathrm{Coker}(SK_1(f)) & \longrightarrow & 1 \\ \downarrow{\scriptstyle SK_1(\alpha_H)} & & \downarrow{\scriptstyle SK_1(\alpha)} & & \downarrow{\scriptstyle S(\alpha)} & & \\ SK_1(R[\hat{H}]) & \xrightarrow{SK_1(\hat{f})} & SK_1(R[\hat{G}]) & \longrightarrow & \mathrm{Coker}(SK_1(\hat{f})) & \longrightarrow & 1 \end{array}$$

Since $S(\alpha)([\xi]) = 1$, there exists $\hat{\eta} \in SK_1(R[\hat{H}])$ such that $SK_1(\hat{f})(\eta) = SK_1(\alpha)(\xi)$. By Proposition 8.1, we can lift $\hat{\eta}$ to some $\eta \in Wh(R[H])$ such that $\Gamma_{RH}(\eta) = ra(1-z)g_0$, for some $a \in \mathbb{Z}$ and any desired $g_0 \in H$. In particular, since z is a commutator in G, we may choose g_0 such that g_0 is conjugate in G to zg_0.

Now set

$$\Omega = \{g \in G : g \text{ conjugate } zg\} = \{g \in G : [g,h] = z, \text{ some } h \in G\} \neq \emptyset,$$

and let $\sim$ be the relation on Ω from Theorem 7.1. For any $g \in \Omega$, either $g \in H$, or $[g,h] = z$ for some $h \in H$ (since G/H is cyclic). So each $\sim$-equivalence class of Ω includes elements of H. By Theorem 7.1,

$$\begin{aligned} Ker(SK_1(\alpha)) &= \langle Exp(r(1-z)(g-h)) : g,h \in H \cap \Omega \rangle \\ &\subseteq \langle Wh(f)(\Gamma_{RH}^{-1}(r(g-zg))) : g \in H, \ g \text{ conj. } zg \text{ in } G \rangle. \end{aligned}$$

Since $Wh(f)(\eta) \equiv \xi \pmod{Ker(SK_1(\alpha))}$, this shows that we can write

$$\xi = Wh(f)(\xi_0), \qquad \text{where} \qquad \Gamma_{RH}(\xi_0) = r(1-z)\cdot\sum g_i,$$

and where g_i is conjugate in G to zg_i for all i.

Recall that $x \in G$ generates G/H. Hence, for each i, there is some $r_i \leq p-1$ such that $x^{r_i}g_ix^{-r_i}$ is conjugate in H to zg_i. In particular, $\Gamma_{RH}(\xi_0) = \sum r(g_i - x^{r_i}g_ix^{-r_i}) \in H_0(H;R[H])$. By relabeling, we can find elements $h_1,\ldots,h_n \in H$ such that

$$\Gamma_{RH}(\xi_0) = \sum_{i=1}^{n} r(h_i - xh_ix^{-1}), \quad \text{and} \quad [h_i,x] \in [H,H] \quad (\text{all } i). \qquad (2)$$

<u>Step 2</u> Now set

$$K = Coker(H_2/H_2^{ab}(f)) = H_2(G)/\langle H_2^{ab}(G), Im(H_2(f))\rangle,$$

for short, and fix a central extension $1 \longrightarrow K \longrightarrow \tilde{G} \xrightarrow{\beta} G \longrightarrow 1$ such that

$$\delta^\beta : H_2(G) \twoheadrightarrow K = \operatorname{Coker}(H_2/H_2^{ab}(f))$$

is the projection (use Lemma 8.3(i)). In particular, $K \cap [\tilde{H},\tilde{H}] = 1$ (where $\tilde{H} = \beta^{-1}(H)$), since $\operatorname{Im}(H_2(f)) \subseteq \operatorname{Ker}(\delta^\beta)$.

The Hochschild-Serre spectral sequence for $1 \to H \xrightarrow{f} G \longrightarrow C_p \to 1$ (see Brown [1, Theorem VII.6.3]) induces an exact sequence

$$H_3(C_p) \xrightarrow{\partial} H_1(C_p;H^{ab}) \xrightarrow{\sigma_0} \operatorname{Coker}(H_2(f)) \longrightarrow 0.$$

The usual identification of $H_1(C_p;—)$ with invariant elements modulo norms takes here the form

$$H_1(C_p;H^{ab}) = \frac{\{h \in H^{ab} : xhx^{-1} = h \text{ in } H^{ab}\}}{\langle h \cdot xhx^{-1} \cdots x^{p-1}hx^{1-p} : h \in H\rangle} = \frac{\{h \in H : [h,x] \in [H,H]\}}{[H,H]\cdot\langle (hx)^p x^{-p} : h \in H\rangle}.$$

Under this identification, $\partial(H_3(C_p)) = \langle x^p\rangle$ by naturality (compare this with the corresponding sequence for $1 \to \langle x^p\rangle \to \langle x\rangle \to C_p \to 1$). So there is an isomorphism

$$\sigma_1 : \frac{\{h \in H : [h,x] \in [H,H]\}}{[H,H]\cdot\langle (hx)^p : h \in H\rangle} \xrightarrow{\cong} \operatorname{Coker}(H_2(f)).$$

Furthermore,

$$\sigma_1^{-1}(H_2^{ab}(G)) = \langle h \in H : h \text{ conj. } xhx^{-1} \text{ in } H\rangle \supseteq \langle (hx)^p : h \in H\rangle;$$

and so σ_1 factors through an isomorphism

$$\sigma : \frac{\{h \in H : [h,x] \in [H,H]\}}{[H,H]\cdot\langle h \in H : h \text{ conj. } xhx^{-1} \text{ in } H\rangle} \xrightarrow{\cong} \operatorname{Coker}(H_2/H_2^{ab}(f)) = K. \quad (3)$$

By construction, for each h,

$$\sigma(h) = [\beta^{-1}x,\beta^{-1}h] \pmod{[\tilde{H},\tilde{H}]}. \qquad (4)$$

By (2), $\Gamma_{RH}(\xi_0) = \sum_{i=1}^n r(h_i - xh_ix^{-1})$, where $[h_i,x] \in [H,H]$ for all i. Choose liftings $\tilde{h}_i,\tilde{x} \in \tilde{G}$ of $h_i,x \in G$. Since $K \cap [\tilde{H},\tilde{H}] = 1$, there are unique elements $z_i \in K$ such that

$$[\tilde{h}_i,\tilde{x}] \equiv z_i \pmod{[\tilde{H},\tilde{H}]}. \qquad (5)$$

Fix $u_i \in Wh(R[\tilde{H}])$ such that $\Gamma_{R\tilde{H}}(u_i) = r(\tilde{h}_i - \tilde{x}\tilde{h}_i\tilde{x}^{-1} + (1-z_i))$ (use Theorem 6.6). Then $Wh(\beta|\tilde{H})(\prod_{i=1}^n u_i) \equiv \xi_0 \pmod{SK_1(R[H])}$. Also, $\Gamma_{R\tilde{G}}(Wh(\tilde{f})(u_i)) = r(1-z_i)$; and so by the formula in Proposition 8.4,

$$\Theta_0([\xi]) = \prod_{i=1}^n z_i^{-1} \in K = \operatorname{Coker}(H_2/H_2^{ab}(f))$$

Then by (4) and (5), $\sigma^{-1}(\Theta_0(\xi)) = \prod_{i=1}^n h_i$.

__Step 3__ Now by (3), we can write $\prod_{i=1}^n h_i = \hat{h}\cdot h_1'\cdots h_m'$, where $\hat{h} \in [H,H]$, and where each h_j' is conjugate in H to $xh_j'x^{-1}$. We may assume $n \equiv m \pmod 2$ (otherwise just take $h_{m+1}' = 1$). By Theorem 6.6, we can choose $\xi_1 \in Wh(R[H])$ such that $\Gamma_{RH}(\xi_1) = r\left(\sum_{i=1}^n h_i - \sum_{j=1}^m h_j'\right)$. Then

$$\Gamma_{RH}(\xi_0\cdot x\xi_1x^{-1}\cdot\xi_1^{-1}) = \sum_{i=1}^n r(h_i - xh_ix^{-1}) - \Gamma(\xi_1) + x\cdot\Gamma(\xi_1)\cdot x^{-1}$$

$$= \sum_{i=1}^m r(h_i'-xh_i'x^{-1}) = 0 \in H_0(H;R[H]);$$

and $\xi_0 \equiv [\xi_1,x] \pmod{SK_1(R[H])}$. It follows that $\xi = Wh(f)(\xi_0) \in \operatorname{Im}(SK_1(f))$, and this finishes the proof. □

The proof that Θ_{RG} always is an isomorphism is now just a matter of choosing the right induction argument.

Theorem 8.6 *Let* R *be the ring of integers in any finite unramified extension of* $\hat{\mathbb{Q}}_p$. *Then for any p-group* G,

$$\Theta_{RG} : SK_1(R[G]) \xrightarrow{\cong} H_2(G)/H_2^{ab}(G)$$

is an isomorphism. Furthermore, the standard involution $(g \mapsto g^{-1})$ *acts on* $SK_1(R[G])$ *by negation.*

Proof This will be shown by induction on $|G/Z(G)|$. Fix any nonabelian p-group G ($SK_1(R[G]) = 1$ if G is abelian); and let $H \lhd G$ be any index p subgroup such that $H \supseteq Z(G)$. By Lemma 8.3(ii), there is a surjection $\alpha_1: G_1 \twoheadrightarrow G/Z(G)$ of p-groups, with $H_1 = \alpha_1^{-1}(H/Z(G))$, such that $Ker(\alpha_1) \subseteq Z(H_1)$ and $H_2(\alpha_1|H_1) = 0$. Let $\tilde{G}$ be the pullback

$$\begin{array}{ccc} \tilde{G} & \xrightarrow{\alpha} & G \\ \downarrow & & \downarrow \\ G_1 & \xrightarrow{\alpha_1} & G/Z(G) \end{array}$$

and set $\tilde{H} = \alpha^{-1}(H)$. Then $1 \longrightarrow \tilde{K} \longrightarrow \tilde{H} \xrightarrow{\tilde{\alpha}} H/Z(G) \longrightarrow 1$ is a central extension, so $|H/Z(H)| < |G/Z(G)|$, and $\Theta_{R\tilde{H}}$ is an isomorphism by the induction hypothesis. Also, $H_2(\tilde{\alpha}) = 0$, since α factors through $\alpha_1|H_1$; and so $H_2^{ab}(\tilde{H}) \supseteq Ker(H_2(\tilde{\alpha})) = H_2(\tilde{H})$ by the 6-term exact sequence of Theorem 8.2. This shows that $SK_1(R[\tilde{H}]) \cong H_2(\tilde{H})/H_2^{ab}(\tilde{H}) = 0$; and Lemma 8.5 now applies to show that $\Theta_{R\tilde{G}}$ is an isomorphism. But $\tilde{G}$ surjects onto G, and so Θ_{RG} is an isomorphism by the last statement in Proposition 8.4.

By the description of Θ_{RG} in Proposition 8.4, for any $[u] \in SK_1(R[G])$, $\Theta_{RG}([\bar{u}]) = -\Theta([u])$. Since Θ_{RG} is an isomorphism, this shows that $SK_1(R[G])$ is negated by the standard involution. □

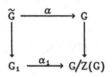

Theorem 8.6 can in fact be extended to include group rings over arbitrary finite extensions of $\hat{\mathbb{Q}}_p$. This does not have the same import-

ance when studying $SK_1(\mathbb{Z}[G])$ as does the case of unramified extensions; but we include the next theorem for the sake of completeness.

Theorem 8.7 *Let* R *be the ring of integers in any finite extension* F *of* $\hat{\mathbb{Q}}_p$. *Then for any p-group* G,

$$SK_1(R[G]) \cong H_2(G)/H_2^{ab}(G).$$

If $E \supseteq F$ *is a finite extension, and if* $S \subseteq E$ *is the ring of integers, then*

(i) $i_* : SK_1(R[G]) \xrightarrow{\cong} SK_1(S[G])$ *(induced by inclusion) is an isomorphism if* E/F *is totally ramified; and*

(ii) $trf : SK_1(S[G]) \xrightarrow{\cong} SK_1(R[G])$ *(the transfer) is an isomorphism if* E/F *is unramified.*

Proof Note first that for any finite extension E of $\hat{\mathbb{Q}}_p$, there is a unique subfield $F \subseteq E$ such that $F/\hat{\mathbb{Q}}_p$ is unramified and E/F is totally ramified. To see this, let $\mathfrak{p} \subseteq S \subseteq E$ be the maximal ideal and ring of integers, and set $m = |(S/\mathfrak{p})^*|$. Let μ_m be the group of m-th roots of unity in E, and set $F = \hat{\mathbb{Q}}_p(\mu_m) \subseteq E$ and $R = \hat{\mathbb{Z}}_p[\mu_m]$. By Theorem 1.10, $F/\hat{\mathbb{Q}}_p$ is unramified, and $R \subseteq F$ is the ring of integers. Also, E/F is totally ramified since $|R/pR| = |S/\mathfrak{p}| = m+1$.

In particular, this shows that it suffices to prove (i) and (ii) under the assumption that F is unramified over $\hat{\mathbb{Q}}_p$. If E is also unramified, then the following triangle commutes by the description of Θ_{SG} and Θ_{RG} in Proposition 8.4:

$$\begin{array}{ccc} SK_1(S[G]) & \xrightarrow{trf} & SK_1(R[G]) \\ {\scriptstyle\Theta_{SG}} \searrow \cong & & \cong \swarrow {\scriptstyle\Theta_{RG}} \\ & H_2(G)/H_2^{ab}(G). & \end{array}$$

So the transfer is an isomorphism in this case.

Now assume that E/F is totally ramified (and $F/\hat{\mathbb{Q}}_p$ is unramified). Let $p \subseteq S$ be the maximal ideal. Then

$$\mathrm{Ker}\left[i_*: SK_1(R[G]) \longrightarrow SK_1(S[G])\right]$$

$$\subseteq \mathrm{Ker}\left[SK_1(R[G]) \longrightarrow K_1(R/p[G]) \cong K_1(S/p[G])\right]$$

$$\subseteq \mathrm{tors}_p \mathrm{Im}\left[K_1(R[G],p) \longrightarrow K_1(R[G])\right].$$

Using the logarithm homomorphism $\log\colon K_1(R[G],p) \longrightarrow H_0(G;pR[G])$ of Theorem 2.8, one checks easily that $K_1(R[G],p)$ is p-torsion free if p is odd, and that the only torsion is $\{\pm 1\}$ if $p = 2$.

Thus, in either case, i_* is injective. The surjectivity of i_* is now shown by induction on $|G|$, using Theorem 7.1 again. For details, see Oliver [2, Proposition 15]. □

We end the section by showing that the isomorphisms Θ_{RG} are natural, not only with respect to group homomorphisms, but also with respect to transfer homomorphisms induced by inclusions of p-groups.

<u>Proposition 8.8</u> *Let R be the ring of integers in any finite extension of $\hat{\mathbb{Q}}_p$. Then for any pair $H \subseteq G$ of p-groups, the square*

$$\begin{array}{ccc} SK_1(R[G]) & \xrightarrow{\Theta_{RG}} & H_2(G)/H_2^{ab}(G) \\ \big\downarrow \mathrm{trf}_{SK} & & \big\downarrow \mathrm{trf}_H \\ SK_1(R[H]) & \xrightarrow{\Theta_{RH}} & H_2(H)/H_2^{ab}(H) \end{array}$$

commutes. Here, trf_{SK} and trf_H are induced by the usual transfer homomorphisms for K_1 and H_2, respectively.

<u>Proof</u> By Theorem 8.7, it suffices to show this when $R = \hat{\mathbb{Z}}_p$. Using Lemma 8.3(ii), choose an extension

$$1 \longrightarrow K \longrightarrow \tilde{G} \xrightarrow{\alpha} G \longrightarrow 1, \qquad \tilde{H} = \alpha^{-1}(H), \qquad \alpha_0 = \alpha|_H$$

of p-groups, such that $K \subseteq Z(\tilde{H})$ and $H_2(\alpha_0) = 0$. Then $\delta^{\alpha_0}: H_2(H) \rightarrowtail K$ is injective, and $SK_1(\hat{\mathbb{Z}}_p[\alpha_0]) = 1$ by Theorem 8.6. By the description of Θ in Proposition 8.4, it will suffice to show that the following squares all commute:

$$\begin{array}{ccccccc}
SK_1(\hat{\mathbb{Z}}_p[G]) & \hookrightarrow & K_1(\hat{\mathbb{Z}}_p[G]) & \xleftarrow{\alpha_*} & K_1(\hat{\mathbb{Z}}_p[\tilde{G}]) & \xrightarrow{\Gamma_G} & H_0(\tilde{G};\hat{\mathbb{Z}}_p[\tilde{G}]) \\
\downarrow{\scriptstyle trf_{SK}} & (1) & \downarrow{\scriptstyle trf_K} & (2) & \downarrow{\scriptstyle trf_{\tilde{K}}} & (3) & \downarrow{\scriptstyle Res^{\tilde{G}}_{\tilde{H}}} \\
SK_1(\hat{\mathbb{Z}}_p[H]) & \hookrightarrow & K_1(\hat{\mathbb{Z}}_p[H]) & \xleftarrow{\alpha_{0*}} & K_1(\hat{\mathbb{Z}}_p[\tilde{H}]) & \xrightarrow{\Gamma_H} & H_0(\tilde{H};\hat{\mathbb{Z}}_p[\tilde{H}])
\end{array}$$

$$\begin{array}{ccccc}
\bar{H}_0(\tilde{G};I_\alpha) & \xrightarrow{\omega_\alpha} & K/\langle [g,h]\in K:\ g,h\in\tilde{G}\rangle & \xleftarrow{\delta^\alpha} & H_2(G)/H_2^{ab}(G) \\
\downarrow{\scriptstyle Res^{\tilde{G}}_{\tilde{H}}} & & \downarrow{\scriptstyle N_{G/H}} & (5) & \downarrow{\scriptstyle trf_H} \\
\bar{H}_0(\tilde{H};I_{\alpha_0}) & \xrightarrow{\omega_{\alpha_0}} & K/\langle [g,h]\in K:\ g,h\in\tilde{H}\rangle & \xleftarrow{\delta^{\alpha_0}} & H_2(H)/H_2^{ab}(H)
\end{array}$$

Here, $Res^{\tilde{G}}_{\tilde{H}}$ is the homomorphism of Theorem 6.8;

$$\bar{H}_0(\tilde{G};I_\alpha) = \operatorname{Ker}\left[H_0(\tilde{G};\hat{\mathbb{Z}}_p[\tilde{G}]) \twoheadrightarrow H_0(G;\hat{\mathbb{Z}}_p[G])\right],$$

$$\omega_\alpha(\textstyle\sum r_i(1-a_i)g_i) = \prod a_i^{r_i} \qquad (r_i \in \hat{\mathbb{Z}}_p,\ \ a_i \in K,\ \ g_i \in \tilde{G}),$$

(and similarly for $\bar{H}_0(\tilde{H};I_{\alpha_0})$ and ω_{α_0}); and $N_{G/H}$ is the norm map for the conjugation action of G/H on K. The commutativity of (1) and (2) is clear, (3) commutes by Theorem 6.8, and (4) by definition of $Res^{\tilde{G}}_{\tilde{H}}$.

The commutativity of (5) follows since trf_H splits as a composite

$$H_2(G) \xrightarrow{f_1} H_2(G;\mathbb{Z}(G/H)) \xleftarrow[\cong]{f_2} H_2(H)$$

(see Brown [1, Section III.9]); and similarly for $N_{G/H}$. Here, f_1 and f_2 are induced by the inclusions $i_1, i_2\colon \mathbb{Z} \hookrightarrow \mathbb{Z}(G/H)$, where $i_1(1) = \sum_{g\in G/H} g$ and $i_2(1) = 1$ (note that i_2 is only $\mathbb{Z}[H]$-linear). □

8c. Examples

It turns out that $H_2(G)$ need not be computed completely in order to describe $H_2(G)/H_2^{ab}(G) \cong SK_1(\hat{\mathbb{Z}}_p[G])$. In practice, the following formula provides the easiest way to make computations and to construct examples.

Lemma 8.9 *Fix a central extension* $1 \longrightarrow K \longrightarrow G \xrightarrow{\alpha} \hat{G} \longrightarrow 1$ *of p-groups, and define*

$$\Lambda(\hat{G}) = \left\{g\wedge h \in H_2(\hat{G}) \;:\; g,h \in \hat{G},\ gh = hg\right\} \subseteq H_2(\hat{G})$$

(a subset of $H_2(\hat{G})$*). Let* $\delta^\alpha\colon H_2(\hat{G}) \longrightarrow K$ *be the boundary map in the 5-term homology exact sequence (Theorem 8.2). Then, if* R *is the ring of integers in any finite extension of* $\hat{\mathbb{Q}}_p$*,*

$$SK_1(R[G]) \cong H_2(G)/H_2^{ab}(G) \cong \mathrm{Ker}(\delta^\alpha)/\langle \Lambda(\hat{G}) \cap \mathrm{Ker}(\delta^\alpha)\rangle.$$

In particular, $SK_1(R[G]) = 1$ *if* $H_2(\alpha) = 0$.

Proof Consider again the 6-term homology exact sequence for a central extension (Theorem 8.2):

$$K\otimes G^{ab} \xrightarrow{\gamma} H_2(G) \xrightarrow{H_2(\alpha)} H_2(\hat{G}) \xrightarrow{\delta^\alpha} K \longrightarrow G^{ab} \longrightarrow \hat{G}^{ab} \longrightarrow 1.$$

Here, $\gamma(x\otimes g) = x\wedge g \in H_2^{ab}(G)$ for any $x \in K$ and any $g \in G$. So $\mathrm{Ker}(H_2(\alpha)) \subseteq H_2^{ab}(G)$. Furthermore,

$$H_2(\alpha)(H_2^{ab}(G)) = \langle g\wedge h \in H_2(\hat{G}) : g,h \text{ lift to commuting elements in } G\rangle$$

$$= \langle \Lambda(\hat{G}) \cap \mathrm{Ker}(\delta^{\alpha})\rangle;$$

and the result follows. □

As a first, simple application of Lemma 8.9, we note the following conditions for $SK_1(\hat{\mathbb{Z}}_p[G])$ to vanish.

<u>Theorem 8.10</u> *Let R be the ring of integers in any finite extension of $\hat{\mathbb{Q}}_p$. Then $SK_1(R[G]) = 1$ if G is a p-group satifying any of the following conditions:*

(i) there exists $H \triangleleft G$ such that H is abelian and G/H is cyclic, or

(ii) $[G,G]$ is central and cyclic, or

(iii) $G/Z(G)$ is abelian of rank ≤ 3.

<u>Proof</u> See Corollary 7.2 and Oliver [2, Proposition 23]. □

The smallest p-groups G with $SK_1(\hat{\mathbb{Z}}_p[G]) \neq 1$ have order 64 if $p=2$, or p^5 if p is odd (see Oliver [2, Proposition 24]). The following examples are larger, but are easier to describe.

<u>Example 8.11</u> *Fix $n \geq 1$, and set*

$$G = \Big\langle a,b,c,d : [G,[G,G]] = 1 = a^{p^n} = b^{p^n} = c^{p^n} = d^{p^n} = [a,b][c,d]\Big\rangle.$$

Then $SK_1(\hat{\mathbb{Z}}_p[G]) \cong \mathbb{Z}/p^n$.

<u>Proof</u> By construction, G sits in a central extension

$$1 \longrightarrow (C_{p^n})^5 \longrightarrow G \xrightarrow{\ \alpha\ } G^{ab} \cong (C_{p^n})^4 \longrightarrow 1;$$

where $\delta^\alpha: H_2(G^{ab}) \longrightarrow (C_{p^n})^5$ is surjective with kernel

$$\mathrm{Ker}(\delta^\alpha) = \langle a\wedge b + c\wedge d\rangle \cong \mathbb{Z}/p^n.$$

Then $\Lambda(G^{ab}) \cap \mathrm{Ker}(\delta^\alpha) = 1$, in the notation of Lemma 8.9, and the result follows. □

Recall that for any G and n, $G\wr C_n$ denotes the wreath product $G^n \rtimes C_n$. The next proposition describes how $H_2(G)/H_2^{ab}(G)$ and $SK_1(\hat{\mathbb{Z}}_p[G])$ act with respect to products and wreath products.

<u>Proposition 8.12</u> *For any finite groups G and H, and any $n > 1$,*

$$H_2(G\times H)/H_2^{ab}(G\times H) \cong H_2(G)/H_2^{ab}(G) \oplus H_2(H)/H_2^{ab}(H), \quad \textit{and}$$

$$H_2(G\wr C_n)/H_2^{ab}(G\wr C_n) \cong H_2(G)/H_2^{ab}(G).$$

In particular, if R is the ring of integers in any finite extension of $\hat{\mathbb{Q}}_p$, and if G and H are p-groups, then

$$SK_1(R[G\times H]) \cong SK_1(R[G]) \oplus SK_1(R[H]) \quad \textit{and} \quad SK_1(R[G\wr C_p]) \cong SK_1(R[G]).$$

Also, $SK_1(R[G]) = 1$ if G is a p-Sylow subgroup in any symmetric group.

<u>Proof</u> See Oliver [2, Proposition 25]. The only point that is at all complicated is that involving $G\wr C_n$.

For $1 \le i \le n$, let $f_i : G \longrightarrow G\wr C_n$ be the inclusion into the i-th factor of G^n. Fix $x \in (G\wr C_n)\smallsetminus G^n$ such that $x^n = 1$, and such that

$$x(g_1,\dots,g_n)x^{-1} = (g_2,\dots,g_n,g_1) \quad (\text{for all } (g_1,\dots,g_n) \in G^n).$$

Define

$$T = \langle f_i(g)\wedge f_j(h) : g,h \in G,\ i\ne j\rangle \subseteq H_2^{ab}(G\wr C_n).$$

A straightforward argument using the Hochschild-Serre spectral sequence (see Brown [1, Theorem VII.6.3]) shows that f_1 induces an isomorphism

$$f_{1*} : H_2(G) \xrightarrow{\cong} H_2(G\wr C_n)/T.$$

Then $H_2^{ab}(G\wr C_n)/T$ is generated by $f_{1*}(H_2^{ab}(G))$ (i. e., elements $g\wedge h$ for commuting $g,h \in G^n$); as well as all $gx\wedge h$ for $g,h \in G^n$ such that $[gx,h] = 1$. For elements of the last type, if $g = (g_1,\ldots,g_n)$ and $h = (h_1,\ldots,h_n)$, then a direct computation shows that

$$gx\wedge h = f_{1*}((g_1\cdots g_n)\wedge h_1) \in f_{1*}(H_2^{ab}(G));$$

and so f_{1*} induces an isomorphism $H_2(G)/H_2^{ab}(G) \cong H_2(G\wr C_n)/H_2^{ab}(G\wr C_n)$. □

By Theorem 7.1, if $1 \longrightarrow \langle z\rangle \longrightarrow \tilde{G} \xrightarrow{\alpha} G \longrightarrow 1$ is any central extension of p-groups such that $|z| = p$, then

$$\mathrm{Ker}(SK_1(\hat{\mathbb{Z}}_p\alpha)) = \Big\langle \mathrm{Exp}((1-z)(g-h)):\ g,h\in\Omega\Big\rangle. \qquad (\Omega = \{g\in\tilde{G}:\ g \text{ conj. } zg\})$$

Also, for any $r\in\hat{\mathbb{Z}}_p$, $[\mathrm{Exp}(r(1-z)(g-h))]$ depends only on $r \pmod{p\hat{\mathbb{Z}}_p}$, and on the classes of g and h modulo a certain equivalence relation $\sim$ in Ω. It is natural now to check where these elements are sent under the isomorphism $\Theta_{\tilde{G}}$. This is done in following theorem, which describes one case where elements in $SK_1(\hat{\mathbb{Z}}_p[\tilde{G}])$ can be constructed or detected directly (in contrast to the very indirect definition of $\Theta_{\tilde{G}}$ in Proposition 8.4).

<u>Theorem 8.13</u> *Fix a p-group* G *and a central commutator* $z\in G$ *of order* p, *and let* $\alpha\colon G \twoheadrightarrow G/z$ *denote the projection. Let* R *be the ring of integers in any finite extension of* $\hat{\mathbb{Q}}_p$, *let* $\mathfrak{p}\subseteq R$ *be the maximal ideal, and let* $\hat{\alpha}\colon R[G] \twoheadrightarrow \mathbb{F}_p[G/z]$ *be the epimorphism induced by* α *and by* $\tau\colon R \twoheadrightarrow R/\mathfrak{p} \xrightarrow{\mathrm{Tr}} \mathbb{F}_p$. *Set*

$$\Omega = \{g\in G : g \textit{ conjugate to } zg\} = \{g\in G : [g,h] = z, \textit{ some } h\in G\}.$$

Define functions

$$\mathbb{F}_p(\Omega/z) \xrightarrow{\ \chi\ } H_2(G/z)/\langle \mathrm{Ker}(\delta^\alpha) \cap \Lambda(G/z)\rangle \xleftarrow{\ \alpha_*\ } H_2(G)/H_2^{ab}(G)$$

where $\chi(g) = \alpha(g)\wedge\alpha(h)$ *for any* $g,h \in \Omega$ *such that* $[g,h] = z$; *and where* α_* *is induced by* $H_2(\alpha)$ *(an injection by Lemma 8.9). Then for any* $[u] = [1+(1-z)\xi] \in \mathrm{Ker}(SK_1(R\alpha))$, *if we set* $\mathrm{Log}(u) = (1-z)\eta$, *then* $\hat{\alpha}(\eta) = \hat{\alpha}(\xi - \xi^p) \in \mathbb{F}_p(\Omega/z)$ *and*

$$\Theta_{RG}([u]) = \alpha_*^{-1} \circ \chi(\hat{\alpha}(\eta)) = \alpha_*^{-1} \circ \chi(\hat{\alpha}(\xi - \xi^p)).$$

<u>Proof</u> See Oliver [2, Proposition 26]. Note that if $\sim$ is the equivalence relation defined in Theorem 7.1, then χ factors through $\mathbb{F}_p(\Omega/\sim)$. This then gives a new interpretation of the inequality

$$\mathrm{rk}_{\mathbb{F}_p}\Big(\mathrm{Ker}(SK_1(R\alpha))\Big) \le |\Omega/\sim| - 1$$

of Theorem 7.1. □

Chapter 9 $Cl_1(\mathbb{Z}[G])$ FOR P-GROUPS

We now turn to the problem of describing $Cl_1(\mathbb{Z}[G])$, when G is a p-group and p is any prime. This question is completely answered for odd p in Theorem 9.5, and partly answered in the case of 2-groups in Theorem 9.6. Conjecture 9.7 then suggests results which would go further towards describing the structure of $Cl_1(\mathbb{Z}[G])$ (and $SK_1(\mathbb{Z}[G])$) in the 2-group case. Some examples of computations of $Cl_1(\mathbb{Z}[G])$ are given at the end of the chapter, in Examples 9.8 and 9.9.

All of these results are based on the localization sequence

$$K_2^c(\hat{\mathbb{Z}}_p[G]) \xrightarrow{\varphi_G} C_p(\mathbb{Q}[G]) \xrightarrow{\partial_G} Cl_1(\mathbb{Z}[G]) \longrightarrow 1$$

of Theorem 3.15. The group $C_p(\mathbb{Q}[G])$ has already been described in Theorem 4.13. So there are two remaining problems to solve before $Cl_1(\mathbb{Z}[G])$ can be computed: a set of generators must be found for $K_2^c(\hat{\mathbb{Z}}_p[G])$, and a simple algorithm is needed for describing their images in $C_p(\mathbb{Q}[G])$. The first problem is solved (in part) in Proposition 9.4, and the second in Proposition 9.3.

If $\mathbb{Q}[G] = \prod_{i=1}^k A_i$, where the A_i are simple, then $C_p(\mathbb{Q}[G]) = \prod_{i=1}^k C_p(A_i)$, and the $C_p(A_i)$ have been described in terms of roots of unity in $Z(A_i)$. The following theorem helps to make this more explicit, by listing all of the possible "representation types" which can occur in a group ring of a p-group: i. e., all of the isomorphism types of simple summands. As usual, when p is fixed, then ξ_n (any $n \geq 0$) denotes the root of unity $\xi_n = \exp(2\pi i/p^n) \in \mathbb{C}$.

Theorem 9.1 *Fix a prime* p *and a p-group* G, *and let* A *be any simple summand of* $\mathbb{Q}[G]$. *If* p *is odd, then* A *is isomorphic to a* *matrix algebra over* $\mathbb{Q}(\xi_n)$ *for some* $n \geq 0$; *and* $K_2^c(\hat{A}_p)_{(p)} \cong \langle \xi_n \rangle \cong C_p(A)$. *If* $p = 2$, *then* A *is a matrix algebra over one of the division algebras* D *in the following table:*

D		$K_2^c(\hat{A}_2)$	$C(A) = C_2(A)$
$\mathbb{Q}$		$\{\pm 1\}$	1
$\mathbb{Q}(\xi_n)$	$(n \geq 2)$	$\langle \xi_n \rangle$	$\langle \xi_n \rangle$
$\mathbb{Q}(\xi_n + \xi_n^{-1})$	$(n \geq 3)$	$\{\pm 1\}$	1
$\mathbb{Q}(\xi_n - \xi_n^{-1})$	$(n \geq 3)$	$\{\pm 1\}$	$\{\pm 1\}$
$\mathbb{Q}(\xi_n, j)$ $(\subseteq \mathbb{H})$	$(n \geq 2)$	$\{\pm 1\}$	$\{\pm 1\}$

Proof See Roquette [1]. In Section 2 of [1], Roquette shows that the division algebra for any irreducible representation of G is isomorphic to that of a primitive, faithful representation of some subquotient of G; and in Section 3 he shows that the only p-groups with primitive faithful representations are the cyclic groups; and (if $p = 2$) the dihedral, quaternion, and semidihedral groups. For each such G, $\mathbb{Q}[G]$ has a unique faithful summand A, given by the following table:

G	C_{p^n}	$D(2^{n+1})$	$Q(2^{n+1})$	$SD(2^{n+1})$
A	$\mathbb{Q}(\xi_n)$	$M_2(\mathbb{Q}(\xi_n + \xi_n^{-1}))$	$\mathbb{Q}(\xi_n, j)$	$M_2(\mathbb{Q}(\xi_n - \xi_n^{-1}))$

The computations of $K_2^c(\hat{A}_p)$ and $C_p(A)$ follow immediately from Theorems 4.11 and 4.13. □

We next turn to the problem of describing $\varphi(\{g,u\}) \in C_p(\mathbb{Q}[G])$, for certain Steinberg symbols $\{g,u\} \in K_2^c(\hat{\mathbb{Z}}_p[G])$. The homology group $H_1(G;\mathbb{Z}[G])$, where G acts on $\mathbb{Z}[G]$ via conjugation, provides a useful bookkeeping device for doing this. Note that for any G, if $g_1, \ldots, g_k$ are conjugacy class representatives for elements of G, then

$$H_1(G;\mathbb{Z}[G]) \cong \bigoplus_{i=1}^{k} H_1(C_G(g_i)) \otimes \mathbb{Z}(g_i) \cong \bigoplus_{i=1}^{k} C_G(g_i)^{ab} \otimes \mathbb{Z}(g_i).$$

In particular, $H_1(G;\mathbb{Z}[G])$ is generated by elements $g\otimes h$, for commuting $g,h \in G$.

Let

$$\sigma_G : C_p(\mathbb{Q}[G]) \xrightarrow{\cong} \prod_{i\in I} (\mu_{K_i})_p$$

be the isomorphism of Theorem 4.13: where $\mathbb{Q}[G] = \prod_{i=1}^k A_i$, $K_i = Z(A_i)$, and $I \subseteq \{1,\ldots,k\}$ is an appropriate subset.

Definition 9.2 *Fix a prime* p *and a p-group* G, *and define a homomorphism*

$$\psi_G : H_1(G;\mathbb{Z}[G]) \cong H_1(G;\hat{\mathbb{Z}}_p[G]) \longrightarrow C_p(\mathbb{Q}[G])$$

as follows. Write $\mathbb{Q}[G] = \prod_{i=1}^k A_i$, *where each* A_i *is simple with irreducible module* V_i *and center* K_i. *Let* $I \subseteq \{1,\ldots,k\}$ *be the set of all* i *such that* $C_p(A_i) \neq 1$; *i. e., such that* $D_i = \mathrm{End}_{A_i}(V_i) \not\subseteq \mathbb{R}$. *For each* $i \in I$, *set*

$$\epsilon_i = \begin{cases} 2^{r-1}+1 & \text{if } p = 2 \text{ and } K_i \cong \mathbb{Q}(\xi_r) \\ 1 & \text{otherwise.} \end{cases}$$

Then, for any commuting pair $g,h \in G$, *set*

$$\psi_G(g\otimes h) = \sigma_G^{-1}\Big(\big(\det_{K_i}(g,V_i^h)^{\epsilon_i}\big)_{i\in I}\Big) \in C_p(\mathbb{Q}[G]). \qquad (V_i^h = \{x \in V_i\colon hx = x\})$$

Note in particular the form taken by ψ_G when G is abelian. Fix such a G, write $\mathbb{Q}[G] = \prod_{i=1}^k K_i$ where the K_i are fields, and let $\chi_i\colon G \twoheadrightarrow \mu_{K_i}$ be the corresponding character. Let ϵ_i be defined as in Definition 9.2, and set $I = \{i\colon K_i \not\subseteq \mathbb{R}\}$. Then

$$\sigma_G\circ\psi_G(g\otimes h) = (\psi_i(g\otimes h))_{i\in I} \in \prod_{i\in I}(\mu_{K_i})_p$$

where $$\psi_i(g\otimes h) = \begin{cases} \chi_i(g)^{\epsilon_i} & \text{if } \chi_i(h) = 1 \\ 1 & \text{if } \chi_i(h) \neq 1. \end{cases}$$

When p is odd, the ψ_G are easily seen to be natural with respect to homomorphisms between p-groups. This is *not* the case for 2-groups: naturality fails even for the inclusion $C_2 \hookrightarrow C_4$.

We now focus attention on certain Steinberg symbols in $K_2^c(\hat{\mathbb{Z}}_p[G])$: symbols of the form $\{g,u\}$, where $g \in G$ and $u \in (\hat{\mathbb{Z}}_p[C_G(g)])^*$ (i. e., each term in u commutes with g). The next proposition describes how ψ_G is used to compute the images in $C(\mathbb{Q}[G])$ of the $\{g,u\}$. Afterwards, Proposition 9.4 will show that when p is odd, $\operatorname{Im}(\partial_G)$ is generated by the images of such symbols.

Proposition 9.3 *Fix a prime* p *and a p-group* G. *If* $p = 2$, *then let* $A_1,\ldots,A_\ell$ *be the distinct quaternionic simple summands of* $\mathbb{Q}[G]$; *i. e., those simple summands which are matrix algebras over* $\mathbb{Q}(\xi_m,j)$ *for some* m. *Define* $C_p^Q(\mathbb{Q}[G]) \subseteq C_p(\mathbb{Q}[G])$ *by setting*

$$C_p^Q(\mathbb{Q}[G]) = \begin{cases} 1 & \text{if } p \text{ is odd} \\ \prod_{i=1}^{\ell} C_p(A_i) & \text{if } p = 2. \end{cases}$$

Then, for any $g \in G$, *any* $H \subseteq G$ *such that* $[g,H] = 1$, *and any* $u \in (\hat{\mathbb{Z}}_p[H])^*$,

$$\varphi_G(\{g,u\}) \equiv \psi_G(g\otimes\Gamma_H(u)) \pmod{C_p^Q(\mathbb{Q}[G])}.$$

Proof This is a direct application of the symbol formulas of Artin and Hasse (see Theorem 4.7(ii)).

Fix a simple summand A of $\mathbb{Q}[G]$, let $\chi\colon \mathbb{Q}[G] \longrightarrow A$ be the projection, and let V be the irreducible A-module. Let $K = Z(A)$ be the center, and assume that K has no real imbeddings. In other words, $K \cong \mathbb{Q}(\xi_r)$ $(p^r > 2)$ or $\mathbb{Q}(\xi_r - \xi_r^{-1})$ $(p=2,\ r \geq 3)$; where $\xi_r = \exp(2\pi i/p^r)$ as usual. Let φ_A and ψ_A denote the composites

$$\varphi_A : K_2^c(\hat{\mathbb{Z}}_p[G]) \xrightarrow{\varphi_G} C_p(\mathbb{Q}[G]) \xrightarrow{C_p(\chi)} C_p(A), \quad \text{and}$$

$$\psi_A : H_1(G;\mathbb{Z}[G]) \xrightarrow{\psi_G} C_p(\mathbb{Q}[G]) \xrightarrow{C_p(\chi)} C_p(A).$$

We must show that $\varphi_A(\{g,u\}) = \psi_A(g\otimes\Gamma_H(u))$ for any g,u as above.

Set $p^n = \exp(G)$, and let $L = K(\xi_n) \cong \mathbb{Q}(\xi_n)$. Define

$$\epsilon_K = \begin{cases} 1 & \text{if } p > 2, \text{ or } p = 2 \text{ and } K \cong \mathbb{Q}(\xi_r - \xi_r^{-1}) \quad (r \geq 3) \\ 1+2^{r-1} & \text{if } p = 2 \text{ and } K \cong \mathbb{Q}(\xi_r) \quad (r \geq 2) \end{cases}$$

and similarly for ϵ_L. Set $W = L\otimes_K V$, and let $\eta_1,\ldots,\eta_m \in \langle\xi_n\rangle$ be the distinct eigenvalues of g on W. Write $W = W_1\oplus\ldots\oplus W_m$, where W_j is the eigenspace for η_j. Then, for each $h \in H \subseteq C_G(g)$, the action of h on W leaves each W_j invariant.

Write

$$\mathrm{Log}(u) = \sum_{i=1}^k a_i h_i, \qquad \Gamma(u) = \sum_{i=1}^k a_i(h_i - \tfrac{1}{p}\cdot h_i^p). \qquad (h_i \in H,\ a_i \in \hat{\mathbb{Q}}_p)$$

Then by definition of ψ,

$$\sigma_A \circ \psi_A(g\otimes\Gamma_H(u)) = \prod_{j=1}^m (\eta_j)^{T_j\epsilon_K}, \tag{1}$$

where $\sigma_A\colon C_p(A) \xrightarrow{\cong} \mu_K$ is the norm residue symbol isomorphism, and where for each j,

$$T_j = \sum_{i=1}^k a_i\Big[\dim_L((W_j)^{h_i}) - \tfrac{1}{p}\cdot\dim_L((W_j)^{h_i^p})\Big]. \tag{2}$$

Note that $T_j\in\hat{\mathbb{Z}}_p$ for all j, since $\Gamma_H(u) \in \hat{\mathbb{Z}}_p[H]$ (modulo conjugacy).

Now let φ_A^L denote the composite

$$\varphi_A^L : K_2^c(\hat{\mathbb{Z}}_p[G]) \xrightarrow{\varphi_A} C_p(A) \xrightarrow{1\otimes} C_p(L\otimes_K A).$$

Then

$$\sigma_{L\otimes A}\circ\varphi_A^L(\{g,u\}) = \prod_{j=1}^{m} (\eta_j, \det_L(u,W_j))_L.$$

The Artin-Hasse formula (Theorem 4.7(ii)) takes here the form

$$\sigma_{L\otimes A} \circ \varphi_A^L(\{g,u\}) = \prod_{j=1}^{m} (\eta_j)^{S_j \epsilon_L}, \tag{3}$$

where for each j,

$$S_j = \frac{1}{p^n}\cdot Tr_{L/\mathbb{Q}}(\log(\det_L(u,W_j))) = \frac{1}{p^n}\cdot Tr_{L/\mathbb{Q}}(Tr_j(\log(u)))$$

$$= \frac{1}{p^n}\cdot Tr_{L/\mathbb{Q}}(Tr_j(\sum_{i=1}^{k} a_i h_i)) = \frac{1}{p^n}\cdot Tr_{L/\mathbb{Q}}(\sum_{i=1}^{k} a_i\cdot\chi_j(h_i)). \tag{4}$$

Here, $Tr_j\colon End_L(W_j) \longrightarrow L$ is the trace map, and $\chi_j(h)$ is the character (in L) of h on W_j.

For each $\zeta \in (\mu_L)_p = \langle \xi_n \rangle$,

$$\frac{1}{p^n}\cdot Tr_{L/\mathbb{Q}}(\zeta) = \begin{cases} 1 - \frac{1}{p} & \text{if } \zeta = 1 \\ -\frac{1}{p} & \text{if } |\zeta| = p \\ 0 & \text{if } |\zeta| \ge p^2. \end{cases}$$

In particular, for each j and each $h \in C_G(g)$,

$$\frac{1}{p^n}\cdot Tr_{L/\mathbb{Q}}(\chi_j(h)) = \dim_L(W_j^h) - \frac{1}{p}\cdot\dim_L((W_j)^{h^p}).$$

Substituting this into (4) and comparing with (2) now gives

$$S_j = \sum_{i=1}^{k} a_i\Big[\dim_L((W_j)^{h_i}) - \frac{1}{p}\cdot\dim_L((W_j)^{h_i^p})\Big] = T_j.$$

Now consider the diagram

$$\begin{array}{ccccccc}
K_2^c(\hat{\mathbb{Z}}_p[G]) & \xrightarrow{\varphi_A} & C_p(A) & \xrightarrow[\cong]{\sigma_A} & (\mu_K)_p & \xleftarrow[\cong]{\sigma_K} & C_p(K) \\
& \searrow^{\varphi_A^L} & \downarrow{\scriptstyle 1\otimes} & (5a) & \downarrow{\scriptstyle \iota} & (5b) & \downarrow{\scriptstyle \mathrm{incl}} \\
& & C_p(L\otimes_K A) & \xrightarrow[\cong]{\sigma_{L\otimes A}} & (\mu_L)_p & \xleftarrow[\cong]{\sigma_L} & C_p(L).
\end{array} \tag{5}$$

Here, (5a+5b) commutes by Proposition 4.8(ii), and so there is a homomorphism ι which makes each square commute. By (1) and (3),

$$\sigma_A\circ\varphi_A(\{g,u\}) = \prod_{j=1}^{m}(\eta_j)^{T_j\epsilon_K} \quad\text{and}\quad \iota\circ\sigma_A\circ\psi_A(g\otimes\Gamma_H(u)) = \prod_{j=1}^{m}(\eta_j)^{S_j\epsilon_L};$$

and $S_j = T_j$ for all j. So the relation $\varphi_A(\{g,u\}) = \psi_A(g\otimes\Gamma_H(u))$ will follow, once we show that $\iota(\zeta^{\epsilon_K}) = \zeta^{\epsilon_L}$ for any $\zeta\in(\mu_K)_p$.

It suffices to do this when $[L:K] = p$; i. e., when $L = \mathbb{Q}(\xi_n)$, and $K = \mathbb{Q}(\xi_{n-1})$ or $\mathbb{Q}(\xi_n - \xi_n^{-1})$. Consider the following diagram:

$$\begin{array}{ccccc}
C_p(L) & \xrightarrow{\mathrm{trf}} & C_p(K) & \xrightarrow{\mathrm{incl}} & C_p(L) \\
\cong\downarrow{\scriptstyle \sigma_L} & & \cong\downarrow{\scriptstyle \sigma_K} & & \cong\downarrow{\scriptstyle \sigma_L} \\
(\mu_L)_p & \xrightarrow{\tau} & (\mu_K)_p & \xrightarrow{\iota} & (\mu_L)_p;
\end{array}$$

where $\tau(\xi_n) = (\xi_n)^q$ if $q = [(\mu_L)_p:(\mu_K)_p]$. The left-hand square commutes by Theorem 4.6. The composite $\mathrm{incl}\circ\mathrm{trf}$ is induced by the (L,L)-bimodule $L\otimes_K L$ (see Proposition 1.18), and is hence the norm homomorphism for the action of $\mathrm{Gal}(L/K)$. If $K = \mathbb{Q}(\xi_{n-1})$, then

$$\iota(\xi_{n-1})^{\epsilon_K} = \iota\circ\tau(\xi_n)^{\epsilon_K} = \Big(\prod_{i=0}^{p-1}(\xi_n)^{1+ip^{n-1}}\Big)^{\epsilon_K} = \Big((\xi_n)^{\epsilon_K}\Big)^{p+\binom{p}{2}p^{n-1}}$$

$$= (\xi_n)^p = \xi_{n-1} = (\xi_{n-1})^{\epsilon_L} \qquad \text{if } p \text{ is odd}$$

$$= (\xi_n)^{(1+2^{n-2})(2+2^{n-1})} = \xi_{n-1} = (\xi_{n-1})^{\epsilon_L} \qquad \text{if } p = 2.$$

If, on the other hand, $K = \mathbb{Q}(\xi_n - \xi_n^{-1})$, then $\epsilon_K = 1$,

$$\iota(-1)^{\epsilon_K} = \iota\circ\tau(\xi_n) = (\xi_n)\cdot(-\xi_n^{-1}) = -1 = (-1)^{\epsilon_L};$$

and this finishes the proof. □

The remaining problem is to determine to what extent $K_2^c(\hat{\mathbb{Z}}_p[G])$ is generated by symbols $\{g,u\}$ of the type dealt with in Proposition 9.3. When G is abelian, then by Corollary 3.4, $K_2^c(\hat{\mathbb{Z}}_p[G])$ is generated by such symbols (and $\{-1,-1\}$ if $p = 2$). The next proposition gives some partial answers to this in the nonabelian case. Recall that for any ring R and any ideal $I \subseteq R$, we have defined $K_2(R,I) = \operatorname{Ker}[K_2(R) \to K_2(R/I)]$ (and similarly for K_2^c).

<u>Proposition 9.4</u> *Fix a prime p and a p-group G, and let R be the ring of integers in some finite unramified extension $F \supseteq \hat{\mathbb{Q}}_p$. Then*

(i) For any central element $z \in Z(G)$,

$$K_2^c(R[G],(1-z)R[G]) = \Big\langle \{g,1-r(1-z)^i h\} \ :\ g,h \in G,\ gh = hg,\ r \in R,\ i \geq 1 \Big\rangle.$$

(ii) For any $H \triangleleft G$ such that $H \cap [G,G] = 1$, if $\alpha\colon G \twoheadrightarrow G/H$ denotes the projection, and if $I_\alpha = \operatorname{Ker}\big[\hat{\mathbb{Z}}_p[G] \twoheadrightarrow \hat{\mathbb{Z}}_p[G/H]\big]$, then

$$K_2^c(\hat{\mathbb{Z}}_p[G],I_\alpha) = \Big\langle \{g,1-r(1-z)h\} \ :\ g,h \in G,\ gh = hg,\ r \in R,\ z \in H \Big\rangle.$$

(iii) If p is odd, then

$$K_2^c(R[G])^+ = \Big\langle \{g,u\} \ :\ g \in G,\ u \in K_1(R[C_G(g)])^+ \Big\rangle.$$

Here, $K_2^c(R[G])^+$ is the group of elements in $K_2^c(R[G])$ fixed under the standard involution.

__Proof__ The most important result here is the first point; all of the others are easy consequences of that. The main idea when finding generators for $K_2^c(R[G],(1-z)R[G])$ is to construct a filtration

$$(1-z)R[G] = I_0 \supseteq I_1 \supseteq I_2 \supseteq \cdots,$$

where $\bigcap_{k=1}^{\infty} I_k = 0$; and such that for each k, Corollary 3.4 applies to give generators for $K_2(\hat{\mathbb{Z}}_p[G]/I_k, I_{k-1}/I_k)$. These generators are then lifted in several stages to $K_2^c(\hat{\mathbb{Z}}_p[G])$. The exact sequences for pairs of ideals are used to show at each stage that all elements which can be lifted are products of liftable Steinberg symbols; and that the given symbols are the only ones which survive. The complete proof is given in Oliver [7, Theorem 1.4].

(ii) If $H \triangleleft G$ and $H \cap [G,G] = 1$, then a pair of elements $g,h \in G$ commutes in G if and only if it commutes in G/H. Hence, if

$$1 = H_0 \subseteq H_1 \subseteq \cdots \subseteq H_k = H$$

is any sequence such that $H_i \triangleleft G$ and $|H_i| = p^i$ for all i; then all of the symbol generators given by (i) for each group

$$\mathrm{Ker}\Big[K_2^c(\hat{\mathbb{Z}}_p[G/H_i]) \longrightarrow K_2^c(\hat{\mathbb{Z}}_p[G/H_{i+1}])\Big]$$

lift to symbols in $K_2^c(\hat{\mathbb{Z}}_p[G])$.

(iii) Now assume p is odd. For each $h \in G$ and each $r \in R$, define $u(rh) \in K_1(R[\langle h\rangle])^+$ such that $\Gamma_{\langle h\rangle}(u(rh)) = \frac{1}{2}r\cdot(h+h^{-1})$ (see Theorem 6.6). Note that this element is unique, since $K_1(R[\langle h\rangle])^+_{(p)}$ is torsion free by Theorem 7.3.

Recall the formula for the action of the standard involution on a symbol in Lemma 5.10(i). In particular, $\overline{\{g,u\}} = \{g,\bar{u}\}$ for any commuting $g \in G$ and $u \in (R[G])^*$. So $\{g,u(rh)\} \in K_2^c(R[G])^+$ for any commuting $g,h \in G$, and the homomorphism

$$\theta_G : H_1(G;R[G]^+) \longrightarrow K_2^c(R[G])^+,$$

defined by setting $\theta_G(g \otimes r\cdot\frac{h+h^{-1}}{2}) = \{g,u(rh)\}$ for any commuting $g,h \in G$ and any $r \in R$, is uniquely defined and natural in G.

We claim that θ_G is surjective. This is clear if $G = 1$: $K_2^c(R) = 1$ since $\mu_p \not\subseteq R$. If $|G| > 1$, then fix a central element $z \in Z(G)$ of order p, set $\hat{G} = G/z$, and assume inductively that $\theta_{\hat{G}}$ is onto. Set $I_z = (1-z)R[G]$, and consider the following diagram:

$$\begin{array}{ccccccc}
 & & H_1(G;R[G]^+) & \longrightarrow & H_1(\hat{G};R[\hat{G}]^+) & \xrightarrow{\partial_H} & H_0(\hat{G};I_z) \\
 & & \Big\downarrow \theta_G & (1a) & \Big\downarrow \theta_{\hat{G}} & (1b) & \Big\uparrow L \qquad (1) \\
K_2^c(R[G],I_z)^+ & \longrightarrow & K_2^c(R[G])^+ & \longrightarrow & K_2^c(R[\hat{G}])^+ & \xrightarrow{\partial_K} & K_1(R[G],I_z)
\end{array}$$

where L is the logarithm homomorphism constructed in Theorem 2.8 ($(I_z{}^p) \subseteq pI_z$). Square (1a) commutes by the naturality of θ. The bottom row is part of the relative exact sequence for the ideal I_z (see Theorem 1.13). The upper row is part of the homology sequence induced by the conjugation $\hat{G}$-action on the short exact sequence

$$0 \longrightarrow I_z \longrightarrow R[G] \longrightarrow R[\hat{G}] \longrightarrow 0$$

(and note that $H_1(G;R[G])$ surjects onto $H_1(\hat{G};R[G])$).

To see that square (1b) commutes, fix any commuting $\hat{g},\hat{h} \in \hat{G}$ together with liftings $g,h \in G$. Then, for any $r \in R$,

$$L\circ\partial_K\circ\theta_{\hat{G}}\Big(\hat{g} \otimes r\cdot\frac{\hat{h}+\hat{h}^{-1}}{2}\Big) = L\circ\partial_K(\{\hat{g},u(r\hat{h})\}) = \operatorname{Log}([g,u(rh)]).$$

Set $H = \langle z,h\rangle$, an abelian group. Then, in $(1-z)R[H]$,

$$\operatorname{Log}([g,u(rh)]) = \operatorname{Log}(g\cdot u(rh)\cdot g^{-1}) - \operatorname{Log}(u(rh))$$

$$= g\cdot\operatorname{Log}(u(rh))\cdot g^{-1} - \operatorname{Log}(u(rh))$$

$$= \left(1 - \tfrac{1}{p}\cdot\Phi\right)\left(g\cdot \mathrm{Log}(u(rh))\cdot g^{-1} - \mathrm{Log}(u(rh))\right) \qquad (\Phi(1-z) = 1-z^p = 0)$$

$$= g\cdot\Gamma(u(rh))\cdot g^{-1} - \Gamma(u(rh))$$

$$= g\cdot r\cdot\frac{h+h^{-1}}{2}\cdot g^{-1} - r\cdot\frac{h+h^{-1}}{2} = \partial_H\left(\hat{g}\otimes r\cdot\frac{\hat{h}+\hat{h}^{-1}}{2}\right).$$

The surjectivity of θ_G now follows from the commutativity of (1), together with the fact that $K_2^c(R[G],I_z)^+ \subseteq \mathrm{Im}(\theta_G)$ (by (i)). □

The proof of Proposition 9.4(i) can also be adapted to show that for any prime p, any p-group G, and any $i \geq 1$,

$$K_2^c(\hat{\mathbb{Z}}_p[G],p^i) = \begin{cases} \langle\{g,1+p^ih\} : g,h\in G,\ gh=hg\rangle & \text{if } p^i>2 \\ \langle\{-1,-1\},\ \{g,1+2h\} : g,h\in G,\ gh=hg\rangle & \text{if } p^i=2. \end{cases}$$

The description of $Cl_1(\mathbb{Z}[G])$ when G is an odd p-group is now immediate.

<u>Theorem 9.5</u> *For any odd prime* p *and any p-group* G, *the sequence*

$$H_1(G;\mathbb{Z}[G]) \xrightarrow{\psi_G} C_p(\mathbb{Q}[G]) \xrightarrow{\partial_G} Cl_1(\mathbb{Z}[G]) \longrightarrow 1$$

is exact. In other words, if $\mathbb{Q}[G] = \prod_{i=1}^k A_i$, *where each* A_i *is simple with center* K_i, *then*

$$Cl_1(\mathbb{Z}[G]) \cong \mathrm{Coker}\left[\sigma_G\circ\psi_G : H_1(G;\mathbb{Z}[G]) \longrightarrow \prod_{i=1}^k (\mu_{K_i})_p\right]$$

$$\cong \left[\prod_{i=1}^k (\mu_{K_i})_p\right]/\langle\sigma_G\circ\psi_G(g\otimes h) : g,h\in G,\ gh=hg\rangle.$$

<u>Proof</u> Consider the localization sequence

$$K_2^c(\hat{\mathbb{Z}}_p[G]) \xrightarrow{\varphi_G} C_p(\mathbb{Q}[G]) \xrightarrow{\partial_G} Cl_1(\mathbb{Z}[G]) \longrightarrow 1$$

of Theorem 3.15. By Proposition 5.11(i), φ_G and ∂_G both commute with the standard involution, and $C_p(\mathbb{Q}[G])$ is fixed by the involution by Theorem 5.12. Hence

$$\begin{aligned}
\operatorname{Im}(\varphi_G) &= \varphi_G(K_2^c(\hat{\mathbb{Z}}_p[G])^+) \\
&= \langle \varphi_G(\{g,u\}) : g \in G,\ u \in K_1(\hat{\mathbb{Z}}_p[C_G(g)])^+ \rangle && \text{(Prop. 9.4(iii))} \\
&= \langle \psi_G(g\otimes x) : g \in G,\ x \in \Gamma(K_1(\hat{\mathbb{Z}}_p[C_G(g)])^+) \rangle && \text{(Prop. 9.3)} \\
&= \psi_G(H_1(G;\hat{\mathbb{Z}}_p[G]^+)). && \text{(Theorem 6.6)}
\end{aligned}$$

But $\psi_G(g\otimes h) = \psi_G(g\otimes h^{-1})$ by definition, and so $\operatorname{Im}(\varphi_G) = \operatorname{Im}(\psi_G)$. □

Note in particular that by Theorem 9.5, for any odd prime p and any p-group G, the kernel of $\partial_G\colon C_p(\mathbb{Q}[G]) \twoheadrightarrow Cl_1(\mathbb{Z}[G])$ is generated by elements which come from rank 2 abelian subgroups of G. In other words, if $\mathcal{A}$ denotes the set of rank 2 abelian subgroups $H \subseteq G$, then there is a pushout square

$$\begin{array}{ccc}
\bigoplus_{H\in\mathcal{A}} C_p(\mathbb{Q}[H]) & \xrightarrow{\quad\partial\quad} & \bigoplus_{H\in\mathcal{A}} Cl_1(\mathbb{Z}[H]) \\
\downarrow & & \downarrow \\
C_p(\mathbb{Q}[G]) & \longrightarrow & Cl_1(\mathbb{Z}[G]).
\end{array}$$

If, furthermore, $Cl_1(\mathbb{Z}[H]) = 1$ for all $H \in \mathcal{A}$ (and by Example 9.8 below this is the case whenever $C_{p^2}\times C_{p^2} \not\subseteq G$), then

$$Cl_1(\mathbb{Z}[G]) \cong \operatorname{Coker}\Big[\bigoplus_{H\in\mathcal{A}} C_p(\mathbb{Q}[H]) \longrightarrow C_p(\mathbb{Q}[G])\Big].$$

In the case of 2-groups, the situation is more complicated. The following theorem gives an algorithm which completely describes $Cl_1(\mathbb{Z}[G])$ when G is abelian, but which only gives a lower bound in the nonabelian case.

Theorem 9.6 *Fix a 2-group* G, *and set*

$$X = \langle 2g,\ (1-g)(1-h):\ g,h \in G\rangle \subseteq \mathbb{Z}[G].$$

Write $\mathbb{Q}[G^{ab}] = \prod_{i=1}^{k} K_i$ *where the* K_i *are fields, set*

$$\mathcal{I} = \{1 \le i \le k : K_i \not\subseteq \mathbb{R}\} = \{1 \le i \le k : K_i \neq \mathbb{Q}\};$$

and for each i *let* $\chi_i: G \longrightarrow (\mu_{K_i})_2$ *be the character induced by projection. Then if* G *is abelian, the sequence*

$$G \otimes X \xrightarrow{\ \psi_G\ } C(\mathbb{Q}[G]) \xrightarrow{\ \partial_G\ } Cl_1(\mathbb{Z}[G]) \longrightarrow 1$$

is exact; and so

$$SK_1(\mathbb{Z}[G]) = Cl_1(\mathbb{Z}[G]) \cong [\prod_{i \in \mathcal{I}} \mu_{K_i}]/\langle \sigma_G \circ \psi_G(g \otimes x):\ g \in G,\ x \in X\rangle.$$

Otherwise, ∂_G *induces a surjection*

$$Cl_1(\mathbb{Z}[G]) \twoheadrightarrow \mathrm{Coker}\Big[H_1(G;\mathbb{Z}[G]) \xrightarrow{\ \psi_G\ } C(\mathbb{Q}[G]) \xrightarrow{\ \mathrm{proj}\ } C(\mathbb{Q}[G^{ab}])\Big]$$

$$\cong \mathrm{Coker}\Big[\sigma_{G^{ab}} \circ \psi : H_1(G;\mathbb{Z}[G]) \longrightarrow \prod_{i \in \mathcal{I}} \mu_{K_i}\Big].$$

Proof If G is abelian, then by Corollary 3.4, applied to the augmentation ideal $I = \langle g-1:\ g \in G\rangle \subseteq \hat{\mathbb{Z}}_2[G]$,

$$K_2^c(\hat{\mathbb{Z}}_2[G]) = K_2^c(\hat{\mathbb{Z}}_2) \oplus K_2^c(\hat{\mathbb{Z}}_2[G],I) = \langle\{-1,-1\}\rangle \oplus \langle\{g,u\} : g \in G,\ u \in 1+I\rangle.$$

Also, by Theorem 6.6, since $(1-g)+(1-h) \equiv (1-gh) \pmod{I^2}$ for $g,h \in G$,

$$\Gamma(1+I) = \mathrm{Ker}\Big[\omega: H_0(G;I) \longrightarrow G^{ab}\Big] = \Big\{\textstyle\sum a_i g_i:\ \sum a_i = 0,\ \prod g_i^{a_i} = 1\Big\} = I^2.$$

Hence, by Proposition 9.3,

$$\mathrm{Im}\Big[\varphi_G\colon K_2^c(\hat{\mathbb{Z}}_2[G]) \longrightarrow C(\mathbb{Q}[G])\Big] = \langle \psi_G(g\otimes\Gamma(u)) : g\in G,\ u\in 1+I\rangle = \psi_G(G\otimes I^2).$$

The relations $\psi_G(g\otimes g) = 1$ and $\psi_G(g\otimes h) = \psi_G(g\otimes h^{-1})$ show that $\psi_G(G\otimes I^2) = \psi_G(G\otimes X)$; and hence that

$$Cl_1(\mathbb{Z}[G]) \cong \mathrm{Coker}(\varphi_G) \cong C(\mathbb{Q}[G])/\psi_G(G\otimes X) \cong [\prod_{i\in\mathcal{I}}(\mu_{K_i})]/\sigma_G\circ\psi_G(G\otimes X).$$

If G is nonabelian, let $\alpha\colon G \longrightarrow G^{ab}$ denote the projection, and set $I_\alpha = \mathrm{Ker}(\hat{\mathbb{Z}}_2[G] \longrightarrow \hat{\mathbb{Z}}_2[G^{ab}])$. Consider the following homomorphisms:

$$\begin{array}{ccccc}
K_2^c(\hat{\mathbb{Z}}_2[G]) & \xrightarrow{K_2^c(\alpha)} & K_2^c(\hat{\mathbb{Z}}_2[G^{ab}]) & \xrightarrow{\partial_\alpha} & K_1(\hat{\mathbb{Z}}_2[G],I_\alpha) \\
 & & & & \big\downarrow \Gamma_\alpha \\
H_1(G;\hat{\mathbb{Z}}_2[G]) & \longrightarrow & H_1(G^{ab};\hat{\mathbb{Z}}_2[G^{ab}]) & \xrightarrow{\partial_H} & H_0(G;I_\alpha).
\end{array}$$

Both rows are exact, and Γ_α is the homomorphism of Theorem 6.9. For any $g\in G^{ab}$ and $u\in\hat{\mathbb{Z}}_2[G^{ab}]$, and any liftings to $\tilde{g}\in G$ and $\tilde{u}\in\hat{\mathbb{Z}}_2[G]$,

$$\begin{aligned}
\Gamma_\alpha(\partial_\alpha(\{g,u\})) &= \Gamma_\alpha([\tilde{g},\tilde{u}]) \\
&= \tilde{g}\cdot\Gamma_G(\tilde{u})\cdot\tilde{g}^{-1} - \Gamma_G(\tilde{u}) \qquad \text{(by definition of } I_\alpha\text{)} \\
&= \partial_H(g\otimes\Gamma(u)).
\end{aligned}$$

Hence, for any $x\in K_2^c(\hat{\mathbb{Z}}_2[G])$, if we write

$$K_2^c(\alpha)(x) = \{-1,-1\}^r\cdot\prod_{i=1}^k\{g_i,u_i\} \qquad (g_i\in G^{ab},\ u_i\in(\hat{\mathbb{Z}}_2[G^{ab}])^*)$$

(using Corollary 3.4), then

$$\partial_H\Big(\sum_{i=1}^k g_i\otimes\Gamma(u_i)\Big) = \Gamma_\alpha\circ\partial_\alpha\Big(\prod_{i=1}^k\{g_i,u_i\}\Big) = \Gamma_\alpha\circ\partial_\alpha\circ K_2^c(\alpha)(x) = 1.$$

In other words,

$$C(\alpha)\circ\varphi_G(x) = \varphi_{G^{ab}}(K_2^c(\alpha)(x)) = \psi_{G^{ab}}\Big(\sum_{i=1}^{k} g_i \otimes \Gamma(u_i)\Big) \qquad \text{(Proposition 9.3)}$$

$$\in \psi_{G^{ab}}(\mathrm{Ker}(\partial_H)) = C(\alpha)\circ\psi_G(H_1(G;\hat{\mathbb{Z}}_2[G])).$$

So there is a surjection

$$Cl_1(\mathbb{Z}[G]) \twoheadrightarrow \mathrm{Coker}\Big[K_2^c(\hat{\mathbb{Z}}_2[G]) \xrightarrow{\varphi_G} C_2(\mathbb{Q}[G]) \xrightarrow{C(\alpha)} C_2(\mathbb{Q}[G^{ab}])\Big]$$

$$\twoheadrightarrow \mathrm{Coker}\Big[H_1(G;\hat{\mathbb{Z}}_2[G]) \xrightarrow{\psi_G} C_2(\mathbb{Q}[G]) \xrightarrow{C(\alpha)} C_2(\mathbb{Q}[G^{ab}])\Big];$$

and this finishes the proof. □

Recall Conjecture 6.13: that for any p-group G, there should be an exact sequence

$$H_3(G) \longrightarrow Wh_2^c(\hat{\mathbb{Z}}_p[G]) \xrightarrow{\Gamma_2} H_1(G;\hat{\mathbb{Z}}_p[G])/\langle g\otimes g\rangle \xrightarrow{\omega_2} \tilde{H}_2(G);$$

which is natural with respect to group homomorphisms. This would still not be enough to give a general formula for $Cl_1(\mathbb{Z}[G])$ in the 2-group case (for reasons discussed below), but it does at least suggest the following approximation formula:

<u>Conjecture 9.7</u> *Fix a 2-group* G, *and write* $\mathbb{Q}[G] = \prod_{i=1}^{k} A_i$, *where each* A_i *is a matrix algebra over a division algebra* D_i *with center* K_i. *Set*

$$\mathcal{I} = \{i : D_i \not\subseteq \mathbb{R}\} \quad \textit{and} \quad \mathcal{J} = \{i : K_i \not\subseteq \mathbb{R}\}$$

(so $i \in \mathcal{I}\smallsetminus\mathcal{J}$ *if and only if* D_i *is a quaternion algebra). Define*

$$C_2^Q(\mathbb{Q}[G]) = \prod_{i\in\mathcal{I}\smallsetminus\mathcal{J}} C_2(A_i) \subseteq C_2(\mathbb{Q}[G]); \quad Cl_1^Q(\mathbb{Z}[G]) = \partial_G(C_2^Q(\mathbb{Q}[G])) \subseteq Cl_1(\mathbb{Z}[G]);$$

and let

$$H_1(G;\mathbb{Z}[G]) \xrightarrow{\psi'_G} C_2(\mathbb{Q}[G])/C_2^Q(\mathbb{Q}[G]) \xrightarrow{\partial'_G} Cl_1(\mathbb{Z}[G])/Cl_1^Q(\mathbb{Z}[G])$$

be the homomorphisms induced by ψ_G *and* ∂_G, *respectively. Then there are homomorphisms*

$$\theta^{ab}: H_2^{ab}(G) \longrightarrow Cl_1(\mathbb{Z}[G])/Cl_1^Q(\mathbb{Z}[G]), \quad \tilde{\theta}: H_2(G) \longrightarrow SK_1(\mathbb{Z}[G])/Cl_1^Q(\mathbb{Z}[G]),$$

such that the following are pushout squares:

$$\begin{array}{ccccc}
H_1(G;\mathbb{Z}[G]) & \xrightarrow{\omega_2} & H_2^{ab}(G) & \longleftarrow & H_2(G) \\
\downarrow{\scriptstyle\psi'_G} & & \downarrow{\scriptstyle\theta^{ab}} & & \downarrow{\scriptstyle\tilde{\theta}} \\
C_2(\mathbb{Q}[G])/C_2^Q(\mathbb{Q}[G]) & \xrightarrow{\partial'_G} & Cl_1(\mathbb{Z}[G])/Cl_1^Q(\mathbb{Z}[G]) & \hookrightarrow & SK_1(\mathbb{Z}[G])/Cl_1^Q(\mathbb{Z}[G]).
\end{array}$$

To see the connection between Conjectures 9.7 and 6.13, assume that Conjecture 6.13 holds, and consider the following diagram:

$$\begin{array}{ccccccc}
K_2^c(\hat{\mathbb{Z}}_2[G]) & \xrightarrow{\Gamma_2} & H_1(G;\hat{\mathbb{Z}}_2[G])/\langle g\otimes g\rangle & \xrightarrow{\omega_2} & \tilde{H}_2^{ab}(G) & \longrightarrow & 0 \\
\cong\downarrow{\scriptstyle id} & & \downarrow{\scriptstyle\psi'_G} & & \downarrow{\scriptstyle\theta_G^{ab}} & & \\
K_2^c(\hat{\mathbb{Z}}_2[G]) & \xrightarrow{\varphi'_G} & C_2(\mathbb{Q}[G])/C_2^Q(\mathbb{Q}[G]) & \xrightarrow{\partial'_G} & Cl_1(\mathbb{Z}[G])/Cl_1^Q(\mathbb{Z}[G]) & \longrightarrow & 1.
\end{array}$$

Both rows are exact; and the left-hand square commutes on symbols $\{g,u\}$, when $u \in (\hat{\mathbb{Z}}_2[C_G(g)])^*$, by Proposition 9.3 (note that $\Gamma_2(\{g,u\}) = g\otimes\Gamma(u)$). If Γ_2 is also natural with respect to transfer homomorphisms (see Oliver [6, Conjecture 5.1] for details), then the relation $\varphi'_G = \psi'_G\circ\Gamma_2$ can be reduced to the case where G is cyclic or semidihedral; and this is easily checked. The first part of the conjecture would then follow immediately.

The second part of the conjecture (the existence of $\tilde{\theta}$ defined on

$H_2(G))$ is motivated partly by the isomorphism

$$SK_1(\mathbb{Z}[G])/Cl_1(\mathbb{Z}[G]) \cong H_2(G)/H_2^{ab}(G)$$

of Theorem 8.6; and partly by the existence of homomorphisms

$$H_2(G) \longrightarrow L_0^s(\mathbb{Z}[G]) \longleftarrow \hat{H}^1(\mathbb{Z}/2;SK_1(\mathbb{Z}[G]))$$

defined via surgery. There is some reason to think that this surgery defined map can be used to show that Θ^{ab}, at least, is well defined. This conjecture seems at present to be the best chance for getting information about the extension

$$1 \longrightarrow Cl_1(\mathbb{Z}[G]) \longrightarrow SK_1(\mathbb{Z}[G]) \longrightarrow SK_1(\hat{\mathbb{Z}}_2[G]) \longrightarrow 1$$

when G is a 2-group. In fact, if the conjecture can be proven, it should then be easy to construct examples of G where this extension does not split. In contrast, it will be shown in Section 13c that this extension always splits when G is a p-group and p is odd.

There seems to be no obvious conjecture which would describe $Cl_1(\mathbb{Z}[G])$ or $SK_1(\mathbb{Z}[G])$ completely. The problem with including quaternionic components in the above diagram is that when $G = \langle a,b\rangle \cong Q(8)$, for example, the element $x = \{a^2, \Gamma^{-1}(1+a+b+ab)\} \in K_2^c(\hat{\mathbb{Z}}_2[G])$ has the property that $\Gamma_2(x) = 0$, but $\varphi_G(x) \neq 1$.

There are, however, some other cases which can be handled with the present techniques. For example, if G is a 2-group such that $[G,G]$ is central and cyclic, then $K_2^c(\hat{\mathbb{Z}}_2[G])$ can be shown to be generated by $\{-1,-1\}$, and symbols $\{g,u\}$ for $g \in G$ and $u \in (\hat{\mathbb{Z}}_2[C_G(g)])^*$. Using this, the image of $K_2^c(\hat{\mathbb{Z}}_2[G])$ in $C_2(\mathbb{Q}[G])$ can be described — in principal, at least — also when $\mathbb{Q}[G]$ contains quaternionic components.

Another class of nonabelian 2-groups for which $Cl_1(\mathbb{Z}[G])$ can be computed using Proposition 9.4 is that of products $G \times H$, where H is abelian and $Cl_1(\mathbb{Z}[G])$ is already known. Fix such G and H, and set

$$I = \mathrm{Ker}\Big[\mathbb{Z}[G\times H] \longrightarrow \mathbb{Z}[G]\Big] \quad \text{and} \quad I_{\mathbb{Q}} = \mathrm{Ker}\Big[\mathbb{Q}[G\times H] \longrightarrow \mathbb{Q}[G]\Big].$$

Then

$$Cl_1(\mathbb{Z}[G\times H]) = Cl_1(\mathbb{Z}[G]) \oplus Cl_1(\mathbb{Z}[G\times H], I);$$

and using Proposition 9.4(ii):

$$Cl_1(\mathbb{Z}[G\times H], I) \cong \mathrm{Coker}\Big[K_2^c(\hat{\mathbb{Z}}_2[G\times H], \hat{I}_2) \longrightarrow C(\mathbb{Q}[G\times H], I_{\mathbb{Q}})\Big]$$

$$= C(\mathbb{Q}[G\times H], I_{\mathbb{Q}})/\langle \varphi_G(\{g, 1+(1-z)h\}) : z\in H,\ g,h\in G\times H,\ gh=hg\rangle.$$

A special case of this will be shown in Example 9.10 below.

We now look at some more specific examples of computations. The case of abelian p-groups will first be considered.

It will sometimes be convenient to describe elements in $C(\mathbb{Q}[G])$ using the epimorphism $\tilde{\mathfrak{J}}_G\colon R_{\mathbb{C}}(G) \twoheadrightarrow C(\mathbb{Q}[G])$ of Section 5b — or rather its projection $\tilde{\mathfrak{J}}_{G,p}\colon R_{\mathbb{C}}(G) \twoheadrightarrow C_p(\mathbb{Q}[G])$ to p-torsion. Recall the description of $\tilde{\mathfrak{J}}_G$ (but adapted to $\tilde{\mathfrak{J}}_{G,p}$) given in Lemma 5.9(ii). For any irreducible $\mathbb{C}[G]$-representation V, let A be the unique simple summand of $\mathbb{Q}[G]$, and let $\alpha\colon K = Z(A) \hookrightarrow \mathbb{C}$ be the unique embedding, such that V is the irreducible $\mathbb{C}\otimes_{\alpha K} A$-module. Then $\tilde{\mathfrak{J}}_{G,p}([V]) \in C_p(A)$. If $C_p(A) \not\cong 1$, if $\sigma_A\colon C_p(A) \xrightarrow{\cong} (\mu_K)_p$ is the norm residue symbol isomorphism, and if $p^n = |(\mu_K)_p|$, then

$$\tilde{\mathfrak{J}}_{G,p}([V]) = \sigma_A^{-1}\circ\alpha^{-1}(\xi_n) \in C_p(A). \qquad (\xi_n = \exp(2\pi i/p^n))$$

<u>Example 9.8</u> *Fix any prime* p. *Then*

(*i*) $SK_1(\mathbb{Z}[C_{p^n}\times C_p]) = 1$ *for any* $n\geq 0$, *and*

(*ii*) $SK_1(\mathbb{Z}[C_{p^2}\times C_{p^2}]) \cong (\mathbb{Z}/p)^{p-1}$.

Proof The two computations will be carried out separately. To simplify the notation, the groups $C_p(\mathbb{Q}[G])$ are written additively here.

Step 1 For each n, write $G_n = C_{p^n} \times C_p = \langle g,h\rangle$, where $|g| = p^n$ and $|h| = p$. We identify $G_{n-1} = G_n/\langle g^{p^{n-1}}\rangle$ for each n. Then $SK_1(\mathbb{Z}[G_0]) = 1$ by Theorem 5.6. Also, if $p = 2$, then $SK_1(\mathbb{Z}[G_1]) = 1$ by Theorem 5.4 $(C(\mathbb{Q}[G_1]) = C(\mathbb{Q}[C_2{\times}C_2]) = 1)$. If p is odd, then $C_p(\mathbb{Q}[G_1]) \cong (\mathbb{Z}/p)^{p+1}$ is easily seen to be generated by the elements

$$\psi(h\otimes gh^i) \quad (0 \le i \le p-1) \qquad \text{and} \qquad \psi(g\otimes h).$$

Now fix $n \ge 2$, and assume inductively that $SK_1(\mathbb{Z}[G_{n-1}]) = 1$. Set

$$X = \begin{cases} \langle 2a,\ (1-a)(1-b):\ a,b \in G_n\rangle \subseteq \mathbb{Z}[G_n] & \text{if } p = 2 \\ \mathbb{Z}[G_n] & \text{if } p \text{ is odd;} \end{cases}$$

so that $SK_1(\mathbb{Z}[G_n]) \cong C_p(\mathbb{Q}[G_n])/\psi(G_n\otimes X)$ by Theorem 9.5 or 9.6. Write $\mathbb{Q}[G_n] = \mathbb{Q}[G_{n-1}] \times A$, where A is the product of those simple summands upon which g acts with order p^n. For each $r = 0,\ldots,p-1$, let V_r denote the $\mathbb{C}[G_n]$-representation with character $\chi_{V_r}(g) = \xi_n$, $\chi_{V_r}(h) = \xi_1^r = (\xi_n)^{rp^{n-1}}$ $(\xi_n = \exp(2\pi i/p^n))$. Then $C_p(A)$ is generated by the elements $\tilde{\mathfrak{z}}(V_r)$, each of which has order p^n.

Since $SK_1(\mathbb{Z}[G_{n-1}]) = 1$, we have

$$C_p(\mathbb{Q}[G_n]) = \langle C_p(A),\ \psi(G_n\otimes X)\rangle. \tag{1}$$

Also, a direct computation shows that for each $0 \le r \le p-1$,

$$\psi\Big(g \otimes \Big(\sum_{i=0}^{p-1} g^{ip^{n-2}}\Big)\Big(1 + (p-1)g^{-rp^{n-1}}h - \sum_{i=0}^{p-1} g^{ip^{n-1}}\Big)\Big)$$

$$\in \begin{cases} \tilde{\mathfrak{z}}((p-1)\cdot V_r) + p\cdot C_p(\mathbb{Q}[G_{n-1}]) & \text{if } p \text{ is odd} \\ \tilde{\mathfrak{z}}((2^{n-1}+1)\cdot V_r) + 2\cdot C(\mathbb{Q}[G_{n-1}]) & \text{if } p = 2. \end{cases}$$

Together with (1), this shows that for each r, there is some $\eta_r \in G_n \otimes X$ such that

$$\psi(\eta_r) \in \tilde{\mathfrak{z}}(V_r) + p\cdot C_p(A).$$

The elements $\psi(\eta_r)$ then generate $C_p(A)$. Together with (1), this shows that ψ is onto, and hence that $SK_1(\mathbb{Z}[G_n]) = 1$.

<u>Step 2</u> The proof that $SK_1(\mathbb{Z}[C_4\times C_4]) \cong \mathbb{Z}/2$ is very similar to the proof of Example 5.1, and we leave this as an exercise. So assume p is odd. Write $G = C_{p^2}\times C_{p^2}$ for short, fix generators $g,h \in G$, and set $H = \langle g^p,h^p\rangle$.

Let $\mathfrak{R}_1$ and $\mathfrak{R}_2$ denote the sets of irreducible $\mathbb{C}[G]$-representations upon which G acts with order p and p^2, respectively. Define

$$\alpha : \mathfrak{R}_2 \longrightarrow \mathfrak{R}_1$$

by letting $\alpha(V)$, for any $V \in \mathfrak{R}_2$, be the representation whose character satisfies $\chi_{\alpha(V)} = (\chi_V)^p$. Then by Definition 9.2, for any generating pair $a,b \in G$, $\psi(a\otimes b) = \tilde{\mathfrak{z}}(V \oplus \alpha(V))$, where $V \in \mathfrak{R}_2$ is the unique representation such that $\chi_V(a) = \xi_2$ and $\chi_V(b) = 1$.

Now define an epimorphism

$$\beta : C_p(\mathbb{Q}[G]) \longrightarrow\!\!\!\!\rightarrow C_p(\mathbb{Q}[G/H]) \cong C_p(\mathbb{Q}[C_p\times C_p]) \cong (\mathbb{Z}/p)^{p+1}$$

by setting $\beta(\tilde{\mathfrak{z}}(V)) = \tilde{\mathfrak{z}}(V)$ for $V \in \mathfrak{R}_1$; $\beta(\tilde{\mathfrak{z}}(V)) = -\tilde{\mathfrak{z}}(\alpha(V))$ for $V \in \mathfrak{R}_2$.

We have just seen that $Ker(\beta) \subseteq \psi(G\otimes\mathbb{Z}[G])$; and that $\beta\circ\psi(a\otimes b) = 1$ if $\langle a,b\rangle = G$. Also, $\beta\circ\psi(a\otimes b) = 1$ if $a \in H$ (since $C_p(\mathbb{Q}[G/H])$ has exponent p); and $\beta\circ\psi(a\otimes b) = \beta\circ\psi(a\otimes 1)$ if $b \in H$. Since $\psi(a\otimes a) = 1$ for all a, it now follows that

$$SK_1(\mathbb{Z}[G]) \cong C_p(\mathbb{Q}[G/H])/\langle\beta\circ\psi(g\otimes 1),\ \beta\circ\psi(h\otimes 1)\rangle \cong (\mathbb{Z}/p)^{p-1}. \qquad \square$$

Some more complicated examples of computations of $SK_1(\mathbb{Z}[G])$ for abelian p-groups G can be found in Alperin et al [3, Section 5]. Some of these are listed in Example 6 at the end of the introduction.

The next example illustrates some of the techniques for computing $Cl_1(\mathbb{Z}[G])$ for nonabelian p-groups G using Theorems 9.5 and 9.6. We already have seen one example of this: $Cl_1(\mathbb{Z}[G]) = 1$ for any dihedral, quaternion, or semidihedral (2-)group by Example 5.8. Note that for groups of the same size, it is often easier to compute $Cl_1(\mathbb{Z}[G])$ when G is nonabelian — $C(\mathbb{Q}[G])$ is smaller in this case, and computations can frequently be carried out via comparison with proper subgroups $H \subsetneq G$ for which $Cl_1(\mathbb{Z}[H])$ is already known.

<u>Example 9.9</u> *Fix a prime* p, *and let* G *be a nonabelian p-group. Then* $Cl_1(\mathbb{Z}[G]) \neq 1$, *unless (possibly)* $p = 2$ *and* G^{ab} *has exponent* 2. *Also,*

(i) $SK_1(\mathbb{Z}[G]) = Cl_1(\mathbb{Z}[G]) \cong (\mathbb{Z}/p)^{p-1}$ *if* p *is odd and* $|G| = p^3$;

and

(ii) *if* $p = 2$ *and* $|G| = 16$, *then*

$$SK_1(\mathbb{Z}[G]) = Cl_1(\mathbb{Z}[G]) \cong \begin{cases} 1 & \text{if } G^{ab} \cong (C_2)^2 \text{ or } (C_2)^3 \\ \mathbb{Z}/2 & \text{if } G^{ab} \cong C_4\times C_2. \end{cases}$$

<u>Proof</u> The proof will be split into two cases, depending on whether p is odd or $p = 2$. Note first that all of the groups G in (i) and (ii) have abelian subgroups of index p. Hence $SK_1(\hat{\mathbb{Z}}_p[G]) = 1$ for these

G by Corollary 7.2, and $SK_1(\mathbb{Z}[G]) = Cl_1(\mathbb{Z}[G])$.

<u>Case 1</u> Assume p is odd, and fix a nonabelian p-group G. Set $H_0 = [G,G] \triangleleft G$. Then $[G,H_0] \subsetneq H_0$ (G being nilpotent); and $G/[G,H_0]$ is also nonabelian. So $\delta \neq 1$ in the five term homology exact sequence

$$H_2(G) \longrightarrow H_2(G^{ab}) \xrightarrow{\delta} H_0/[G,H_0] \longrightarrow G^{ab} \xrightarrow{\cong} (G/H_0)^{ab} \longrightarrow 1$$

of Theorem 8.2. Write $G/H_0 = G^{ab} = \langle g_1\rangle \times \ldots \times \langle g_k\rangle$, where the g_i are ordered so that $\delta(g_1 \wedge g_2) \neq 1$; and let $H \triangleleft G$ be the subgroup such that $H/H_0 = \langle (g_1)^p, (g_2)^p, g_3, \ldots, g_k\rangle \triangleleft G/H_0$. Then H has the property that any commuting pair $g,h \in G$ generates a cyclic subgroup in G/H.

Now consider the composite

$$\psi' : H_1(G;\mathbb{Z}[G]) \xrightarrow{\psi_G} C_p(\mathbb{Q}[G]) \xrightarrow{\alpha_*} C_p(\mathbb{Q}[G/H]) \cong C_p(\mathbb{Q}[C_p \times C_p]) \cong (\mathbb{Z}/p)^{p+1}. \tag{1}$$

By the construction of H, we see that $\mathrm{Im}(\psi')$ is generated by $\psi'(g_1\otimes 1)$ and $\psi'(g_2\otimes 1)$ ($\psi(a\otimes a) = 1$ for all $a \in G$). Hence, there is a surjection

$$Cl_1(\mathbb{Z}[G]) \xrightarrow{\cong} \mathrm{Coker}(\psi_G) \xrightarrow{\hat{\alpha}} \mathrm{Coker}(\psi') \cong (\mathbb{Z}/p)^{p-1}; \tag{2}$$

and $Cl_1(\mathbb{Z}[G]) \neq 1$.

If $|G| = p^3$, so that $H = [G,G] \cong C_p$, then all nonabelian $\mathbb{C}[G]$-representations are induced up from proper subgroups $K \subsetneq G$, for which $Cl_1(\mathbb{Z}[K]) = 1$. So the $\mathrm{Ker}(\alpha_*) \subseteq \mathrm{Im}(\psi_G)$ in (1) above, and $\hat{\alpha}$ is an isomorphism in (2).

<u>Case 2</u> Now assume that $p = 2$, and that G is a nonabelian p-group such that G^{ab} is not elementary abelian. Set $H_0 = [G,G]$, as in Case 1, and write $G/H_0 = G^{ab} = \langle g_1\rangle \times \ldots \times \langle g_k\rangle$ such that $\delta(g_1 \wedge g_2) \neq 1$ (i.e., g_1, g_2 lift to noncommuting elements of $G/[G,H_0]$); but this time arrange the g_i so that $|g_1| \geq 4$. Let $H \triangleleft G$ be such that $H/H_0 = \langle (g_1)^4, (g_2)^2, g_3, \ldots, g_k\rangle$. Then $G/H \cong C_4 \times C_2$, and no abelian subgroup of G surjects onto G/H.

Define

$$\psi' : H_1(G;\mathbb{Z}[G]) \xrightarrow{\psi_G} C(\mathbb{Q}[G]) \twoheadrightarrow C(\mathbb{Q}[G/H]) \cong C(\mathbb{Q}[C_4\times C_2]) \cong (\mathbb{Z}/4)^2,$$

as before; so that there by Theorem 9.6 is a surjection

$$Cl_1(\mathbb{Z}[G]) \twoheadrightarrow \operatorname{Coker}(\psi').$$

Then in this case,

$$\operatorname{Im}(\psi') = \langle \psi'(g_1\otimes 1),\ \psi'(g_2\otimes 1) = \psi'(g_2\otimes g_1^2 g_2),\ \psi'(g_1^2\otimes g_2)\rangle,$$

and this has index 2 in $C(\mathbb{Q}[G/H])$.

If G is any nonabelian group of order 16, then $Cl_1(\mathbb{Z}[K]) = 1$ for all proper subgroups $K \subsetneq G$ (see Examples 5.8 and 9.8, and Theorems 5.4 and 5.6). So by Proposition 5.2, there is a commutative square

$$\begin{array}{ccc}
\bigoplus_{K\subsetneq G} R_{\mathbb{C}}(K) & \longrightarrow & \bigoplus_{K\subsetneq G} Cl_1(\mathbb{Z}[K]) = 1 \\
{\scriptstyle f\,=\,\oplus \operatorname{Ind}_K^G}\big\downarrow & & \big\downarrow \\
R_{\mathbb{C}/\mathbb{R}}(G) & \xrightarrow{\ \tilde{\vartheta}_G\ } & Cl_1(\mathbb{Z}[G]);
\end{array}$$

and hence $|Cl_1(\mathbb{Z}[G])| \le |\operatorname{Coker}(f)|$. $\operatorname{Coker}(f)$ is easily checked to have order 2 if $G^{ab} \cong C_4\times C_2$, and order 1 otherwise (note, for example, that G always has an abelian subgroup K of index 2, and that all nonabelian irreducible $\mathbb{C}[G]$-representations are induced up from $\mathbb{C}[K]$-representations). We have seen that $Cl_1(\mathbb{Z}[G])$ has order at least 2 if $G^{ab} \cong C_4\times C_2$, and this completes the computation. □

As has been mentioned above, Proposition 9.4 can be used to calculate $Cl_1(\mathbb{Z}[G\times H])$, for any abelian 2-group H, and any 2-group G for which $Cl_1(\mathbb{Z}[G])$ is already known. The last example illustrates a special case of this.

Example 9.10 *Let* G *be any 2-group. Then, for any* k,

$$Cl_1(\mathbb{Z}[G\times(C_2)^k]) \cong \bigoplus_{i=0}^{k} \binom{k}{i}\cdot Cl_1(\mathbb{Z}[G],2^i);$$

where for each $i \geq 1$,

$$Cl_1(\mathbb{Z}[G],2^i) \cong C_2(\mathbb{Q}[G])/\langle\varphi(\{g,1+2^ih\}): g,h \in G,\ gh = hg\rangle.$$

In particular, if G *is any quaternion or semidihedral 2-group, then*

$$Cl_1(\mathbb{Z}[G\times(C_2)^k]) \cong (\mathbb{Z}/2)^{2^k-k-1}.$$

Proof For abelian G, this is shown in Alperin et al [3, Theorems 1.10 and 1.11]. The proof in the nonabelian case is almost identical; except that Proposition 9.4(i) is now used to construct generators for $\varphi(K_2^c(\hat{\mathbb{Z}}_2[G],2^i)) \subseteq C_2(\mathbb{Q}[G])$. The last formula (when G is quaternion or semidihedral) is an easy exercise. □

Chapter 10 THE TORSION FREE PART OF Wh(G)

So far, all of the results on $K_1(\mathbb{Z}[G])$ and $Wh(G)$ presented here have dealt with either their torsion subgroups or their ranks; and that suffices when trying to detect whether or not any given $x \in Wh(G)$ vanishes. For many problems, however, it is necessary to know specific generators for $Wh'(G) = Wh(G)/SK_1(\mathbb{Z}[G])$; or to know generators p-locally for some prime p. In general, this problem seems quite difficult, since it depends on knowing generators for the units in rings of integers in global cyclotomic fields, and this is in turn closely related to class numbers.

One case which partly avoids these problems is that of p-groups for regular primes p (including the case $p = 2$). For such G, $Wh'(\hat{\mathbb{Z}}_p[G])$ is a free $\hat{\mathbb{Z}}_p$-module by Theorems 2.10(i) and 7.3; and so the inclusion $\mathbb{Z}[G] \subseteq \hat{\mathbb{Z}}_p[G]$ induces a homomorphism $\hat{\mathbb{Z}}_p \otimes Wh'(G) \longrightarrow Wh'(\hat{\mathbb{Z}}_p[G])$. This is a monomorphism (Theorem 10.3 below); and the image of the composite

$$\hat{\Gamma}_G : \hat{\mathbb{Z}}_p \otimes Wh'(G) \rightarrowtail Wh'(\hat{\mathbb{Z}}_p[G]) \xrightarrow{\Gamma_G} H_0(G;\hat{\mathbb{Z}}_p[G])$$

will be described in Theorems 10.3 and 10.4. One consequence of these results (Theorem 10.5) is a description of the behavior of $Wh'(G)$ under surjections, and under induction from cyclic subgroups of G.

In the last part of the chapter, we turn to the problem of determining which elements of $Wh'(G)$ (or of $Wh(G)$) are representable by units. Theorem 10.7 gives some applications of logarithmic methods to this problem in the case of 2-groups. For example, it is shown that not all elements in $Wh'(Q(32) \times C_2 \times C_2)$ are represented by units in the group ring. In addition, some of the results in Magurn et al [1] are listed: these include examples (Theorem 10.8) of quaternion groups for which $Wh'(G)$ is or is not generated by units.

The first step towards obtaining these results is to establish an upper bound for the image of $\hat{\Gamma}_G$ in $H_0(G;\hat{\mathbb{Z}}_p[G])$. This is based on a simple symmetry argument, and applies in fact to arbitrary finite G.

<u>Lemma 10.1</u> *Fix a prime* p, *and let* G *be any finite group. Let* $\hat{\Gamma}_G\colon \hat{\mathbb{Z}}_p \otimes Wh'(G) \longrightarrow H_0(G;\hat{\mathbb{Z}}_p[G])$ *be defined as above, and set*

$$Y(G) = \Big\langle g + g^{-1} - g^{n} - g^{-n},\ h - h^{m} : g,h \in G,\ \ h \text{ conj. } h^{-1},\ (n,|g|)=1,\ \ (m,|h|)=1 \Big\rangle \subseteq H_0(G;\hat{\mathbb{Z}}_p[G]).$$

Then $\hat{\Gamma}_G(\hat{\mathbb{Z}}_p \otimes Wh'(G)) \subseteq Y(G)$.

<u>Proof</u> Set $n = \exp(G)$, $K = \mathbb{Q}\zeta_n$ $(\zeta_n = \exp(2\pi i/n))$, and $R = \mathbb{Z}\zeta_n$. By Theorem 1.5, R is the ring of integers in K, and K is a splitting field for G. In particular, we can write

$$K[G] \cong \prod_{i=1}^{k} M_{m_i}(K)$$

for some m_i. Consider the following commutative diagram:

$$\begin{array}{ccccc}
K_1(\mathbb{Z}[G]) & \xrightarrow{\ \mathbb{Z} \hookrightarrow R\ } & K_1(R[G]) & \xrightarrow{\ \prod \det \circ \mathrm{pr}_i\ } & \prod_{i=1}^{k} R^* \\
\big\downarrow \log & & \big\downarrow \log & & \big\downarrow \prod \log \\
H_0(G;\hat{\mathbb{Q}}_p[G]) & \xrightarrow{\ \mathbb{Q} \hookrightarrow K\ } & H_0(G;\hat{K}_p[G]) & \xrightarrow[\cong]{\ \prod \mathrm{Tr} \circ \mathrm{pr}_i\ } & \prod_{i=1}^{k} \hat{K}_p;
\end{array}$$

where $\mathrm{pr}_i\colon K[G] \twoheadrightarrow M_{m_i}(K)$ denotes the projection onto the i-th component.

By Theorem 1.5(i) again, for any $a \in (\mathbb{Z}/n)^*$, there is an element $\gamma_a \in \mathrm{Gal}(K/\mathbb{Q}) = \mathrm{Gal}(\mathbb{Q}\zeta_n/\mathbb{Q})$ such that $\gamma_a(\zeta_n) = (\zeta_n)^a$. Also, $(\mathbb{Z}/n)^*$ acts on $H_0(G;\hat{\mathbb{Q}}_p[G])$ and $H_0(G;\hat{K}_p[G])$ via the action $\gamma_a(\sum r_i g_i) = \sum \gamma_a(r_i)\cdot g_i^a$. Then $\prod \mathrm{Tr}\circ\mathrm{pr}_i$ commutes with the $(\mathbb{Z}/n)^*$-actions on $H_0(G;\hat{K}_p[G])$ and $\prod \hat{K}_p$ (note that each matrix $\mathrm{pr}_i(g) \in M_{r_i}(K)$, for any $g \in G$, can be diagonalized).

Write $T = (\mathbb{Z}/n)^*$, for short. For any $u \in R^*$,

$$\bar{u}/u = \gamma_{-1}(u)/u \in \langle \pm\zeta_n \rangle \qquad \text{and} \qquad \prod_{a\in T} \gamma_a(u) = N_{K/\mathbb{Q}}(u) \in \mathbb{Z}^* = \{\pm 1\}.$$

So by the commutativity of (1), for any $u \in K_1(\mathbb{Z}[G])$ and any $1 \le i \le k$,

$$\gamma_{-1}(\mathrm{Tropr}_i \circ \log(u)) = \mathrm{Tropr}_i \circ \log(u) \qquad \text{and} \qquad \sum_{a\in T} \gamma_a(\mathrm{Tropr}_i \circ \log(u)) = 0.$$

Since $\prod \mathrm{Tropr}_i$ is a T-linear isomorphism,

$$\gamma_{-1}(\log(x)) = \log(x) \qquad \text{and} \qquad \sum_{a\in T} \gamma_a(\log(x)) = 0 \qquad (\text{in } H_0(G;\hat{\mathbb{Q}}_p[G]))$$

for all $x \in K_1(\mathbb{Z}[G])$. Also, $\Gamma_G = (1 - \frac{1}{p}\cdot\Phi)\circ\log$, and Φ commutes with the γ_a $(\Phi(\sum r_i g_i) = \sum r_i g_i^p)$. It follows that

$$\Gamma_G(K_1(\mathbb{Z}[G])) \subseteq H_0(G;\hat{\mathbb{Z}}_p[G]) \cap \left\{x \in H_0(G;\hat{\mathbb{Q}}_p[G]) \,:\, \gamma_{-1}(x) = x,\ \sum_{a\in T} \gamma_a(x) = 0\right\}$$

$$= Y(G). \qquad \square$$

We now restrict to the case where G is a p-group. The goal is to show that $\hat{\Gamma}_G(\hat{\mathbb{Z}}_p \otimes \mathrm{Wh}'(G)) = Y(G)$ whenever p is an odd regular prime, and to describe $\mathrm{Im}(\hat{\Gamma}_G)$ when $p = 2$. The key to these results is the following proposition, due to Weber for $p = 2$, and to Hilbert and Iwasawa for odd p.

<u>Proposition 10.2</u> *Fix a prime* p *and a number field* K *such that* $K \subseteq \mathbb{Q}(\xi_n)$ *for some* $n \ge 1$ $(\xi_n = \exp(2\pi i/p^n))$. *Let* $R = \mathbb{Z}[\xi_n] \cap K$ *be the ring of integers. Then the homomorphism*

$$\iota_K : \hat{\mathbb{Z}}_p \otimes_{\mathbb{Z}} R^* \longrightarrow (\hat{R}_p)^*,$$

induced by the inclusion $R \subseteq \hat{R}_p$ *(and by the* $\hat{\mathbb{Z}}_p$*-module structure on*

$(\hat{R}_p)^*_{(p)})$ *is injective. If* p *is regular (possibly* $p = 2$*), then* $\mathrm{Coker}(\iota_K)$ *is p-torsion free. If* $p = 2$*, and if* $K = \mathbb{Q}(\xi_n+\xi_n^{-1})$ *(and* $R = \mathbb{Z}[\xi_n+\xi_n^{-1}]$*), then*

$$\left\{u \in R^* : v(u) > 0, \text{ all } v\colon K \hookrightarrow \mathbb{R}\right\} = \{u^2 : u \in R^*\}. \tag{1}$$

<u>Proof</u> The injectivity of ι_K is a special case of Leopoldt's conjecture. For a proof, see, e. g., Washington [1,Corollary 5.32].

We next show that $\mathrm{Coker}(\iota_K)$ is torsion free whenever p is regular. If $L \subseteq K$ is any subfield, then $\mathrm{Coker}(\iota_L)$ is a subgroup of $\mathrm{Coker}(\iota_K)$; it thus suffices to consider the case $K = \mathbb{Q}(\xi_n)$. Set $\xi = \xi_n$, for short, and let $\mathfrak{p} = (1-\xi)R \subseteq R$ denote the maximal ideal.

Define indexing sets

$$I = \{i : 1 \le i \le p^n,\ p\nmid i \text{ or } i = p^n\}; \qquad J = \{i : 1 \le i \le (p^n-3)/2,\ p\nmid i\}.$$

Assume that $\{x_i\}_{i\in I}$ is any set of units in $(\hat{R}_p)^*$ satisfying

$$x_1 = \xi; \qquad x_i \equiv 1 + a_i(1-\xi)^i \pmod{\mathfrak{p}^{i+1}} \quad (\text{some } a_i \in \mathbb{Z}\smallsetminus p\mathbb{Z}) \tag{2}$$

for all $i \in I$. We can then define x_i inductively for all $1 \le i \notin I$ by setting

$$x_i = \begin{cases} (x_{i/p})^p & \text{if } p \mid i < p^n \\ (x_{i-v(p)})^p & \text{if } i > p^n; \end{cases}$$

where $v(p) = (p-1)p^{n-1} = [K : \mathbb{Q}]$ is the p-adic valuation of p. One easily checks that (2) is satisfied for all $i \ge 1$; and hence that the x_i generate $1+\mathfrak{p}$ as a $\hat{\mathbb{Z}}_p$-module. Since $\mathrm{rk}_{\hat{\mathbb{Z}}_p}(1+\mathfrak{p}) = (p-1)p^{n-1} = |I| - 1$, and since $x_1 = \xi$, this shows that the elements x_i for $2 \le i \in I$ are a basis for the torsion free part of $(\hat{R}_p)^*$.

Assume now that there exist real global units $\{u_j\}_{j\in J}$

($u_j \in (\mathbb{Z}[\xi+\xi^{-1}])^*$) which satisfy

$$u_j \equiv 1 + b_j(1-\xi)^j(1-\xi^{-1})^j \pmod{p^{2j+1}} \qquad (\text{some } b_i \in \mathbb{Z}\smallsetminus p\mathbb{Z}) \tag{3}$$

for all $j \in J$. If p is odd, then we can set $x_{2j} = u_j$ for $j \in J$, and extend this to some set $\{x_i\}_{i\in I}$ satisfying (2). If $p = 2$, a set $\{x_i\}_{i\in I}$ can be chosen to satisfy (2) and such that $x_{2j+1} = u_j\cdot(x_j)^{-2}$ for all $j \in J$: note that $2 \in p^{j+2}$ by assumption, and hence that

$$u_j\cdot x_j^{-2} \equiv \left(1 + (1-\xi)^j(1-\xi^{-1})^j\right)\left(1 + (1-\xi)^j\right)^{-2} \equiv 1 + (1-\xi)^j(1-\xi^{-1})^j - (1-\xi)^{2j}$$

$$= 1 + (1-\xi)^{2j}(-\xi^{j}-1) \equiv 1 + (1-\xi)^{2j+1} \pmod{p^{2j+2}}.$$

In either case, since $\mathrm{rk}_{\mathbb{Z}}(R^*) = |J|$ by Dirichlet's unit theorem (see Janusz [1, Theorem I.11.19]), this shows that there is a $\hat{\mathbb{Z}}_p$-basis for $(\hat{R}_p)^*/\langle\pm\xi\rangle$ which includes a $\mathbb{Z}$-basis for $R^*/\langle\pm\xi\rangle$, and hence that $\mathrm{Coker}(\iota_K)$ is torsion free. Also, once the u_j have been constructed, this gives a second (and more elementary) proof that ι_K is injective.

If p is odd and regular, then global units $\{u_j\}_{j\in J}$ satisfying (3) are constructed by Hilbert [1, §138, Hilfsatz 29] when $n = 1$, and by Galovich [1, Proposition 2.5] when $n > 1$. We include here a construction of the u_j when $p = 2$, due to Hambleton & Milgram [1].

We may assume that $n \geq 3$ (otherwise $J = \emptyset$). Set

$$\lambda = \xi+\xi^{-1} = 2 - (1-\xi)(1-\xi^{-1}) \qquad (\text{so } \lambda R = (1-\xi)(1-\xi^{-1})R = p^2).$$

Then $(1+\lambda)(1-\lambda) = -(1+\xi^2+\xi^{-2})$; and an easy induction shows that

$$N_{K_0/\mathbb{Q}}(1+\lambda) = -1. \qquad (K_0 = \mathbb{Q}(\xi)\cap\mathbb{R} = \mathbb{Q}(\lambda)) \tag{4}$$

Let $g \in \mathrm{Gal}(K/\mathbb{Q})$ be the element $g(\xi) = \xi^3$. Then $g(\lambda) = \lambda^3 - 3\lambda$; and so for all $i \geq 1$,

$$g(\lambda^i) - \lambda^i = (\lambda^3 - 4\lambda)\Big(\sum_{\ell=0}^{i-1} g(\lambda)^{\ell}\cdot\lambda^{i-\ell-1}\Big) \in \lambda^{i+2}R = p^{2i+4}. \tag{5}$$

Set $u_1 = 1+\lambda$ ($u_1 \in R^*$ by (4)); and define inductively $u_j = u_{j-2}^{-1}\cdot g(u_{j-2})$ for all odd $j \geq 3$. If congruence (3) holds for u_{j-2}, then

$$\begin{aligned} u_j &= (1+\lambda^{j-2}+a)^{-1}\cdot(1+g(\lambda^{j-2})+g(a)) \qquad (\text{some } a \in p^{2j-3}\cap\mathbb{R} = \lambda^{j-1}R) \\ &= 1 + (g(\lambda^{j-2}) - \lambda^{j-2} + g(a) - a)\cdot(1+\lambda^{j-2}+a)^{-1} \\ &\equiv 1 + (g(\lambda^{j-2}) - \lambda^{j-2}) \qquad (\text{mod } \lambda^{j+1}R + \lambda^{2j-2}R \subseteq \lambda^{j+1}R) \qquad (\text{by } (5)) \\ &= 1 + (\lambda^3-3\lambda)^{j-2} - \lambda^{j-2} \equiv 1 + \lambda^j \qquad (\text{mod } \lambda^{j+1}R = p^{2j+2}). \end{aligned}$$

In other words, congruences (3) are satisfied for all j; and this finishes the proof that $\text{Coker}(\iota_K)$ is torsion free.

It remains to prove (1): the description of strictly positive units in $K = \mathbb{Q}(\xi_n+\xi_n^{-1})$. Set $G = \text{Gal}(K/\mathbb{Q})$. Let V be the set of all real places of K; i. e., all embeddings $v\colon K \hookrightarrow \mathbb{R}$. Define

$$\lambda = \oplus\lambda_v : R^* \longrightarrow \bigoplus_{v\in V}\mathbb{Z}/2: \quad \text{where} \quad \lambda_v(u) = \begin{cases} 0 & \text{if } v(u) > 0 \\ 1 & \text{if } v(u) < 0. \end{cases}$$

Regard $M = \oplus_{v\in V}\mathbb{Z}/2$ as a free $\mathbb{Z}/2[G]$-module of rank 1, so that $\text{Im}(\lambda)$ is a $\mathbb{Z}/2[G]$-submodule of M. By (4), $N_{K/\mathbb{Q}}(1+\xi+\xi^{-1}) = -1$; and hence $\sum_{v\in V}\lambda_v(1+\xi+\xi^{-1}) = 1$. Then by Example 1.12, $\text{Im}(\lambda) \not\subseteq J(\mathbb{Z}/2[G])\cdot M$, the unique maximal proper submodule of M; and so λ is surjective. Also,

$$\text{rk}_{\mathbb{Z}}(R^*) = [K:\mathbb{Q}] - 1 = |V| - 1 \quad \text{and} \quad R^* \cong \mathbb{Z}^{|V|-1} \times \mathbb{Z}/2$$

by Dirichlet's unit theorem (Janusz [1, Theorem I.11.19]); and this implies that

$$\{u \in R^*\colon v(u) > 0, \text{ all } v\colon K \hookrightarrow \mathbb{R}\} = \text{Ker}(\lambda) = \{u^2 : u \in R^*\}. \qquad \square$$

Proposition 10.2 will now be combined with a Mayer-Vietoris sequence to give information about $Wh'(G)$.

For any prime p and any p-group G, define

$$Y_0(G) = \langle g + g^{-1} - g^n - g^{-n} \;:\; g \in G,\ p\nmid n\rangle \subseteq H_0(G;\hat{\mathbb{Z}}_p[G]);$$

$$Y(G) = Y_0(G) + \langle g - g^n \;:\; g \in G,\ g \text{ conj. } g^{-1},\ p\nmid n\rangle \subseteq H_0(G;\hat{\mathbb{Z}}_p[G]).$$

Note that $2\cdot Y(G) \subseteq Y_0(G)$, that $Y(G) = Y_0(G)$ if p is odd or if G is abelian, and that $H_0(G;\hat{\mathbb{Z}}_p[G])/Y(G)$ is torsion free. As before, $\hat{\Gamma}_G$ denotes the composite

$$\hat{\Gamma}_G : \hat{\mathbb{Z}}_p \otimes Wh'(G) \longrightarrow Wh'(\hat{\mathbb{Z}}_p[G]) \xrightarrow{\Gamma_G} H_0(G;\hat{\mathbb{Z}}_p[G]),$$

where the first map is induced by the inclusion $\mathbb{Z}[G] \subseteq \hat{\mathbb{Z}}_p[G]$ and the $\hat{\mathbb{Z}}_p$-module structure on $Wh'(\hat{\mathbb{Z}}_p[G])$. By Lemma 10.1, $\mathrm{Im}(\hat{\Gamma}_G) \subseteq Y(G)$.

<u>Theorem 10.3</u> *For any prime* p, *for any p-group* G, *and for any maximal order* $\mathfrak{M} \subseteq \mathbb{Q}[G]$, *there is an exact sequence*

$$1 \longrightarrow \hat{\mathbb{Z}}_p \otimes Wh'(G) \xrightarrow{\hat{\Gamma}_G} Y(G) \xrightarrow{\theta} \text{tors Coker}\Big[\hat{\mathbb{Z}}_p \otimes K_1'(\mathfrak{M}) \longrightarrow K_1'(\hat{\mathfrak{M}}_p)\Big].$$

Furthermore,

(i) $\mathrm{Im}(\hat{\Gamma}_G) = Y(G)$ *if* p *is an odd regular prime, or if* $p = 2$ *and* $\mathbb{Q}[G]$ *is a product of matrix algebras over fields; and*

(ii) $Y_0(G) \subseteq \mathrm{Im}(\hat{\Gamma}_G) \subseteq Y(G)$ *if* G *is an arbitrary 2-group.*

<u>Proof</u> Write $\mathbb{Q}[G] = \prod_{i=1}^k A_i$, where each A_i is simple with center K_i, and let $R_i \subseteq K_i$ be the ring of integers. By Theorem 9.1, each K_i is contained in $\mathbb{Q}(\xi_i)$ for some i. In the following diagrams

$$\begin{array}{ccc} \hat{\mathbb{Z}}_p \otimes Wh'(G) & \xrightarrow{\iota} & Wh'(\hat{\mathbb{Z}}_p[G]) \\ \downarrow \prod nr & (1a) & \downarrow \prod nr \\ \prod \hat{\mathbb{Z}}_p \otimes (R_i)^* & \xrightarrow{\prod \iota_{K_i}} & \prod (R_i)_p^{\wedge *} \end{array} \quad \text{and} \quad \begin{array}{ccc} \hat{\mathbb{Z}}_p \otimes K_1(\mathfrak{M}) & \longrightarrow & K_1(\hat{\mathfrak{M}}_p) \\ \downarrow \prod nr & (1b) & \downarrow \prod nr \\ \prod \hat{\mathbb{Z}}_p \otimes (R_i)^* & \xrightarrow{\prod \iota_{K_i}} & \prod (R_i)_p^{\wedge *} \end{array} \qquad (1)$$

the reduced norm homomorphisms have finite kernel and cokernel (Theorem 2.3 and Lemma 2.4), and the ι_{K_i} are injective by Proposition 10.2. So $Ker(\iota)$ is finite. But $Wh'(G)$ is torsion free, and hence ι and $\hat{\Gamma}_G = \Gamma_G \circ \iota$ are both injective. Also, $rk_{\mathbb{Z}}(Wh(G)) = rk_{\hat{\mathbb{Z}}_p}(Y(G))$ by Theorem 2.6, and so $[Y(G):Im(\hat{\Gamma}_G)]$ is finite.

By Milnor [2, Theorem 3.3], for each $n \geq 1$ such that $p^n\mathfrak{M} \subseteq \mathbb{Z}[G]$, there is a Mayer-Vietoris exact sequence

$$K_1(\mathbb{Z}[G]) \longrightarrow K_1(\mathfrak{M}) \oplus K_1(\mathbb{Z}[G]/p^n\mathfrak{M}) \longrightarrow K_1(\mathfrak{M}/p^n\mathfrak{M}).$$

The group $Ker\Big[K_1(\mathbb{Z}[G]) \longrightarrow K_1(\mathfrak{M})\Big] \subseteq SK_1(\mathbb{Z}[G])$ is finite, by Theorem 2.5(i), and so this sequence remains exact after taking the inverse limit over n. Since

$$\varprojlim_n K_1(\mathbb{Z}[G]/p^n\mathfrak{M}) \cong K_1(\hat{\mathbb{Z}}_p[G]) \quad \text{and} \quad \varprojlim_n K_1(\mathfrak{M}/p^n\mathfrak{M}) \cong K_1(\hat{\mathfrak{M}}_p)$$

by Theorem 2.10(iii), this shows that the sequence

$$K_1(\mathbb{Z}[G]) \longrightarrow K_1(\mathfrak{M}) \oplus K_1(\hat{\mathbb{Z}}_p[G]) \longrightarrow K_1(\hat{\mathfrak{M}}_p)$$

is exact; and remains exact after tensoring by $\hat{\mathbb{Z}}_p$. Also, $SK_1(\mathfrak{M})$ surjects onto $SK_1(\hat{\mathfrak{M}}_p)$ (Theorem 3.9), and so the top row in the following diagram is exact:

$$\begin{array}{ccccc} 1 \to \hat{\mathbb{Z}}_p \otimes Wh'(G) & \xrightarrow{\iota} & Wh'(\hat{\mathbb{Z}}_p[G]) & \xrightarrow{\hat{\theta}} & Coker\Big[\hat{\mathbb{Z}}_p \otimes K_1'(\mathfrak{M}) \to K_1'(\hat{\mathfrak{M}}_p)\Big]_{(p)}. \\ \downarrow \hat{\Gamma}_G & & \downarrow \Gamma_G & & \\ Y(G) & \hookrightarrow & H_0(G;\hat{\mathbb{Z}}_p[G]) & & \end{array} \qquad (2)$$

By Theorem 6.6, $Y(G) \subseteq \mathrm{Im}(\Gamma_G)$. In particular, we can identify $Y(G)$ with $\Gamma_G^{-1}(Y(G)) \subseteq Wh'(\hat{\mathbb{Z}}_p[G])$. Since $Y(G)/\mathrm{Im}(\hat{\Gamma}_G)$ is finite, the top row in (2) now restricts to an exact sequence

$$1 \longrightarrow \hat{\mathbb{Z}}_p \otimes Wh'(G) \xrightarrow{\hat{\Gamma}_G} Y(G) \xrightarrow{\theta} \text{tors Coker}\Big[\hat{\mathbb{Z}}_p \otimes K_1'(\mathfrak{M}) \xrightarrow{\iota_{\mathfrak{M}}} K_1'(\hat{\mathfrak{M}}_p)\Big],$$

where $\theta = \hat{\theta} \circ (\Gamma_G^{-1}|Y(G))$.

If p is regular, and if $\mathbb{Q}[G]$ is a product of matrix algebras over fields, then the reduced norm homomorphisms in (1b) are isomorphisms, and so $\mathrm{Coker}(\iota_{\mathfrak{M}})$ is torsion free by Proposition 10.2. So $\mathrm{Im}(\hat{\Gamma}_G) = Y(G)$ in this case. In particular, by Theorem 9.1, this always applies if p is odd (and regular), or if $p = 2$ and G is abelian. If G is an arbitrary 2-group, then $\mathrm{Im}(\hat{\Gamma}_G)$ contains the image of $Y(H) = Y_o(H)$ for all cyclic $H \subseteq G$; and hence $\mathrm{Im}(\hat{\Gamma}_G) \supseteq Y_o(G)$. □

In principle, it should be possible to use these methods to get information about $Wh'(G)$ when G is a p-group and p an irregular prime. With certain conditions on p (see Ullom [1]), the p-power torsion in $\mathrm{Coker}[\hat{\mathbb{Z}}_p \otimes (\mathbb{Z}[\xi_k])^* \longrightarrow (\hat{\mathbb{Z}}_p[\xi_k])^*]$ is understood (see also Washington [1, Theorem 13.56]). But most results which we know of, shown using Theorem 10.3, seem either to be obtainable by simpler methods (as in Ullom [1]); or to be quite technical.

The next theorem gives a precise description of $\hat{\Gamma}_G(\hat{\mathbb{Z}}_2 \otimes Wh'(G))$ when G is a nonabelian 2-group. Recall (Theorem 9.1) that if $\exp(G) = 2^n$, then $\mathbb{Q}[G]$ is isomorphic to a product of matrix algebras over subfields of $\mathbb{Q}(\xi_n)$ $(\xi_n = \exp(2\pi i/2^n))$, and over division algebras $\mathbb{Q}(\xi_k, j)$ $(\subseteq \mathbb{H})$ for $k \leq n$.

<u>**Theorem 10.4**</u> *Let* G *be a 2-group, and let* $Y(G)$ *be as in Theorem 10.3. Then*

$$\hat{\Gamma}_G(\hat{\mathbb{Z}}_2 \otimes Wh'(G)) = \mathrm{Ker}\Big[\theta_G' = \prod_{i=1}^{k} \theta_{G,i}' : Y(G) \longrightarrow \prod_{i=1}^{k} (\mathbb{Z}/2)\Big],$$

where $V_1,\dots,V_k$ *are the distinct irreducible* $\mathbb{C}[G]$*-representations which are quaternionic, and where*

$$\theta'_{G,i} : Y(G) \subseteq H_0(G;\hat{\mathbb{Z}}_2[G]) \longrightarrow \mathbb{Z}/2$$

is defined by setting

$$\theta'_{G,i}(g) = \sum_{r\geq 0} \dim_{\mathbb{C}}\Bigl(\xi_r\text{-eigenspace of } (g: V_i \to V_i)\Bigr). \qquad (\text{for } g \in G)$$

Proof This is based on the exact sequence

$$1 \to \hat{\mathbb{Z}}_2 \otimes Wh'(G) \xrightarrow{\hat{\Gamma}_G} Y(G) \xrightarrow{\theta} \text{tors Coker}\Bigl[\hat{\mathbb{Z}}_2 \otimes K'_1(\mathfrak{M}) \longrightarrow K'_1(\hat{\mathfrak{M}}_2)\Bigr] \quad (1)$$

of Theorem 10.3; where $\mathfrak{M} \subseteq \mathbb{Q}[G]$ is a maximal order containing $\mathbb{Z}[G]$, and where $\theta(\Gamma(u)) = [u] \in K'_1(\hat{\mathfrak{M}}_2)$.

Write $\mathbb{Q}[G] = \prod_{i=1}^m A_i$, where the A_i are simple. Fix i, and set $A = A_i \cong M_r(D)$, where D is a division algebra with center K. Let $\mathfrak{M}_A \subseteq A$ be a maximal $\mathbb{Z}$-order, and let $R \subseteq K$ be the ring of integers. If $D = K$ (i. e., D is a field), then $K \subseteq \mathbb{Q}(\xi_n)$ ($\xi_n = \exp(2\pi i/2^n)$) for some n by Theorem 9.1, and hence $\mathfrak{M}_A$ is Morita equivalent to R by Theorem 1.19. So

$$\text{Coker}\Bigl[\hat{\mathbb{Z}}_2 \otimes K'_1(\mathfrak{M}_A) \longrightarrow K'_1(\hat{\mathfrak{M}}_{A2})\Bigr] \cong \text{Coker}\Bigl[\hat{\mathbb{Z}}_2 \otimes R^* \longrightarrow (\hat{R}_2)^*\Bigr]$$

is torsion free in this case by Proposition 10.2.

By Theorem 9.1 again, the only other possibility is that $D \cong \mathbb{Q}(\xi_n, j)$ ($\subseteq \mathbb{H}$) for some $n \geq 2$ (so $K = \mathbb{Q}(\xi_n + \xi_n^{-1})$). In this case, consider the commutative diagram

$$\begin{array}{ccccc} 1 \longrightarrow \hat{\mathbb{Z}}_2 \otimes K_1(\mathfrak{M}_A) & \xrightarrow{\text{nr}} & \hat{\mathbb{Z}}_2 \otimes R^* & \xrightarrow{\lambda} & \bigoplus_{v\in V} \mathbb{Z}/2 \\ \Big\downarrow \iota_A & & \Big\downarrow \iota_K & & \\ K_1(\hat{\mathfrak{M}}_{A2}) & \xrightarrow[\cong]{\text{nr}_2} & (\hat{R}_2)^* & & \end{array} \tag{2}$$

where $\lambda = \oplus\lambda_v$ is defined by setting $\lambda_v(u) = 0$ if $v(u) > 0$; $\lambda_v(u) = 1$ if $v(u) < 0$. By Theorem 2.3, nr_2 is an isomorphism, and the top row in (2) is exact. By Proposition 10.2, λ is onto and $\text{Coker}(\iota_K)$ is torsion free; and so by (2),

$$\text{tors Coker}\Big[\hat{\mathbb{Z}}_2 \otimes K_1'(\mathfrak{M}_A) \longrightarrow K_1'(\hat{\mathfrak{M}}_{A2})\Big] \cong (\mathbb{Z}/2)^{|V|}.$$

In other words, $\text{tors}(\text{Coker}(\iota_A))$ includes one copy of $\mathbb{Z}/2$ for each quaternion representation of $\mathbb{R} \otimes_{\mathbb{Q}} A$. Sequence (1) now takes the form

$$1 \longrightarrow \hat{\mathbb{Z}}_2 \otimes \text{Wh}'(G) \xrightarrow{\hat{\Gamma}_G} Y(G) \xrightarrow{\theta} (\mathbb{Z}/2)^k;$$

where k is the number of quaternion components in $\mathbb{R}[G]$. The details of identifying θ with θ_G' as defined above are shown in Oliver & Taylor [1, Section 3]. □

A second description of $\text{Im}(\hat{\Gamma}_G)$, when G is a 2-group, is given in Oliver & Taylor [1, Propositions 4.4 and 4.5].

The next result is an easy application of Theorem 10.3.

<u>Theorem 10.5</u> *Fix a regular prime* p *and a* p*-group* G.

(i) For any surjection $\alpha: \tilde{G} \twoheadrightarrow G$ *of* p*-groups,*

$$\text{Wh}'(\alpha) : \text{Wh}'(\tilde{G}) \longrightarrow \text{Wh}'(G)$$

is surjective if p *is odd or if* G *is abelian; and* $\text{Coker}(\text{Wh}'(\alpha))$ *has exponent at most 2 otherwise.*

(ii) For any $x \in Wh'(G)$, x *is a product of elements induced up from cyclic subgroups of* G *if* p *is odd or if* G *is abelian; and* x^2 *is a product of such elements otherwise.*

Proof Fix a p-group G, and consider the group

$$C = \mathrm{Coker}\Big[\textstyle\sum\{Wh'(H) : H \subseteq G,\ H \text{ cyclic}\} \longrightarrow Wh'(G)\Big].$$

By a result of Lam [1, Section 4.2] (see also Theorem 11.2 below), C is a finite p-group. So $C \cong \hat{\mathbb{Z}}_p \otimes C$ is isomorphic to a subgroup of $Y(G)/Y_0(G)$ by Theorem 10.3. By definition, $Y(G)/Y_0(G)$ is trivial if p is odd or if G is abelian, and has exponent at most 2 otherwise.

This proves (ii). To prove (i), it now suffices to consider the case where $\tilde{G}$ and G are both cyclic. By Theorem 10.3, $\hat{\mathbb{Z}}_p \otimes Wh'(\tilde{G})$ surjects onto $\hat{\mathbb{Z}}_p \otimes Wh'(G)$ in this case, and so $\mathrm{Coker}(Wh'(\alpha))$ is finite of order prime to p. It thus suffices to show, for any $u \in (\mathbb{Z}[G])^*$, that $u^{p^k} \in \mathrm{Im}(Wh'(\alpha))$ for some k.

Assume that $\tilde{G} \cong C_{p^n}$ and $G \cong C_{p^{n-1}}$; and consider the pullback square

$$\begin{array}{ccc} \mathbb{Z}[C_{p^n}] & \longrightarrow & \mathbb{Z}[\xi_n] \\ \alpha\big\downarrow & & \big\downarrow \\ \mathbb{Z}[C_{p^{n-1}}] & \xrightarrow{\ \beta\ } & \mathbb{Z}/p[C_{p^{n-1}}]. \end{array} \qquad (\xi_n = \exp(2\pi i/p^n))$$

This induces a Mayer-Vietoris exact sequence

$$K_1(\mathbb{Z}[C_{p^n}]) \longrightarrow K_1(\mathbb{Z}[C_{p^{n-1}}]) \oplus K_1(\mathbb{Z}[\xi_n]) \longrightarrow (\mathbb{Z}/p[C_{p^{n-1}}]).$$

Set $I = \mathrm{Ker}\Big[\mathbb{Z}/p[C_{p^{n-1}}] \longrightarrow \mathbb{Z}/p\Big]$, the augmentation ideal. Then for any $u \in K_1(\mathbb{Z}[C_{p^{n-1}}])$, either $\beta_*(u)$ or $\beta_*(-u)$ lies in $1+I$, and this is a group of p-power order. In other words, $u^{p^k} \in \langle -1, \mathrm{Ker}(\beta_*)\rangle \subseteq \mathrm{Im}(\alpha_*)$ for some k; and we are done. □

The result that $Wh'(\alpha)$ is onto whenever α is a surjection of cyclic p-groups (for regular p) is due to Kervaire & Murthy [1].

We now turn to the problem of determining, for a given finite group G, which elements of $K_1(\mathbb{Z}[G])$ or $Wh(G)$ can be represented by units in $\mathbb{Z}[G]$. This was studied in detail by Magurn, Oliver and Vaserstein in [1]. The main general results in that paper are summarized in the following theorem.

A simple $\mathbb{Q}$-algebra A with center K is called *Eichler* if there is an embedding $v\colon K \hookrightarrow \mathbb{C}$ such that either $v(K) \not\subseteq \mathbb{R}$, or $v(K) \subseteq \mathbb{R}$ and $\mathbb{R}\otimes_{vK} A \not\cong \mathbb{H}$. Note that A is always Eichler if $[A:K] \neq 4$. A semisimple $\mathbb{Q}$-algebra is called Eichler if all of its simple components are Eichler.

<u>Theorem 10.6</u> *Let* $A = V\times B$ *be any semisimple* $\mathbb{Q}$*-algebra, where* B *is the product of all commutative and all non-Eichler simple components in* A. *Then for any* $\mathbb{Z}$*-order* $\mathfrak{A}$ *in* A, *if* $\mathfrak{B} \subseteq B$ *is the image of* $\mathfrak{A}$ *under projection to* B, *an element* $x \in K_1(\mathfrak{A})$ *can be represented by a unit if and only if its image in* $K_1(\mathfrak{B})$ *can be represented by a unit. In particular, if* A *is Eichler — i. e., if* B *is commutative — then there is an exact sequence*

$$\mathfrak{A}^* \longrightarrow K_1(\mathfrak{A}) \xrightarrow{(\mathfrak{A}\twoheadrightarrow\mathfrak{B})} SK_1(\mathfrak{B}) \longrightarrow 1.$$

<u>Proof</u> See Magurn et al [1, Theorems 6.2 and 6.3]. □

We now list two results containing examples of finite groups G where $Wh'(G)$ is or is not generated by units. The first theorem involves 2-groups, and is an application of Theorems 10.4 and 10.5 above. The second theorem will deal with generalized quaternion groups, and is proven using Theorem 10.6.

<u>Theorem 10.7</u> (i) *For any 2-group* G *and any* $x \in Wh'(G)$, x^2 *is represented by some unit* $u \in (\mathbb{Z}[G])^*$.

(ii) *Set* $G = Q(32)\times C_2\times C_2$, *where* $Q(32)$ *is quaternionic of order* 32. *Then* $Wh'(G)$ *contains elements not represented by units in* $\mathbb{Z}[G]$.

<u>Proof</u> Point (i) is clear: x^2 is a product of elements induced from cyclic subgroups of G by Theorem 10.5(ii); and $K_1'(\mathbb{Z}[G]) \cong (\mathbb{Z}[G])^*$ by definition if G is abelian.

To prove (ii), fix any element $a \in Q(32)$ of order 16, and let t_1, t_2 generate the two factors C_2 in $G = Q(32) \times C_2 \times C_2$. Set

$$x = (1-t_1)(1-t_2)(a-a^3) \in H_0(G;\hat{\mathbb{Z}}_2[G]).$$

A straightforward application of Theorem 10.4 (in fact, of Theorem 10.3) shows that $x \in \hat{\Gamma}_G(\hat{\mathbb{Z}}_2 \otimes Wh'(G))$. Thus, if all elements of $Wh'(G)$ are represented by units, then there must be a unit $u \in (\mathbb{Z}[G])^*$ such that $\Gamma_G(u) \equiv x \pmod{64}$. But using the relation $(\mathbb{Z}[\xi_4, j])^* = \langle(\mathbb{Z}[\xi_4])^*, j\rangle$ (Magurn et al [1, Lemma 7.5(b)]), it can be shown that no such u exists. See Oliver & Taylor [1, Theorem 4.7] for details. □

The following results are similar to those in Theorem 10.7, but for generalized quaternion groups instead of 2-groups. Recall that for any $n \geq 2$, $Q(4n)$ denotes the quaternion group of order $4n$.

<u>Theorem 10.8</u> *For any* $n \geq 2$, *and any* $x \in Wh(Q(4n))$, x^2 *is represented by a unit in* $\mathbb{Z}[Q(4n)]$. *Furthermore:*

(i) If n *is a power of* 2, *then all elements of* $Wh(Q(4n))$ *can be represented by units.*

(ii) If p *is an odd prime, then the elements of* $Wh(Q(4p))$ *can all be represented by units, if and only if the class number* h_p *is odd.*

(iii) For any prime $p \equiv -1 \pmod 8$, $Wh(Q(16p))$ *contains elements not represented by units.*

<u>Proof</u> See Magurn et al [1, Theorems 7.15, 7.16, 7.18, and 7.22]. □

An obvious question now is whether, for *any* finite group G and any $x \in Wh'(G)$, x^2 is represented by a unit in $\mathbb{Z}[G]$.

PART III: GENERAL FINITE GROUPS

One of the standard procedures when working with almost any K-theoretic functor defined on group rings of finite groups, is to reduce problems involving arbitrary groups to problems involving hyperelementary groups: i. e., groups containing a normal cyclic subgroup of prime power index. For most of the functors dealt with here, one can go even further. The main idea, when dealing with $SK_1(\hat{\mathbb{Z}}_p[G])$, $Cl_1(\mathbb{Z}[G])_{(p)}$, etc., is to reduce computations first to the case where G is p-elementary (i. e., a product of a cyclic group with a p-group); and then from that to the case where G is a p-group.

The formal machinery for the reduction to p-elementary groups is set up in Chapter 11. The actual reductions to p-elementary groups, and then to p-groups, are carried out in Chapters 12 (for $SK_1(\hat{\mathbb{Z}}_p[G])$) and 13 (for $Cl_1(\mathbb{Z}[G])$). The inclusion $Cl_1(\mathbb{Z}[G]) \subseteq SK_1(\mathbb{Z}[G])$ is then shown in Section 13c to be split in odd torsion. Finally, in Chapter 14, some applications of these results are listed.

Since much of the philosophy behind the reductions in Chapters 12 and 13 is similar, it seems appropriate to outline it here. The main tool used in the reduction to p-elementary groups is induction theory as formulated by Dress [2]. This sets up conditions for when $\mathcal{M}(G)$, $\mathcal{M}$ being a functor defined on finite groups, can be completely completely computed as the direct or inverse limit of the groups $\mathcal{M}(H)$ for subgroups $H \subseteq G$ lying in some family. The main general results on this subject are Theorem 11.1 (Dress' theorem), Theorem 11.8 (a decomposition formula for certain functors defined on $\mathbb{Z}$- or $\hat{\mathbb{Z}}_p$-orders), and Theorem 11.9 (conditions for computability with respect to p-elementary subgroups).

Using these results, $SK_1(\mathbb{Z}[G])_{(p)}$ is shown in Chapters 12 and 13 to be p-elementary computable for odd p, and 2-$\mathbb{R}$-elementary comput-able when $p = 2$ (Theorems 12.4 and 13.5). In particular, for odd p,

$$SK_1(\mathbb{Z}[G])_{(p)} \cong \varinjlim_{H\in\mathcal{E}} SK_1(\mathbb{Z}[H])_{(p)};$$

where $\mathcal{E}$ is the set of p-elementary subgroups of G, and the limits are taken with respect to inclusions of subgroups and conjugation by elements of G. When $p = 2$, the connection between $SK_1(\mathbb{Z}[G])_{(2)}$ and 2-elementary subgroups is described by a pushout square (Theorem 13.5 again).

The process of reduction from p-elementary groups to p-groups is simpler. Let G be a p-elementary group: $G = C_n \times \pi$, where $p \nmid n$ and π is a p-group. Write $\hat{\mathbb{Q}}_p[C_n] = \prod_{i=1}^k F_i$, where the F_i are fields, and let $R_i \subseteq F_i$ be the ring of integers. Then

$$SK_1(\hat{\mathbb{Z}}_p[G]) \cong \bigoplus_{i=1}^{k} SK_1(R_i[\pi]) \qquad \text{and} \qquad Cl_1(\mathbb{Z}[G]) \cong \bigoplus_{d|n} Cl_1(\mathbb{Z}\zeta_d[\pi]):$$

the first isomorphism is induced by an isomorphism of rings (Theorem 1.10(i)), and the second by an inclusion $\mathbb{Z}[G] \subseteq \prod_{d|n} \mathbb{Z}\zeta_d[\pi]$ of orders of index prime to p (Example 1.2, Theorem 1.4(v), and Corollary 3.10). The groups $SK_1(R_i[\pi])$ have already been described in Theorem 8.6, and the $Cl_1(\mathbb{Z}\zeta_d[\pi])$ are studied in Section 13b by comparing them with $Cl_1(\mathbb{Z}[\pi])$.

These results then lead to explicit descriptions of $SK_1(\hat{\mathbb{Z}}_p[G])$ for arbitrary p and G (Theorems 12.5 and 12.10), and of $Cl_1(\mathbb{Z}[G])_{(p)}$ when p is odd (Theorem 13.9) or G is abelian (Theorem 13.13). For nonabelian G, the situation in 2-torsion is as usual incomplete, but partial descriptions of $Cl_1(\mathbb{Z}[G])_{(2)}$ in terms of $Cl_1(\mathbb{Z}[\pi])$ for 2-subgroups π can be pulled out of Theorems 13.5 and 13.12.

Chapter 11 A QUICK SURVEY OF INDUCTION THEORY

The term "induction theory" refers here to techniques used to get information about $\mathcal{M}(G)$ in terms of the groups $\mathcal{M}(H)$ for certain $H \subseteq G$, when $\mathcal{M}$ is a functor defined on finite groups. Such methods were first applied to K-theoretic functors by Swan [1], when studying the groups $K_0(\mathbb{Z}[G])$ and $G_0(\mathbb{Z}[G])$ for finite G. Swan's techniques were systematized by Lam [1]; whose Frobenius functors gave very general conditions for $\mathcal{M}(G)$ to be generated by induction from subgroups of G lying in some family $\mathcal{F}$, or to be detected by restriction to subgroups in $\mathcal{F}$. Later, Lam's ideas were developed further by Dress [2], who gave conditions for when $\mathcal{M}(G)$ can be completely computed in terms of $\mathcal{M}(H)$ for subgroups $H \subseteq G$ in $\mathcal{F}$.

The results of Dress are based on the concepts of Mackey functors, and Green rings and modules, whose general definitions and properties are summarized in Section 11a. The central theorem, Theorem 11.1, gives conditions for a Green module to be "computable" with respect to a certain family of finite groups. Two examples of Green modules are then given: functors defined on a certain category of R-orders (when R is any Dedekind domain of characteristic zero) are shown to induce Green modules over the Green ring $G_0(R[-])$ (Theorem 11.2), and Mackey functors are shown to be Green modules over the Burnside ring (Proposition 11.3).

In Section 11b, attention is focused on p-local Mackey functors: i. e., Mackey functors which take values in $\hat{\mathbb{Z}}_p$-modules. A decomposition formula is obtained in Theorem 11.8, using idempotents in the localized Burnside ring $\Omega(G)_{(p)}$; and this reduces the computation of $Cl_1(\mathbb{Z}[G])_{(p)}$, $SK_1(\mathbb{Z}[G])_{(p)}$, $SK_1(\hat{\mathbb{Z}}_p[G])$, etc. to that of certain twisted group rings over p-groups. This is the first step toward results in Chapters 12 and 13, which reduce the computation of $SK_1(\mathbb{Z}[G])_{(p)}$ (at least for odd p) to the case where G is a p-group.

11a. Induction properties for Mackey functors and Green modules

The following definitions are all due to Dress [2, Section 1].

(A) A *Mackey functor* is a bifunctor $\mathcal{M} = (\mathcal{M}^*, \mathcal{M}_*)$ from the category of finite groups with monomorphisms to the category of abelian groups, such that $\mathcal{M}^*$ is contravariant, $\mathcal{M}_*$ is covariant,

$$\mathcal{M}^*(G) = \mathcal{M}_*(G) = \mathcal{M}(G)$$

for all G, and the following conditions are satisfied:

(i) $\mathcal{M}^*$ and $\mathcal{M}_*$ send inner automorphisms to the identity.

(ii) For any isomorphism $\alpha\colon G \xrightarrow{\cong} G'$, $\mathcal{M}^*(\alpha) = \mathcal{M}_*(\alpha)^{-1}$.

(iii) The Mackey subgroup property holds for $\mathcal{M}^*$ and $\mathcal{M}_*$: for any G, and any pair $H,K \subseteq G$, the composite

$$\mathcal{M}(H) \xrightarrow{\mathcal{M}_*} \mathcal{M}(G) \xrightarrow{\mathcal{M}^*} \mathcal{M}(K)$$

is equal to the sum, over all double cosets $KgH \subseteq G$, of the composites

$$\mathcal{M}(H) \xrightarrow{\mathcal{M}^*} \mathcal{M}(g^{-1}Kg \cap H) \xrightarrow{\mathcal{M}_*(c_g)} \mathcal{M}(K \cap gHg^{-1}) \xrightarrow{\mathcal{M}_*} \mathcal{M}(K).$$

Here, c_g denotes conjugation by g.

(B) A *Green ring* $\mathcal{G}$ is a Mackey functor together with a commutative ring structure on $\mathcal{G}(G)$ for all G, and satisfying the Frobenius reciprocity conditions. More precisely, for any inclusion $\alpha\colon H \hookrightarrow G$,

$$\alpha^*(xy) = \alpha^*(x)\cdot\alpha^*(y) \qquad \text{for } x,y \in \mathcal{G}(G)$$

$$x\cdot\alpha_*(y) = \alpha_*(\alpha^*(x)\cdot y) \qquad \text{for } x \in \mathcal{G}(G),\ y \in \mathcal{G}(H)$$

$$\alpha_*(x)\cdot y = \alpha_*(x\cdot\alpha^*(y)) \qquad \text{for } x \in \mathcal{G}(H),\ y \in \mathcal{G}(G)$$

(where $\alpha^* = \mathcal{G}^*(\alpha)$, $\alpha_* = \mathcal{G}_*(\alpha)$).

(C) A *Green module* over a Green ring $\mathcal{G}$ is a Mackey functor $\mathcal{M}$, together with a $\mathcal{G}(G)$-module structure on $\mathcal{M}(G)$ for all G, such that the same Frobenius relations hold as in (B), but with $y \in \mathcal{M}(G)$ or $y \in \mathcal{M}(H)$ instead.

(D) Let $\mathcal{C}$ be any class of finite groups closed under subgroups. For each G, set $\mathcal{C}(G) = \{H \subseteq G: H \in \mathcal{C}\}$. Then a Mackey functor $\mathcal{M}$ is called *$\mathcal{C}$-generated* if, for any finite G,

$$\bigoplus_{H\in\mathcal{C}(G)} \mathcal{M}(H) \xrightarrow{\ \mathcal{M}_*\ } \mathcal{M}(G)$$

is onto; $\mathcal{M}$ is called *$\mathcal{C}$-computable* (with respect to induction) if, for any G, $\mathcal{M}_*$ induces an isomorphism

$$\mathcal{M}(G) \cong \varinjlim_{H\in\mathcal{C}(G)} \mathcal{M}(H).$$

Here, the limit is taken with respect to all maps between subgroups induced by inclusions, or by conjugation by elements of G. Similarly, $\mathcal{M}$ is *$\mathcal{C}$-detected*, or *$\mathcal{C}$-computable* with respect to restriction, if for all finite G the homomorphism

$$\mathcal{M}(G) \longrightarrow \varprojlim_{H\in\mathcal{C}(G)} \mathcal{M}(H)$$

(induced by $\mathcal{M}^*$) is a monomorphism or isomorphism, respectively.

For convenience of notation, if $H \subseteq G$ is any pair of finite groups and $i: H \longrightarrow G$ is the inclusion map, we usually write

$$\mathrm{Ind}_H^G = \mathcal{M}_*(i) : \mathcal{M}(H) \longrightarrow \mathcal{M}(G), \qquad \mathrm{Res}_H^G = \mathcal{M}^*(i) : \mathcal{M}(G) \longrightarrow \mathcal{M}(H)$$

to denote the induced homomorphisms.

The first theorem can be thought of as the "fundamental theorem" of induction theory for Green modules.

Theorem 11.1 (Dress [2, Propositions 1.1′ and 1.2]) *Let $\mathcal{M}$ be a Green module over a Green ring $\mathcal{G}$, and let $\mathcal{C}$ be a class of finite groups such that $\mathcal{G}$ is $\mathcal{C}$-generated. Then $\mathcal{M}$ is $\mathcal{C}$-computable for both induction and restriction.*

Proof Fix any G, write $\mathcal{C} = \mathcal{C}(G)$ for short, and let

$$\hat{I} : \varinjlim_{H\in\mathcal{C}} \mathcal{M}(H) \longrightarrow \mathcal{M}(G) \qquad \text{and} \qquad \hat{R} : \mathcal{M}(G) \longrightarrow \varprojlim_{H\in\mathcal{C}} \mathcal{M}(H)$$

be the induced maps. Choose elements $a_H \in \mathcal{G}(H)$, for $H \in \mathcal{C}$, such that

$$\sum_{H\in\mathcal{C}} \mathrm{Ind}_H^G(a_H) = 1 \in \mathcal{G}(G). \tag{1}$$

For any $x \in \mathcal{M}(G)$,

$$x = \sum_{H\in\mathcal{C}} \mathrm{Ind}_H^G(a_H)\cdot x = \sum_{H\in\mathcal{C}} \mathrm{Ind}_H^G(a_H\cdot \mathrm{Res}_H^G(x)) \in \mathrm{Im}(\hat{I})$$

by Frobenius reciprocity. In particular, $x = 0$ if $\mathrm{Res}_H^G(x) = 0$ for all $H \in \mathcal{C}$, i. e., if $\hat{R}(x) = 0$. Thus, $\hat{I}$ is onto and $\hat{R}$ is one-to-one.

To show that $\hat{I}$ is injective and $\hat{R}$ surjective, the Mackey subgroup property is needed. For each pair $H,K \in \mathcal{C}$, let g_{HKi} $(1 \le i \le n_{HK})$ be double coset representatives for $H\backslash G/K$. Consider the maps

$$\mathcal{M}(K) \underset{I'_{HKi}}{\overset{R'_{HKi}}{\rightleftarrows}} \mathcal{M}(g_{HKi}Kg_{HKi}^{-1} \cap H) \underset{R_{HKi}}{\overset{I_{HKi}}{\rightleftarrows}} \mathcal{M}(H)$$

(and similarly for $\mathcal{G}$). Here, I and R denote induction and restriction, while I′ and R′ are the induction and restriction maps composed with conjugation by g_{HKi}.

Fix any $x \in \mathrm{Ker}(\hat{I})$. Write

$$x = \sum_{K\in\mathcal{C}} [x_K, K] \in \varinjlim_{K\in\mathcal{C}} \mathcal{M}(K),$$

where $x_K \in \mathcal{M}(K)$ and $\sum_{K\in\mathcal{C}} \mathrm{Ind}_K^G(x_K) = 0$. Then, in $\varinjlim \mathcal{M}(K)$,

$$x = \sum_{K\in\mathcal{C}} [x_K, K] = \sum_{K\in\mathcal{C}} [\mathrm{Res}_K^G(1)\cdot x_K, K]$$

$$= \sum_{K\in\mathcal{C}} \sum_{H\in\mathcal{C}} [(\mathrm{Res}_K^G \circ \mathrm{Ind}_H^G(a_H))\cdot x_K, K] \qquad \text{(by (1))}$$

$$= \sum_{K\in\mathcal{C}} \sum_{H\in\mathcal{C}} \sum_{i=1}^{n_{HK}} [(I'_{HKi} \circ R_{HKi}(a_H))\cdot x_K, K]$$

$$= \sum_{K\in\mathcal{C}} \sum_{H\in\mathcal{C}} \sum_{i=1}^{n_{HK}} [I'_{HKi}(R_{HKi}(a_H)\cdot R'_{HKi}(x_K)), K]$$

$$= \sum_{H\in\mathcal{C}} \sum_{K\in\mathcal{C}} \sum_{i=1}^{n_{HK}} [I_{HKi}(R_{HKi}(a_H)\cdot R'_{HKi}(x_K)), H] \qquad \text{(by defn. of } \varinjlim)$$

$$= \sum_{H\in\mathcal{C}} \sum_{K\in\mathcal{C}} \sum_{i=1}^{n_{HK}} [a_H\cdot(I_{HKi} \circ R'_{HKi}(x_K)), H]$$

$$= \sum_{H\in\mathcal{C}} \sum_{K\in\mathcal{C}} [a_H\cdot(\mathrm{Res}_H^G \circ \mathrm{Ind}_K^G(x_K)), H]$$

$$= \sum_{H\in\mathcal{C}} [a_H\cdot \mathrm{Res}_H^G(\sum_{K\in\mathcal{C}} \mathrm{Ind}_K^G(x_K)), H] = 0;$$

and so $\hat{I}$ is injective.

Now fix some element $y = (y_K)_{K\in\mathcal{C}}$ in $\varprojlim_{H\in\mathcal{C}} \mathcal{M}(H)$. Set

$$\hat{y} = \sum_{H\in\mathcal{C}} \mathrm{Ind}_H^G(a_H\cdot y_H) \in \mathcal{M}(G).$$

Then, for each $K \in \mathcal{C}$,

$$\mathrm{Res}_K^G(\hat{y}) = \sum_{H\in\mathcal{C}} \mathrm{Res}_K^G \circ \mathrm{Ind}_H^G(a_H\cdot y_H)$$

$$= \sum_{H\in\mathcal{C}} \sum_{i=1}^{n_{HK}} I'_{HKi} \circ R_{HKi}(a_H\cdot y_H)$$

$$= \sum_{H\in\mathscr{C}} \sum_{i=1}^{n_{HK}} I'_{HKi}(R_{HKi}(a_H)\cdot R_{HKi}(y_H))$$

$$= \sum_{H\in\mathscr{C}} \sum_{i=1}^{n_{HK}} I'_{HKi}(R_{HKi}(a_H)\cdot R'_{HKi}(y_K)) \qquad (\text{since } (y_H) \in \varprojlim \mathscr{M}(H))$$

$$= \sum_{H\in\mathscr{C}} \sum_{i=1}^{n_{HK}} (I'_{HKi}\circ R_{HKi}(a_H))\cdot y_K$$

$$= \sum_{H\in\mathscr{C}} (\mathrm{Res}^G_K \circ \mathrm{Ind}^G_H(a_H))\cdot y_K = \mathrm{Res}^G_K(1)\cdot y_K = y_K. \qquad (\text{by } (1))$$

In other words, $y = (y_H)_{H\in\mathscr{C}} = \hat{R}(\hat{y})$; and so $\hat{R}$ is surjective. □

In fact, Dress [2] also proves that a Green module $\mathscr{M}$ as above is "$\mathscr{C}$-acyclic" with respect to induction and restriction, in that the derived functors for the limits in (D) above are all zero. This is important, for example, if one is given a sequence of Mackey functors $\mathscr{M}_i$ which is exact for all $G \in \mathscr{C}$, and which one wants to prove is exact for all finite G.

Recall that for any prime p, a *p-hyperelementary* group is a finite group of the form $C_n \rtimes \pi$, where $p \nmid n$ and π is a p-group. For any field K of characteristic zero, a p-hyperelementary group $G = C_n \rtimes \pi$ is *p-K-elementary* if

$$\mathrm{Im}\Big[\pi \xrightarrow{\text{conj}} \mathrm{Aut}(C_n) \cong (\mathbb{Z}/n)^*\Big] \subseteq \mathrm{Gal}(K\zeta_n/K);$$

where ζ_n is a primitive n-th root of unity, and $\mathrm{Gal}(K\zeta_n/K)$ is regarded as a subgroup of $\mathrm{Aut}(C_n)$ via the action on $\langle\zeta_n\rangle \cong C_n$ (Theorem 1.5). A finite group is *K-elementary* if it is p-K-elementary for some p. Note that hyperelementary is the same as $\mathbb{Q}$-elementary ($\mathrm{Gal}(\mathbb{Q}\zeta_n/\mathbb{Q}) = \mathrm{Aut}(C_n)$ by Theorem 1.5(i)); and that $C_n \rtimes \pi$ is $\mathbb{C}$-elementary only if it is a direct product. A second characterization of p-K-elementary groups will be given in Proposition 11.6 below.

The next theorem gives one way of constructing examples of Green modules. For any ring R, $G_0(R)$ denotes the Grothendieck group on all isomorphism classes of finitely generated R-modules, modulo the relation $[M] = [M'] + [M'']$ for any short exact sequence

$$0 \longrightarrow M' \longrightarrow M \longrightarrow M'' \longrightarrow 0.$$

As defined in Section 1d, the category of "rings with bimodule morphisms" is the category whose objects are rings; and where $Mor(R,S)$, for any R and S, is the Grothendieck group (modulo short exact sequences) of all isomorphism classes of bimodules ${}_SM_R$ such that M is finitely generated and projective as a left S-module.

<u>Theorem 11.2</u> *Let R be a Dedekind domain with field of fractions K of characteristic zero, and let X be an additive functor from the category of R-orders in semisimple K-algebras with bimodule morphisms to the category of abelian groups. Then, for finite G, $\mathcal{M}(G) = X(R[G])$ is a Mackey functor, and is a Green module over the Green ring $G_0(R[G])$. In particular, $\mathcal{M}$ is computable with respect to induction from and restriction to K-elementary subgroups; and $\mathcal{M}(G)_{(p)}$ (for any prime p) is computable with respect to induction from and restriction to p-K-elementary subgroups.*

<u>Proof</u> Induction and restriction maps for pairs $H \subseteq G$ are defined using the obvious bimodules ${}_{RG}RG_{RH}$ and ${}_{RH}RG_{RG}$. This makes $\mathcal{M}(G) = X(R[G])$ into a Mackey functor: the properties all follow from easy identities among bimodules. For example, the Mackey subgroup property for a pair $H,K \subseteq G$ follows upon decomposing ${}_{RK}RG_{RH}$ as a sum of bimodules, one for each double coset KgH.

Next consider the $G_0(R[G])$-module structure on $X(R[G])$. For any finitely generated (left) R[G]-module M, make $M \otimes_R R[G]$ into an (R[G],R[G])-bimodule by setting

$$g\cdot(x \otimes y)\cdot h = gx \otimes gyh \quad \text{for} \quad g,h \in G, \ x \in M, \ y \in R[G].$$

Then multiplication in $X(R[G])$ by $[M] \in G_0(R[G])$ is induced by $[M \otimes_R R[G]] \in Mor(R[G], R[G])$. The module relations, and the Frobenius reciprocity relations as well, are again immediate from bimodule identities.

The computability of $X(R[G])$ will now follow from Theorem 11.1, once we have checked that $G_0(R[G])$ is generated by induction from K-elementary subgroups (and $G_0(R[G])_{(p)}$ from p-K-elementary subgroups). For $K_0(K[G])$ and $K_0(K[G])_{(p)}$, this is a theorem of Berman and Witt (see Serre [2, §12.6, Theorems 27 and 28] or Curtis & Reiner [1, Theorem 21.6]). Also, for any maximal ideal $\mathfrak{p} \subseteq R$, $G_0(R/\mathfrak{p}[G])$ is made into a Green module over $K_0(K[G])$ by the "decomposition map"

$$d : K_0(K[G]) \longrightarrow G_0(R/\mathfrak{p}[G]);$$

where $d([V]) = [M/\mathfrak{p}M]$ for any $K[G]$-module V and any G-invariant R-lattice M in V (see Serre [2, §15.2] or Curtis & Reiner [1, Proposition 16.17]). There is an exact localization sequence

$$\bigoplus_{\mathfrak{p}} G_0(R/\mathfrak{p}[G]) \longrightarrow G_0(R[G]) \longrightarrow K_0(K[G]) \longrightarrow 0$$

(see Bass [2, Proposition IX.6.9]); and so $G_0(R[G])$ and $G_0(R[G])_{(p)}$ are also generated by induction from K-elementary and p-K-elementary subgroups, respectively. □

Note in particular that by Proposition 1.18, Theorem 11.2 applies to the functors $K_n(\mathbb{Z}[G])$, $Cl_1(\mathbb{Z}[G])$, $SK_1(\mathbb{Z}[G])$, $SK_1(\hat{\mathbb{Z}}_p[G])$, etc.

The next proposition gives another example of Green modules. For any finite G, $\Omega(G)$ denotes the *Burnside ring*: the Grothendieck group on all finite G-sets (i. e., finite sets with G-action), where addition is induced by disjoint union and multiplication by Cartesian product. Additively, $\Omega(G)$ is a free abelian group with basis the set of all orbits G/H for $H \subseteq G$; where $[G/H_1] = [G/H_2]$ if and only if H_1 and H_2 are conjugate in G.

Proposition 11.3 *The Burnside ring Ω is a Green ring. Any Mackey functor $\mathcal{M}$ is a Green module over Ω, where the $\Omega(G)$-module structure on $\mathcal{M}(G)$ is given by*

$$[G/H]\cdot x = \mathrm{Ind}_H^G \circ \mathrm{Res}_H^G(x) \qquad \textit{for} \quad H \subseteq G, \quad x \in \mathcal{M}(G).$$

<u>Proof</u> For any pair $H \subseteq G$ of finite groups,

$$\mathrm{Ind}_H^G : \Omega(H) \longrightarrow \Omega(G) \qquad \text{and} \qquad \mathrm{Res}_H^G : \Omega(G) \longrightarrow \Omega(H)$$

are defined by setting $\mathrm{Ind}_H^G([S]) = [G \times_H S]$ for any finite H-set S, and $\mathrm{Res}_H^G([T]) = [T|_H]$ for any finite G-set T. The Mackey property and Frobenius reciprocity are easily checked.

Now let $\mathcal{M}$ be an arbitrary Mackey functor. To see that the above definition does make $\mathcal{M}(G)$ into an $\Omega(G)$-module for each G, note that for any pair $H,K \subseteq G$,

$$(\mathrm{Ind}_H^G \circ \mathrm{Res}_H^G)\circ(\mathrm{Ind}_K^G \circ \mathrm{Res}_K^G) = \mathrm{Ind}_H^G \circ \Big(\sum_{g \in H\backslash G/K} \mathrm{Ind}_{H \cap gKg^{-1}}^H \circ (c_g)_* \circ \mathrm{Res}_{g^{-1}Hg\cap K}^K \Big) \circ \mathrm{Res}_K^G$$

$$= \sum_{g \in H\backslash G/K} \mathrm{Ind}_{H\cap gKg^{-1}}^G \circ (c_g)_* \circ \mathrm{Res}_{g^{-1}Hg\cap K}^G = \sum_{g \in H\backslash G/K} \mathrm{Ind}_{H\cap gKg^{-1}}^G \circ \mathrm{Res}_{H\cap gKg^{-1}}^G .$$

Thus, the composite of multiplication first by $[G/K]$ and then by $[G/H]$ is multiplication by

$$\sum_{g \in H\backslash G/K} [G/(H\cap gKg^{-1})] = [(G/H)\times(G/K)].$$

Checking the Frobenius relations for this module structure is straightforward. □

Proposition 11.3 cannot be directly combined with Theorem 11.1 to give general induction properties of Mackey functors: the Burnside ring is not generated by induction from any proper family of finite groups. But often an apparently weak induction property of a Mackey functor $\mathcal{M}$ implies that $\mathcal{M}$ is a Green module over some quotient ring of Ω, which in turn yields stronger induction properties for $\mathcal{M}$. One example of this is seen in Proposition 11.5 below: any Mackey functor which is generated by hyperelementary induction is also hyperelementary computable with respect to both induction and restriction.

11b. Splitting p-local Mackey functors

By a "p-local" Mackey functor, for any prime p, is meant a Mackey functor $\mathcal{M}$ for which $\mathcal{M}(G)$ is a $\mathbb{Z}_{(p)}$-module for all G. In Proposition 11.5 below, we will see that for any p-local Mackey functor $\mathcal{M}$ generated by p-hyperelementary induction, $\mathcal{M}(G)$ has for all G a natural splitting indexed by conjugacy classes of cyclic subgroups of G of order prime to p. Under certain conditions, these summands can be described in terms of functors on twisted group rings (Lemma 11.7 and Theorem 11.8), and this is then used to set up conditions for when $\mathcal{M}$ is computable with respect to p-elementary subgroups.

By Proposition 11.3, any p-local Mackey functor $\mathcal{M}$ is a module over the localized Burnside ring $\Omega(-)_{(p)}$. Hence, splittings of $\mathcal{M}(G)$ are automatically induced by idempotents in $\Omega(G)_{(p)}$. These idempotents were first studied by Dress [1, Proposition 2].

When working with $\Omega(G)$, it is often convenient to use its "character" homomorphism

$$\chi = \prod \chi_H : \Omega(G) \longrightarrow \prod_{H \in \mathcal{S}(G)} \mathbb{Z}.$$

Here, $\mathcal{S}(G)$ denotes the set of conjugacy classes of subgroups of G, and $\chi_H([S]) = |S^H|$ for any finite G-set S and any $H \in \mathcal{S}(G)$. Note that $\Omega(G)$ and $\prod_{H \in \mathcal{S}(G)} \mathbb{Z}$ are free abelian groups of the same rank; and that for $H, K \subseteq G$, $\chi_K([G/H])$ is nonzero if and only if $K \subseteq gHg^{-1}$ for some $g \in G$. In other words, if the elements of $H \in \mathcal{S}(G)$ are ordered according to size, the matrix for χ is triangular with nonzero diagonal entries. So χ is injective and has finite cokernel. In particular, an element $x \in \Omega(G)$ (or $x \in \Omega(G)_{(p)}$) is an idempotent if and only if $\chi_H(x) \in \{0,1\}$ for all $H \subseteq G$.

<u>Lemma 11.4</u> *Fix a prime* p *and a finite group* G. *Then, for any cyclic subgroup* $C \subseteq G$ *of order prime to* p, *there is an idempotent* $E_C = E_C(G) \in \Omega(G)_{(p)}$ *such that for all* $H \subseteq G$,

$$\chi_H(E_C) = \begin{cases} 1 & \textit{if for some } C' \textit{ conj. } C,\ C' \triangleleft H,\ H/C' \textit{ a p-group} \\ 0 & \textit{otherwise.} \end{cases}$$

<u>Proof</u> Fix C. For each $\bar{C} \subseteq C$, set

$$\mathcal{L}(\bar{C}) = \left\{H \subseteq G : g\bar{C}g^{-1} \triangleleft H,\ H/(g\bar{C}g^{-1}) \text{ a p-group, some } g \in G\right\}.$$

We first claim that for any $x \in \Omega(G)_{(p)}$,

$$\chi_H(x) \equiv \chi_{\bar{C}}(x) \pmod{p\mathbb{Z}_{(p)}} \qquad \text{for all } H \in \mathcal{L}(\bar{C}). \tag{1}$$

It suffices to check this when $\bar{C} \triangleleft H$ and $x = [S]$ (some finite G-set S); and in this case the p-group $\bar{C}/H$ acts on $S^{\bar{C}} \smallsetminus S^H$ without fixed points.

Fix some $M \supseteq C$ such that M/C is a p-Sylow subgroup of $N(C)/C$ (so M is maximal in $\mathcal{L}(C)$). For any $k \geq 0$, and any $H \subseteq G$,

$$\chi_H([G/M]^{(p-1)p^k}) \equiv \begin{cases} 1 \pmod{p^{k+1}} & \text{if } |(G/M)^H| \not\equiv 0 \pmod p \\ 0 \pmod{p^{k+1}} & \text{if } |(G/M)^H| \equiv 0 \pmod p). \end{cases}$$

Since $\chi = \prod \chi_H$ has finite cokernel, this shows that there exists $E \in \Omega(G)_{(p)}$ such that

$$\chi_H(E) = \begin{cases} 1 & \text{if } |(G/M)^H| \not\equiv 0 \pmod p \\ 0 & \text{if } |(G/M)^H| \equiv 0 \pmod p). \end{cases}$$

In particular, if $\chi_H(E) = 1$, then $gHg^{-1} \subseteq M$ for some g; so H is p-hyperelementary, and $H \in \mathcal{L}(\bar{C})$ for some $\bar{C} \subseteq C$. Also, by (1), $\chi_H(E) = \chi_{\bar{C}}(E)$ if $H \in \mathcal{L}(\bar{C})$; and $\chi_C(E) = 1$ since by choice of M:

$$|(G/M)^C| = |\{gM : g^{-1}Cg \subseteq M\}| = |N(C)|/|M| \not\equiv 0 \pmod p).$$

We may assume inductively that for each $\bar{C} \subsetneq C$, an idempotent $E_{\bar{C}}$ is defined such that $\chi_H(E_{\bar{C}}) = 1$ if and only if $H \in \mathcal{L}(\bar{C})$. Then E_C can be defined by setting

$$E_C = E - \sum\{E_{\bar{C}} : \bar{C} \subsetneq C,\ \chi_{\bar{C}}(E) = 1\}. \quad \square$$

Lemma 11.4 will now be applied to split p-local Mackey functors. Assume that p is a fixed prime, and that $\mathcal{M}$ is a p-local Mackey functor which is generated by p-hyperelementary induction. For each finite G, let $Cy(G)$ be a set of conjugacy class representatives of cyclic subgroups $C \subseteq G$ of order prime to p. For each $C \in Cy(G)$, let $E_C(G) \in \Omega(G)_{(p)}$ be the idempotent defined in Lemma 11.4, and set

$$\mathcal{M}_C(G) = E_C(G)\cdot\mathcal{M}(G) \subseteq \mathcal{M}(G).$$

If G is p-hyperelementary — if $G = C_n \rtimes \pi$ where $p \nmid n$ and π is a p-group — then for $k|n$ we write $\mathcal{M}_k(G) = \mathcal{M}_C(G)$ when $C \subseteq G$ is the subgroup of order k (and set $\mathcal{M}_k(G) = 0$ if $k \nmid n$).

<u>**Proposition 11.5**</u> *Fix a prime* p, *and let* $\mathcal{M}$ *be any p-local Mackey functor generated by p-hyperelementary induction. Then* $\mathcal{M}$ *is p-hyperelementary computable for induction and restriction. Also:*

(i) $\mathcal{M}(G) = \bigoplus_{C\in Cy(G)} \mathcal{M}_C(G)$ *for any finite* G.

(ii) *For any finite* G *and each* $C \in Cy(G)$,

$$\mathcal{M}_C(G) \cong \varinjlim_{\pi\in\mathcal{P}(N(C))} \mathcal{M}_C(C\rtimes\pi) \cong \varprojlim_{\pi\in\mathcal{P}(N(C))} \mathcal{M}_C(C\rtimes\pi).$$

Here, $\mathcal{P}(-)$ *denotes the set of p-subgroups, and the limits are taken with respect to* $\mathcal{M}_*$ (*or* $\mathcal{M}^*$) *applied to inclusions, and to conjugation by elements in* $N_G(C)$.

(iii) *Assume* $G = C_n \rtimes \pi$ *is p-hyperelementary (where* $p \nmid n$ *and* π *is a p-group). Then for any* $H = C_m \rtimes \pi \subseteq G$ $(m|n)$, $\mathrm{Res}^G_H \circ \mathrm{Ind}^G_H$ *is an automorphism of* $\mathcal{M}(H)$; *and for each* $k|m$ *the induction and restriction maps*

$$ {}^k\mathrm{Ind}_H^G : \mathcal{M}_k(H) \xrightarrow{\cong} \mathcal{M}_k(G) \quad \textit{and} \quad {}^k\mathrm{Res}_H^G : \mathcal{M}_k(G) \xrightarrow{\cong} \mathcal{M}_k(H) $$

are isomorphisms. Furthermore,

$$ \mathcal{M}_n(G) = \mathrm{Ker}\Big[\oplus\,\mathrm{Res} : \mathcal{M}(G) = \mathcal{M}(C_n \rtimes \pi) \longrightarrow \bigoplus_{p|n} \mathcal{M}(C_{n/p} \rtimes \pi)\Big]. $$

Proof (i) Let $\mathcal{H}$ denote the class of p-hyperelementary groups, for short. Define $E_0(G) = \sum_{C\in\mathcal{C}y(G)} E_C(G) \in \Omega(G)_{(p)}$. Then for any $H \subseteq G$,

$$ \chi_H(E_0(G)) = \begin{cases} 1 & \text{if } H \in \mathcal{H}(G) \\ 0 & \text{otherwise.} \end{cases} \tag{1} $$

In particular, $E_0(G)$ is an idempotent. Also, $\mathrm{Res}_H^G(E_0(G)) = 1 \in \Omega(H)$ for any $H \subseteq \mathcal{H}(G)$. Since $\mathcal{M}$ is generated by p-hyperelementary induction, and since

$$ E_0(G)\cdot\mathrm{Ind}_H^G(x) = \mathrm{Ind}_H^G\Big(\mathrm{Res}_H^G(E_0(G))\cdot x\Big) = \mathrm{Ind}_H^G(1\cdot x) = \mathrm{Ind}_H^G(x) $$

for any $H \in \mathcal{H}(G)$ and any $x \in \mathcal{M}(H)$; this shows that

$$ \mathcal{M}(G) = E_0(G)\cdot\mathcal{M}(G) = \bigoplus_{C\in\mathcal{C}y(G)} E_C(G)\cdot\mathcal{M}(G) = \bigoplus_{C\in\mathcal{C}y(G)} \mathcal{M}_C(G). \tag{2} $$

(iii) Now assume that $G = C_n \rtimes \pi$ is p-hyperelementary, and that $H = C_m \rtimes \pi$ for some $m|n$. Consider the maps

$$ \mathcal{M}_k(H) \xrightarrow{{}^k\mathrm{Ind}_H^G} \mathcal{M}_k(G) \xrightarrow{{}^k\mathrm{Res}_H^G} \mathcal{M}_k(H). \tag{3} $$

Choose double coset representatives $g_1,\ldots,g_r$ for $H\backslash G/H$ such that $g_i \in C_n$ for all i. For each i, write

$$ K_i = H \cap g_i H g_i^{-1} = C_m \rtimes \rho_i \qquad \text{where} \qquad \rho_i = \{x \in \pi : g_i x g_i^{-1} \equiv x \pmod{C_m}\}. $$

Then $g_i \in N_G(K_i)$, so ρ_i and $g_i\rho_i g_i^{-1}$ are both p-Sylow subgroups of

K_i, and are therefore conjugate in K_i. It follows that conjugation by g_i is an inner automorphism of K_i. Hence, in (3),

$$ {}^k\mathrm{Res}^G_H \circ {}^k\mathrm{Ind}^G_H = \sum_{i=1}^r {}^k\mathrm{Ind}^H_{K_i} \circ (c_{g_i})_* \circ {}^k\mathrm{Res}^H_{K_i} = \sum_{i=1}^r {}^k\mathrm{Ind}^H_{K_i} \circ {}^k\mathrm{Res}^H_{K_i} ; $$

and this is multiplication by

$$ E_k(H)\cdot \sum_{i=1}^r [H/(H\cap g_iHg_i^{-1})] = E_k(H)\cdot \mathrm{Res}^G_H([G/H]) = \mathrm{Res}^G_H\Big(E_k(G)\cdot[G/H]\Big) \in \Omega_k(H). $$

For any $K \subseteq G$ such that $\chi_K(E_k(G)) = 1$, $C_k \triangleleft K$ and K/C_k is a p-group. Hence, for such K,

$$ \chi_K(G/H) = |(G/H)^K| = |(G/H)^{K/C_k}| \equiv |G/H| \not\equiv 0 \pmod p. $$

This shows that $E_k(G)\cdot[G/H]$ is invertible in the p-local ring $\Omega_k(G)$. We have just seen that ${}^k\mathrm{Res}^G_H \circ {}^k\mathrm{Ind}^G_H$ is multiplication by $\mathrm{Res}^G_H(E_k(G)\cdot[G/H]) \in \Omega_k(H)^*$; and ${}^k\mathrm{Ind}^G_H \circ {}^k\mathrm{Res}^G_H$ is multiplication by $E_k(G)\cdot[G/H] \in \Omega_k(G)^*$ by Proposition 11.3. It follows that ${}^k\mathrm{Ind}^G_H$ and ${}^k\mathrm{Res}^G_H$ are both isomorphisms between $\mathcal{M}_k(H)$ and $\mathcal{M}_k(G)$; and (after summing over all $k|m$) that $\mathrm{Res}^G_H \circ \mathrm{Ind}^G_H$ is an isomorphism of $\mathcal{M}(H)$ to itself.

In particular, this shows that for any prime $p|n$,

$$ \mathrm{Ker}\Big[\mathrm{Res} : \mathcal{M}(C_n \rtimes \pi) \longrightarrow \mathcal{M}(C_{n/p} \rtimes \pi)\Big] = \bigoplus_{\substack{k|n \\ k\nmid n/p}} \mathcal{M}_k(C_n \rtimes \pi); $$

and hence that

$$ \mathcal{M}_n(C_n \rtimes \pi) = \mathrm{Ker}\Big[\oplus \mathrm{Res} : \mathcal{M}(C_n \rtimes \pi) \longrightarrow \bigoplus_{p|n} \mathcal{M}(C_{n/p} \rtimes \pi)\Big]. $$

(ii) Now let G be arbitrary, again, and define

$$\Omega_0(G) = E_0(G)\cdot\Omega(G)_{(p)} = \bigoplus_{C\in Cy(G)} E_C(G)\cdot\Omega(G)_{(p)}.$$

Then $\Omega_0(G)$ is a ring factor of $\Omega(G)_{(p)}$, and $\mathcal{M}(G)$ is an $\Omega_0(G)$-module by (2). Also, since $\mathrm{Res}^G_H(E_0(G)) = E_0(H)$ for all $H \subseteq G$ by (1), the Frobenius reciprocity relations show that Ω_0 is a Green ring, and that $\mathcal{M}$ is a Green module over Ω_0.

If G is not p-hyperelementary, then $\chi_G(E_0(G)) = 0$ by (1), and so the coefficient of $[G/G]$ in $E_0(G) \in \Omega(G)_{(p)}$ is zero. Since multiplication by $[G/H] \in \Omega(G)_{(p)}$ is $\mathrm{Ind}^G_H \circ \mathrm{Res}^G_H$ by definition, this shows that $\Omega_0(G) = E_0(G)\cdot\Omega(G)_{(p)}$ is generated by induction from proper subgroups in this case. In other words, Ω_0 is $\mathcal{H}$-generated; and by Theorem 11.1 $\mathcal{M}$ is $\mathcal{H}$-computable with respect to both induction and restriction.

In particular, for any $C \in Cy(G)$,

$$\mathcal{M}_C(G) = E_C(G)\cdot\mathcal{M}(G) \cong E_C(G)\cdot \varinjlim_{H\in\mathcal{H}(G)} \mathcal{M}(H) \cong \varinjlim_{H\in\mathcal{H}(G)} \mathrm{Res}^G_H(E_C(G))\cdot\mathcal{M}(H). \tag{4}$$

By definition of $E_C(G)$, for $H \in \mathcal{H}(G)$, $\mathrm{Res}^G_H(E_C(G))\cdot\mathcal{M}(H) = \mathcal{M}_{C'}(H)$ if $H \supseteq C'$ for some (unique) C' conjugate to C, and is zero otherwise. Also, if $H = C_n \rtimes \pi$, where $C' \subseteq C_n$, $p\nmid n$, and π is a p-group, then $\mathcal{M}_{C'}(H) \cong \mathcal{M}_{C'}(C' \rtimes \pi)$ by (iii). So (4) now takes the form

$$\mathcal{M}_C(G) \cong \varinjlim_{\substack{H\in\mathcal{H}(G)\\ C\subseteq H}} \mathcal{M}_C(H) \cong \varinjlim_{\pi\in\mathcal{P}(N(C))} \mathcal{M}_C(C\rtimes\pi).$$

where the limits here are taken with respect to conjugation in $N_G(C)$. The proof for inverse limits is similar. □

So far, the results in this section apply to any p-local Mackey functor $\mathcal{M}$. When $\mathcal{M}(G) = X(R[G])$ for some functor X on R-orders, then the summands $\mathcal{M}_C(G)$ and $\mathcal{M}_n(G)$ can sometimes be given a more accessible description in terms of twisted group rings. Such rings arise naturally as summands of (ordinary) group rings $K[G]$ when G is p-hyperelementary. This is explained in the following proposition. As usual, ζ_n denotes a primitive n-th root of unity.

Proposition 11.6 *Fix a prime* p, *a field* K *of characteristic zero, and a p-hyperelementary group* $G = C_n \rtimes \pi$ $(p \nmid n,\ \pi$ *a p-group). Write* $K[C_n] = \prod_{i=1}^m K_i$, *where the* K_i *are fields. Then* G *is p-K-elementary if and only if the conjugation action of* π *on* $K[C_n]$ *leaves each* K_i *invariant. In this case,* K[G] *splits as a product*

$$K[G] = K[C_n \rtimes \pi] \cong \prod_{i=1}^{m} K_i[\pi]^t,$$

where each $K_i[\pi]^t$ *is the twisted group ring with twisting map*

$$t : \pi \longrightarrow \mathrm{Gal}(K_i/K)$$

induced by the conjugation action of π *on* K_i. *If, furthermore,* R *is a Dedekind domain with field of fractions* K, *and if* $R_i \subseteq K_i$ *is the integral closure of* R, *then* $\prod_{i=1}^m R_i[\pi]^t$ *is an* R*-order in* K[G] *and*

$$R[G] = R[C_n \rtimes \pi] \subseteq \prod_{i=1}^{m} R_i[\pi]^t \subseteq \tfrac{1}{n}\cdot R[G].$$

Proof By Example 1.2, we can write $K[C_n] \cong K\otimes_{\mathbb{Q}} \mathbb{Q}[C_n] \cong \prod_{d|n} K\otimes_{\mathbb{Q}} \mathbb{Q}\zeta_d$; and π acts on each $K\otimes_{\mathbb{Q}} \mathbb{Q}\zeta_d$ via the composite

$$\pi \xrightarrow{\text{conj}} \mathrm{Aut}(C_n) \twoheadrightarrow \mathrm{Aut}(C_d) \cong \mathrm{Gal}(\mathbb{Q}\zeta_d/\mathbb{Q}).$$

Then $\mathrm{Gal}(K\zeta_d/K)$ is the subgroup of elements in $\mathrm{Gal}(\mathbb{Q}\zeta_d/\mathbb{Q})$ which leave all field summands of $K\otimes_{\mathbb{Q}} \mathbb{Q}\zeta_d$ invariant. So π leaves the K_i invariant if and only if $\mathrm{Im}[\pi \longrightarrow \mathrm{Aut}(C_d)] \subseteq \mathrm{Gal}(K\zeta_d/K)$ for all $d|n$, if and only if $G = C_n \rtimes \pi$ is p-K-elementary.

The splitting $K[G] \cong \prod_{i=1}^m K_i[\pi]^t$ is immediate. Also, $\prod_{i=1}^m R_i$ is the maximal R-order in $K[C_n] \cong \prod_{i=1}^m K_i$; and so

$$R[C_n] \subseteq \prod_{i=1}^m R_i \subseteq \frac{1}{n}\cdot R[C_n]$$

by Theorem 1.4(v). □

As an example of how this can be used, consider the functor $\mathcal{M}(G) = Cl_1(\mathbb{Z}[G])_{(p)}$. If $G = C_n \rtimes \pi$, where $p \nmid n$ and π is a p-group, then by the proposition, there is an inclusion $\mathbb{Z}[G] = \mathbb{Z}[C_n \rtimes \pi] \subseteq \prod_{d|n} \mathbb{Z}\zeta_d[\pi]^t$ of orders of index prime to p. So by Corollary 3.10,

$$\mathcal{M}(G) = Cl_1(\mathbb{Z}[C_n \rtimes \pi])_{(p)} \cong \bigoplus_{k|n} Cl_1(\mathbb{Z}\zeta_k[\pi]^t)_{(p)}.$$

We thus have two decompositions of $\mathcal{M}(G) \cong \oplus_{k|n} \mathcal{M}_k(G)$, both indexed on divisors of n; and it is natural to expect that $\mathcal{M}_k(G) \cong Cl_1(\mathbb{Z}\zeta_k[\pi]^t)_{(p)}$ for each k. This is, in fact, the case; but the actual isomorphism is fairly complicated.

<u>Lemma 11.7</u> *Fix a prime p, let X be an additive (covariant) functor from the category of $\mathbb{Z}$-orders in semisimple $\mathbb{Q}$-algebras with bimodule morphisms to $\mathbb{Z}_{(p)}$-modules, and write $\mathcal{M}(G) = X(\mathbb{Z}[G])$ for finite G. Assume that for any p-hyperelementary group $G = C_n \rtimes \pi$ ($p \nmid n$, π a p-group), the projections $\mathbb{Z}[C_n] \twoheadrightarrow \mathbb{Z}\zeta_k$ induce an isomorphism*

$$X(\mathbb{Z}[G]) \xrightarrow{\cong} \prod_{k|n} X(\mathbb{Z}\zeta_k[\pi]^t). \tag{1}$$

Then there is an isomorphism

$$\hat{\beta} : \mathcal{M}_n(G) = \mathcal{M}_n(C_n \rtimes \pi) \xrightarrow{\cong} X(\mathbb{Z}\zeta_n[\pi]^t)$$

which is natural with respect to both induction and restriction in π, as *well as to the Galois action of* $(\mathbb{Z}/n)^*$ *on* $\mathbb{Z}\zeta_n$.

Proof The definition of $\hat{\beta}$, as well as the proof that it is an isomorphism, are both fairly long and complicated. The best way to see what is going on is to first read the proof under the assumption that n is a square of a prime, or a product of two distinct primes.

Fix $G = C_n \rtimes \pi$, where $p \nmid n$ and π is a p-group, and let $G \in C_n$ be a generator. For all $m|n$, set $G_m = C_m \rtimes \pi \subseteq G$; i. e., the subgroup generated by $g^{n/m}$ and π. For all $k|m|n$, we fix the following homomorphisms:

(i) $\mathrm{Ind}_k^m : \mathcal{M}(G_k) \longrightarrow \mathcal{M}(G_m)$ and $\mathrm{Res}_k^m : \mathcal{M}(G_m) \longrightarrow \mathcal{M}(G_k)$ are the induction and restriction maps

(ii) $\mathrm{Proj}_k^m : \mathcal{M}(G_m) \longrightarrow \mathcal{M}(G_k)$ is induced by the surjection $C_m \rtimes \pi \twoheadrightarrow C_k \rtimes \pi$ which is the identity on π, and which on the q-Sylow subgroup of C_m (any prime $q|m$) is induced by $a \mapsto a^{q^r}$ for appropriate r (so $\mathrm{Proj}_k^m \circ \mathrm{Ind}_k^m = \mathrm{Id}$ if $(k, \frac{m}{k}) = 1$)

(iii) $\mathrm{Pr}_k^m : \mathcal{M}(G_m) = X(\mathbb{Z}[C_m \rtimes \pi]) \longrightarrow X(\mathbb{Z}\zeta_k[\pi]^t)$ is the composite of Proj_k^m with the map induced by sending $g^{n/k} \in C_k$ to $\zeta_k = \exp(2\pi i/k)$

(iv) $I_k^m : X(\mathbb{Z}\zeta_k[\pi]^t) \longrightarrow X(\mathbb{Z}\zeta_m[\pi]^t)$, $R_k^m : X(\mathbb{Z}\zeta_m[\pi]^t) \longrightarrow X(\mathbb{Z}\zeta_k[\pi]^t)$ are the induction and restriction maps for $\mathbb{Z}\zeta_k[\pi]^t \subseteq \mathbb{Z}\zeta_m[\pi]^t$.

For $k > 0$, let $\mathfrak{p}(k)$ be the number of distinct prime divisors. For each $u|m|n$ such that $(u, \frac{m}{u}) = 1$, set

$$\beta_u^m = \sum_{k|m} (-1)^{\mathfrak{p}([k,u]) - \mathfrak{p}(k)} \cdot \left(I_k^{[k,u]} \circ \mathrm{Pr}_k^m \right) : \mathcal{M}(G_m) \longrightarrow \bigoplus_{u|k|m} X(\mathbb{Z}\zeta_k[\pi]^t)$$

(where $[k,u]$ denotes the least common multiple). We claim that

$$\hat{\beta}^m_u = \beta^m_u \circ \mathrm{incl} : \bigoplus_{u|k|m} \mathcal{M}_k(G_m) \hookrightarrow \mathcal{M}(G_m) \longrightarrow \bigoplus_{u|k|m} X(\mathbb{Z}\zeta_k[\pi]^t)$$

is an isomorphism for all $u|m|n$ such that $(u,\frac{m}{u}) = 1$. In particular, the lemma will follow from the case $\hat{\beta} = \hat{\beta}^n_n$.

To simplify notation, we write

$$X(k) = X(\mathbb{Z}\zeta_k[\pi]^t) \quad (\text{any } k|n); \qquad \mathcal{M}^u(G_m) = \bigoplus_{u|k|m} \mathcal{M}_k(G_m) \quad (\text{any } u|m|n).$$

The following naturality relations will be needed in the proof below. The naturality of Ind^n_m with respect to projections to the $X(k)$ is described by the commutative square

$$\begin{array}{ccc} \mathcal{M}(G_m) & \xrightarrow{\mathrm{Ind}^n_m} & \mathcal{M}(G_n) \\ \cong\Big\downarrow \oplus \mathrm{Pr}^m_k & & \cong\Big\downarrow \oplus \mathrm{Pr}^n_k \\ \bigoplus\limits_{k|m} X(k) & \xrightarrow{\oplus_{k|n} I^k_{\alpha(m,k)}} & \bigoplus\limits_{k|n} X(k) \end{array} \qquad \Big(\alpha(m,k) = k/(k,\tfrac{n}{m})\Big) \qquad (2)$$

(this is induced by a commutative square of rings). No analogous result for Res^n_m seems to hold in general; but if $q^2|n$ for some prime q, then the following square does commute:

$$\begin{array}{ccc} \mathcal{M}(G_n) & \xrightarrow{\mathrm{Res}^n_{n/q}} & \mathcal{M}(G_{n/q}) \\ \cong\Big\downarrow \mathrm{Proj}^n_{n/q} \oplus \bigoplus \mathrm{Pr}^n_k & & \cong\Big\downarrow \mathrm{Proj}^{n/q}_{n/q^2} \oplus \bigoplus \mathrm{Pr}^{n/q}_{k/q} \\ \mathcal{M}(G_{n/q}) \oplus \bigoplus\limits_{\substack{k|n \\ q\nmid n/k}} X(k) & \xrightarrow[\oplus \bigoplus R^k_{k/q}]{\mathrm{Res}^{n/q}_{n/q^2}} & \mathcal{M}(G_{n/q^2}) \oplus \bigoplus\limits_{\substack{k|n \\ q\nmid n/k}} X(k/q). \end{array} \qquad (3)$$

The commutativity of (3) follows upon comparing bimodules, and the vertical maps are isomorphisms by (1). Since $\mathrm{Res}^n_{n/q} \circ \mathrm{Ind}^n_{n/q}$ is an isomorphism (Proposition 11.5(iii)), (2) and (3) combine to show that $R^k_{k/q} \circ I^k_{k/q}$ is an isomorphism whenever $q^2|n$ and $q \nmid (n/k)$. In particular, for each such k,

$$I^k_{k/q} \oplus \text{incl} : X(k/q) \oplus \text{Ker}(R^k_{k/q}) \xrightarrow{\;\cong\;} X(k) \tag{4}$$

is an isomorphism. Finally, the commutativity of the following square is immediate from the definition of Pr^m_k:

$$\begin{array}{ccc} \mathcal{M}(G_n) & \xrightarrow{\text{Proj}^n_m} & \mathcal{M}(G_m) \\ \cong\Big\downarrow \oplus Pr^n_k & & \cong\Big\downarrow \oplus Pr^m_k \\ \bigoplus_{k|n} X(k) & \xrightarrow{\text{proj}} & \bigoplus_{k|m} X(k) \end{array} \qquad (\text{any } m|n) \tag{5}$$

It suffices to prove that $\hat{\beta}^m_u$ is an isomorphism in the case $m = n$. Fix u, where $(u,\frac{n}{u}) = 1$. If $u = 1$, then $\hat{\beta}^n_u$ is an isomorphism by (1). Otherwise, let $q|u$ be any prime divisor, and define v, m, and r to satisfy

$$u = q^r v, \qquad n = q^r m, \qquad q\nmid v, \quad q\nmid m.$$

We assume inductively that $\hat{\beta}^n_v$ and $\hat{\beta}^{n/q}_{u/q}$ both are isomorphisms.

<u>Case 1</u> Assume first that $r = 1$; i. e., that $q^2\nmid n$, $m = n/q$, and $v = u/q$. Consider the following diagram:

$$\begin{array}{ccccc} \mathcal{M}^v(G_m) & \xrightarrow{f_2} & \mathcal{M}(G_m) & \xrightarrow{\beta^m_v} & \bigoplus_{v|k|m} X(k) \\ \Big\downarrow {}^v\text{Ind}^n_m & & \Big\downarrow \text{Ind}^n_m & & \Big\downarrow \bigoplus_{v|k|n} I^k_{(m,k)} = \bigoplus_{v|k|m} (I^k_k + I^{qk}_k) \\ \mathcal{M}^v(G_n) & \xrightarrow{f_3} & \mathcal{M}(G_n) & \xrightarrow{\beta^n_v} & \bigoplus_{v|k|n} X(k) \\ \Big\uparrow f_1 & & & & \Big\downarrow \bigoplus_{v|k|m} (-I^{qk}_k + I^{qk}_{qk}) \\ \mathcal{M}^u(G) & & \xrightarrow{\hat{\beta}^n_u} & & \bigoplus_{u|k|n} X(k) \end{array}$$

where ${}^v\text{Ind}^n_m$ is the restriction of Ind^n_m, and the f_i are inclusion maps. The two small squares commute — the right-hand square by (2) —

and the lower rectangle commutes by definition of β and $\hat{\beta}$. Also, $f_1 \oplus {}^v\mathrm{Ind}_m^n$ is an isomorphism (recall $\mathcal{M}_k(G_m) \cong \mathcal{M}_k(G_n)$ for all $k|m$, by Proposition 11.5(iii)); and the right-hand column is short exact. Since $\beta_v^m \circ f_2 = \hat{\beta}_v^m$ and $\beta_v^n \circ f_3 = \hat{\beta}_v^n$ are isomorphisms, by assumption, this shows that $\hat{\beta}_u^n$ is also an isomorphism.

<u>Case 2</u> Now assume that $q^2|n$, and consider the following diagram:

$$\begin{array}{ccc} \mathcal{M}^v(G_n) & \xrightarrow{\ {}^v\mathrm{Res}^n_{n/q}\ } & \mathcal{M}^v(G_{n/q}) \\ \cong\Big\downarrow f_1 & & \cong\Big\downarrow f_2 \\ \mathcal{M}^v(G_{n/q}) \oplus \bigoplus\limits_{u|k|n} X(k) & \xrightarrow[\ \oplus \bigoplus R^k_{k/q}\ ]{\ {}^v\mathrm{Res}^{n/q}_{n/q^2}\ } & \mathcal{M}^v(G_{n/q^2}) \oplus \bigoplus\limits_{u|k|n} X(k/q) \end{array} \tag{6}$$

where

$$f_1 = {}^v\mathrm{Proj}^n_{n/q} \oplus (\mathrm{proj} \circ \hat{\beta}_v^n) \qquad \text{and} \qquad f_2 = {}^v\mathrm{Proj}^{n/q}_{n/q^2} \oplus (\mathrm{proj} \circ \hat{\beta}_v^{n/q}).$$

Diagram (6) commutes by (3), and the relations

$$I_{k/q}^{[k/q,v]} \circ R^k_{k/q} = R^{[k,v]}_{[k,v]/q} \circ I_k^{[k,v]}$$

when $q^r|k|n$ (i. e., $q \nmid \frac{n}{k}$). The maps f_1 and f_2 are isomorphisms:

$$\hat{\beta}_v^n = (\hat{\beta}_v^{n/q} \oplus \mathrm{Id}) \circ f_1 \ : \ \mathcal{M}^v(G_n) \longrightarrow \mathcal{M}^v(G_{n/q}) \oplus \bigoplus_{u|k|n} X(k) \longrightarrow \bigoplus_{v|k|n} X(k),$$

where $\hat{\beta}_v^n$ and $\hat{\beta}_v^{n/q}$ are isomorphisms by assumption; and similarly for f_2. But now $\hat{\beta}_u^n$ is the composite

$$\mathcal{M}^u(G_n) = \mathrm{Ker}({}^v\mathrm{Res}^n_{n/q}) \qquad \text{(Proposition 11.5(iii))}$$

$$\xrightarrow[\cong]{f_1} \mathrm{Ker}({}^v\mathrm{Res}^{n/q}_{n/q^2}) \oplus \bigoplus_{u|k|n} \mathrm{Ker}(R^k_{k/q}) \qquad \text{(by (6))}$$

$$= \mathcal{M}^{u/q}(G_{n/q}) \oplus \bigoplus_{u|k|n} \mathrm{Ker}(R^k_{k/q})$$

$$\xrightarrow[\cong]{\hat{\beta}^{n/q}_{u/q} \oplus \mathrm{Id}} \bigoplus_{u|k|n} \Big(X(k/q) \oplus \mathrm{Ker}(R^k_{k/q}) \Big)$$

$$\xrightarrow[\cong]{I^k_{k/q} \oplus \mathrm{incl}} \bigoplus_{u|k|n} X(k); \qquad \text{(by (4))}$$

and is hence an isomorphism. □

Proposition 11.5 and Lemma 11.7 now lead to the following theorem, which greatly simplifies the limits involved when applying Theorem 11.1 to calculate $SK_1(\mathbb{Z}[G])$, $Cl_1(\mathbb{Z}[G])$, $SK_1(\hat{\mathbb{Z}}_p[G])$, etc., in terms of hyperelementary subgroups. Recall (Theorem 1.6) that if G is any finite group, and if K is a field of characteristic zero, then two elements $g,h \in G$ are called *K-conjugate* if h is conjugate to g^a for some $a \in \mathrm{Gal}(K\zeta_n/K)$, where $n = |g|$. Also, for any cyclic $\sigma = \langle g \rangle \subseteq G$, with $n = |g| = |\sigma|$, we define

$$N^K_G(\sigma) = N^K_G(g) = \{x \in G : xgx^{-1} = g^a, \ \text{some } a \in \mathrm{Gal}(K\zeta_n/K)\}.$$

<u>Theorem 11.8</u> *Fix a prime* p *and a Dedekind domain* R *with field of fractions* K *of characteristic zero. Let* X *be an additive functor from the category of* R*-orders in semisimple* K*-algebras with bimodule morphisms to the category of* $\mathbb{Z}_{(p)}$*-modules. Assume that any inclusion* $\mathfrak{A} \subseteq \mathfrak{B}$ *of orders, such that* $n\mathfrak{B} \subseteq \mathfrak{A}$ *for some* n *prime to* p*, induces an isomorphism* $X(\mathfrak{A}) \xrightarrow{\cong} X(\mathfrak{B})$*. Then, for any finite* G*, if* $g_1,\ldots,g_k \in G$ *are* K*-conjugacy class representatives for elements of order prime to* p*, where* $n_i = |g_i|$*, there are isomorphisms:*

$$X(R[G]) \cong \bigoplus_{i=1}^{k} \varinjlim \{X(R\zeta_{n_i}[\pi]^t) : \pi \in \mathcal{P}(N_G^K(g_i))\} \tag{1}$$

(where $\mathcal{P}(-)$ denotes the set of p-subgroups), and

$$X(R[G]) \cong \bigoplus_{i=1}^{k} \varprojlim \{X(R\zeta_{n_i}[\pi]^t) : \pi \in \mathcal{P}(N_G^K(g_i))\}. \tag{2}$$

Here, the limits are taken with respect to inclusion of subgroups, and conjugation by elements of $N_G^K(g_i)$. For all n, $R\zeta_n$ denotes the integral closure of R in $K\zeta_n$. The first isomorphism is natural with respect to induction, and the second with respect to restriction maps.

<u>Proof</u> Write $\mathcal{M} = X(R[-])$, for convenience. Fix G, and let $Cy(G)$ be a set of conjugacy class representatives for cyclic subgroups $C \subseteq G$ of order prime to p. By Proposition 11.5,

$$\mathcal{M}(G) = \bigoplus_{C \in Cy(G)} \mathcal{M}_C(G); \tag{3}$$

where for each C, if $n = |C|$, then

$$\mathcal{M}_C(G) \cong \varinjlim_{\pi \in \mathcal{P}(N(C))} \mathcal{M}_n(C \rtimes \pi). \tag{4}$$

Fix $C \in Cy(G)$, and set $n = |C|$. By Theorem 11.2, $\mathcal{M}$ is computable with respect to p-K-elementary subgroups. In particular, for any $\pi \in \mathcal{P}(N(C))$,

$$\mathcal{M}_n(C \rtimes \pi) \cong \varinjlim \{\mathcal{M}_n(C \rtimes \rho) : \rho \subseteq \pi \cap N_G^K(C)\}.$$

Using this, the limit in (4) takes the form

$$\mathcal{M}_C(G) \cong H_0\Big(N(C)/N_G^K(C);\ \varinjlim_{\pi \in \mathcal{P}(N_G^K(C))} \mathcal{M}_n(C \rtimes \pi)\Big). \tag{5}$$

This time, the limit is taken with respect to inclusion, and conjugation by elements in $N_G^K(C)$.

Now write $K\otimes_{\mathbb{Q}}\mathbb{Q}\zeta_n = \prod_{i=1}^r K_i$, where $K_i \cong K\zeta_n$ for each i; and let $R_i \subseteq K_i$ be the integral closure of R. By Proposition 11.6,

$$R[C\rtimes\pi] \subseteq \prod_{k|n} R\otimes_{\mathbb{Z}}\mathbb{Z}\zeta_k[\pi]^t \subseteq \tfrac{1}{n}\cdot R[C\rtimes\pi], \quad \text{and}$$

$$R\otimes_{\mathbb{Z}}\mathbb{Z}\zeta_n[\pi]^t \subseteq \prod_{i=1}^r R_i[\pi]^t \subseteq \tfrac{1}{n}\cdot R\otimes_{\mathbb{Z}}\mathbb{Z}\zeta_n[\pi]^t.$$

So by Lemma 11.7 (applied to the functor $\mathfrak{A} \longmapsto X(R\otimes_{\mathbb{Z}}\mathfrak{A})$ on $\mathbb{Z}$-orders), for each $\pi \subseteq \mathcal{P}(N_G^K(C))$,

$$\begin{aligned} \mathcal{M}_n(C\rtimes\pi) &\cong X(R\otimes_{\mathbb{Z}}\mathbb{Z}\zeta_n[\pi]^t) \cong X\Big(\prod_{i=1}^r R_i[\pi]^t\Big) \\ &\cong \bigoplus_{i=1}^r X(R_i[\pi]^t) \cong \bigoplus_{i=1}^r X(R\zeta_n[\pi]^t). \end{aligned} \tag{6}$$

Note that r, the number of field summands of $\mathbb{Q}\zeta_n\otimes_{\mathbb{Q}}K$, is equal to the number of equivalence classes of generators of C under the relation $g\sim g^a$ if $a\in \mathrm{Gal}(K\zeta_n/K)$. The factors $X(R_i[\pi]^t)$ are permuted, under conjugation by $N_G(C)$, in the same way that these equivalence classes are permuted in G. Thus, if there are m K-conjugacy classes (in G) of generators of C, then (5) and (6) combine to give an isomorphism

$$\mathcal{M}_C(G) \cong \bigoplus^m \varinjlim \{X(R\zeta_n[\pi]^t) : \pi\in\mathcal{P}(N_G^K(C))\}, \tag{7}$$

where again the limit is taken with respect to inclusion, and conjugation by elements of $N_G^K(C)$. Formula (1) now follows upon combining (3) and (7); and formula (2) (for restriction) is shown in a similar fashion. □

Induction properties with respect to *p-elementary groups* — i. e., subgroups of the form $C_n\times\pi$ where $p\nmid n$ and π is a p-group — will play an important role in Chapters 12 and 13. The next theorem gives a simple criterion, in terms of twisted group rings, for checking them.

__Theorem 11.9__ *Fix a prime p, and a Dedekind domain R with field of fractions K. Let X be an additive functor on R-orders with bimodule morphisms satisfying the hypotheses of Theorem 11.8. For any n, let* ζ_n *be a primitive n-th root of unity, and let* $R\zeta_n$ *denote the integral closure of R in* $K\zeta_n$. *Then*

(i) $X(R[G])$ *is generated by (computable for) induction from p-elementary subgroups if and only if for any n with* $p\nmid n$, *any p-group* π, *and any* $t: \pi \longrightarrow Gal(K\zeta_n/K)$ *with* $\rho = Ker(t)$, *the induction map*

$$ind : H_0(\pi/\rho;\ X(R\zeta_n[\rho])) \longrightarrow X(R\zeta_n[\pi]^t)$$

is surjective (bijective).

(ii) $X(R[G])$ *is detected by (computable for) restriction to p-elementary subgroups if and only if for any n with* $p\nmid n$, *any p-group* π, *and any* $t: \pi \longrightarrow Gal(K\zeta_n/K)$ *with* $\rho = Ker(t)$, *the restriction map*

$$res : X(R\zeta_n[\pi]^t) \longrightarrow H^0(\pi/\rho;\ X(R\zeta_n[\rho]))$$

is injective (bijective).

__Proof__ We prove here point (i) for computability; the other claims are shown similarly. Write $\mathcal{M}(G) = X(R[G])$. Then $\mathcal{M}$ is p-K-elementary computable by Theorem 11.2; and $\mathcal{M}_k(C_n \rtimes \pi) \cong \mathcal{M}_k(C_k \rtimes \pi)$ if $k|n$ by Proposition 11.5(iii). So $\mathcal{M}$ is p-elementary computable if and only if

$$\mathcal{M}_n(C_n \rtimes \pi) \cong \varinjlim_{H\in\mathcal{E}} \mathcal{M}_n(H) \qquad (1)$$

for any p-K-elementary group of the form $G = C_n \rtimes \pi$; where $p\nmid n$, π is a p-group, and $\mathcal{E}$ is the set of p-elementary subgroups of G.

For $H \subseteq C_n \rtimes \pi$, $\mathcal{M}_n(H) = 0$ unless $n \mid |H|$; i. e., unless $C_n \subseteq H$.

Hence, if $\rho = \mathrm{Ker}[t\colon \pi \longrightarrow \mathrm{Gal}(K\zeta_n/K)]$, then

$$\varinjlim_{H\in\mathscr{E}} \mathcal{M}_n(H) \cong \varinjlim_{\sigma\subseteq\rho} \mathcal{M}_n(C_n \times \sigma)$$

where the limit is taken with respect to inclusion and conjugation in G. In other words,

$$\varinjlim_{H\in\mathscr{E}} \mathcal{M}_n(H) \cong \varinjlim_{\sigma\subseteq\rho} \mathcal{M}_n(C_n \times \sigma) \cong H_0(\pi/\rho;\, \mathcal{M}_n(C_n \times \rho));$$

and the result follows from the isomorphisms

$$\mathcal{M}_n(C_n \rtimes \pi) \cong X(R \otimes_{\mathbb{Z}} \mathbb{Z}\zeta_n[\pi]^t) \cong X(R\zeta_n[\pi]^t)^N$$

$$\mathcal{M}_n(C_n \times \rho) \cong X(R \otimes_{\mathbb{Z}} \mathbb{Z}\zeta_n[\rho]) \cong X(R\zeta_n[\rho])^N$$

(where $N = \varphi(n)/[K\zeta_n:K]$) of Lemma 11.7. □

Finally, we list some specific applications of Theorem 11.8, which will be used in later chapters. For technical reasons, a new functor $SK_1^{[p]}$ will be needed. If $\mathfrak{A}$ is any $\mathbb{Z}$-order in a semisimple $\mathbb{Q}$-algebra, and p is a prime, set

$$SK_1^{[p]}(\mathfrak{A}) = \mathrm{Ker}\Big[SK_1(\mathfrak{A}) \longrightarrow \prod_{q\neq p} SK_1(\hat{\mathfrak{A}}_q)\Big]_{(p)}.$$

In particular, there is a short exact sequence

$$1 \longrightarrow Cl_1(\mathfrak{A})_{(p)} \longrightarrow SK_1^{[p]}(\mathfrak{A}) \longrightarrow SK_1(\hat{\mathfrak{A}}_p)_{(p)} \longrightarrow 1.$$

By Theorem 3.14, for any finite G, $SK_1(\hat{\mathbb{Z}}_q[G])_{(p)} = 1$ for all primes $q \neq p$; and so $SK_1^{[p]}(\mathbb{Z}[G]) = SK_1(\mathbb{Z}[G])_{(p)}$ in this case. This is, however, not always the case for twisted group rings. For example, if q is any odd prime, and $C_2 \subseteq \mathrm{Gal}(\mathbb{Q}\zeta_q/\mathbb{Q})$, then it is not hard to show using Theorems 2.5 and 2.10 that $SK_1(\hat{\mathbb{Z}}_q\zeta_q[C_2]^t) \cong \mathbb{Z}/(q-1)$. So in this case,

$SK_1^{[2]}(\mathbb{Z}\zeta_q[C_2]^t) \subsetneqq SK_1(\mathbb{Z}\zeta_q[C_2]^t)_{(2)}$.

<u>Theorem 11.10</u> *Fix a prime* p *and a finite group* G. *For any* $H \subseteq G$, $\mathcal{P}(H)$ *denotes the set of* p-*subgroups. Let* $\sigma_1,\ldots,\sigma_k$ *be a set of conjugacy classes of cyclic subgroups of* G *of order prime to* p. *Set* $n_i = |\sigma_i|$ *and* $N_i = N_G(\sigma_i)$. *Then*

(1) $$Cl_1(\mathbb{Z}[G])_{(p)} \cong \bigoplus_{i=1}^{k} \varinjlim_{\pi\in\mathcal{P}(N_i)} Cl_1(\mathbb{Z}\zeta_{n_i}[\pi]^t)_{(p)},$$

(2) $$SK_1(\mathbb{Z}[G])_{(p)} \cong \bigoplus_{i=1}^{k} \varinjlim_{\pi\in\mathcal{P}(N_i)} SK_1^{[p]}(\mathbb{Z}\zeta_{n_i}[\pi]^t), \text{ and}$$

(3) $$C_p(\mathbb{Q}[G]) \cong \bigoplus_{i=1}^{k} \varinjlim_{\pi\in\mathcal{P}(N_i)} C_p(\mathbb{Q}\zeta_{n_i}[\pi]^t).$$

<u>Proof</u> By Proposition 1.18, $Cl_1(-)_{(p)}$, $SK_1^{[p]}$, and $C_p(\mathbb{Q}\otimes_{\mathbb{Z}}-)$ are all functors on the category of $\mathbb{Z}$-orders with bimodule morphisms. Also, elements $g,h \in G$ are $\mathbb{Q}$-conjugate if and only if they generate conjugate subgroups (by Theorem 1.5(i)). The condition that $X(\mathfrak{A}) \cong X(\mathfrak{B})$ whenever $p\nmid[\mathfrak{B}:\mathfrak{A}]$, is trivial for $C_p(\mathbb{Q}\otimes_{\mathbb{Z}}-)$; and holds for $Cl_1(-)_{(p)}$ and $SK_1^{[p]}$ by Corollary 3.10. So the above decomposition formulas follow from Theorem 11.8. □

Chapter 12 THE P-ADIC QUOTIENT OF $SK_1(\mathbb{Z}[G])$: FINITE GROUPS

The induction techniques of Chapter 11 will first be applied to describe $SK_1(\hat{\mathbb{Z}}_p[G]) = SK_1(\mathbb{Z}[G])/Cl_1(\mathbb{Z}[G])$, as well as $K_1(\hat{\mathbb{Z}}_p[G])_{(p)}$ and $K_1'(\hat{\mathbb{Z}}_p[G])_{(p)}$, for finite groups G. In particular, all three of these functors are shown to be computable for induction from p-elementary subgroups. A detection theorem would be still more useful; but in Example 12.6 a group G is constructed for which $SK_1(\hat{\mathbb{Z}}_p[G])$ is *not* detected by restriction to p-elementary subgroups.

These results lead to two sets of formulas for $SK_1(\hat{\mathbb{Z}}_p[G])$ and $tors_p K_1'(\hat{\mathbb{Z}}_p[G])$. The formulas in Theorem 12.5 are based on the direct sum decompositions of Theorem 11.8, and involve only the functors H_2/H_2^{ab} and $(-)^{ab}$. They are the easiest to use when describing either $SK_1(\hat{\mathbb{Z}}_p[G])$ or $tors_p K_1'(\hat{\mathbb{Z}}_p[G])$ as abstract groups. As applications of these formulas, we show, for example, that $SK_1(\hat{\mathbb{Z}}_p[G]) = 1$ if $S_p(G)$ contains a normal abelian subgroup with cyclic quotient (Proposition 12.7), or if G is any symmetric or alternating group (Example 12.8).

In Theorems 12.9 and 12.10, alternative descriptions of the groups $K_1'(\hat{\mathbb{Z}}_p[G])$, $SK_1(\hat{\mathbb{Z}}_p[G])$ and $tors_p K_1'(\hat{\mathbb{Z}}_p[G])$ are derived, in terms of homology groups of the form $H_n(G;\hat{\mathbb{Z}}_p(G_r))$, where $G_r = \{g \in G: p \nmid |g|\}$. The formula for $SK_1(\hat{\mathbb{Z}}_p[G])$, for example, can be applied directly to determine whether a given element vanishes. The new formula for $tors_p K_1'(\hat{\mathbb{Z}}_p[G])$ is derived from two exact sequences which describe the kernel and cokernel of

$$\Gamma_G: K_1'(\hat{\mathbb{Z}}_p[G]) \longrightarrow H_0(G;\hat{\mathbb{Z}}_p[G])$$

for arbitrary finite G, and which generalize the exact sequences of Theorems 6.6 and 6.7.

As was seen in Chapter 11, results on p-elementary induction are obtained by studying twisted group rings. This is the subject of the first two technical lemmas. Each is stated in two parts: part (i) contains (most of) what will be needed in this chapter, while part (ii) in each lemma will be needed in Chapter 13 (in the proof of Lemma 13.1).

For convenience, for any finite extension F of $\hat{\mathbb{Q}}_p$, $\mathrm{Gal}(F/\hat{\mathbb{Q}}_p)$ will be used to denote the group of all automorphisms of F fixing $\hat{\mathbb{Q}}_p$ — whether or not the extension is Galois. In this situation, for any $\pi \subseteq \mathrm{Gal}(F/\hat{\mathbb{Q}}_p)$, F^π will denote the fixed field.

<u>Lemma 12.1</u> *Fix a prime* p, *let* F *be any finite extension of* $\hat{\mathbb{Q}}_p$, *and let* $R \subseteq F$ *be the ring of integers. Let* $t\colon \pi \longrightarrow \mathrm{Gal}(F/\hat{\mathbb{Q}}_p)$ *be any homomorphism such that* π *is a p-group, and such that the extension* F/F^π *is unramified. Let* $R[\pi]^t$ *denote the induced twisted group ring; and set* $\rho = \mathrm{Ker}(t)$. *Then the following hold.*

(i) *The inclusion* $R[\rho] \subseteq R[\pi]^t$ *induces a surjection*

$$\mathrm{ind} : K_1(R[\rho]) \twoheadrightarrow K_1(R[\pi]^t).$$

(ii) *For any π-invariant radical ideal* $I \subseteq R[\rho]$ *(i. e.,* $gIg^{-1} = I$ *for all* $g \in \pi$*), set* $\bar{I} = \sum_{g\in\pi} I\cdot g \subseteq R[\pi]^t$. *Then*

$$\mathrm{ind}_I : K_1(R[\rho],I) \twoheadrightarrow K_1(R[\pi]^t,\bar{I})$$

is onto.

<u>**Proof**</u> Set $S = R^\pi$, the ring of integers in F^π, and let $\mathfrak{p} \subseteq R$ and $\mathfrak{q} \subseteq S$ be the maximal ideals. Then $\mathfrak{p} = \mathfrak{q}R$, since F/F^π is unramified, and

$$\pi/\rho \cong \mathrm{Gal}(F/F^\pi) \cong \mathrm{Gal}((R/\mathfrak{p})/(S/\mathfrak{q})). \tag{1}$$

We first prove point (ii). Choose some $r \in R$ whose image $\bar{r} \in R/\mathfrak{p}$ generates $(R/\mathfrak{p})^*$. Fix coset representatives $1 = g_0, g_1, g_2, \ldots, g_{m-1}$ for ρ in π, and set

$$s_i = r^{-1}\cdot t(g_i)(r) - 1 \in R^*$$

for each $i \geq 1$. Here, $s_i \in R^*$ since $t(g_i)(\bar{r}) \neq \bar{r}$ in R/p by (1).

Now fix any π-invariant radical ideal $I \subseteq R[\rho]$, and any element $[u] \in K_1(R[\pi]^t, \bar{I})$ $(u \in 1+\bar{I})$. We want to construct a convergent sequence $u=u_1, u_2, u_3, \ldots$, such that for each $k \geq 1$,

$$u_k \in 1 + I + \bar{I}^k \qquad \text{and} \qquad [u] = [u_k] \in K_1(R[\pi]^t, \bar{I}).$$

To do this, assume that

$$u_k = 1 + \sum_{i=0}^{m-1} x_i g_i \qquad (x_0 \in I,\ \ x_i \in I^k \ \text{ for } \ 1 \leq k \leq m-1).$$

has been constructed. Then

$$u_k \equiv \prod_{i=0}^{m-1}(1 + x_i g_i) = (1 + x_0)\cdot \prod_{i=1}^{m-1}\Bigl(1 + x_i s_i^{-1}(r^{-1}\cdot t(g_i)(r) - 1)g_i\Bigr)$$

$$= (1 + x_0)\cdot \prod_{i=1}^{m-1}\Bigl(1 + r^{-1}(x_i s_i^{-1} g_i)r - (x_i s_i^{-1} g_i)\Bigr)$$

$$\equiv (1 + x_0)\cdot \prod_{i=1}^{m-1}[r^{-1}, 1 + x_i s_i^{-1} g_i]. \qquad (\text{mod } \bar{I}^{k+1})$$

Thus, $[u_k] = [u_{k+1}]$ for some $u_{k+1} \in 1 + I + \bar{I}^{k+1}$ with $u_k \equiv u_{k+1}$ (mod $\bar{I}^k$). Hence, since $(1+\bar{I}) \cap E(R[\pi]^t, \bar{I})$ is p-adically closed (Theorem 2.9), and since $\bar{I}^k \to 0$ as $k \to \infty$ ($\bar{I}$ is radical),

$$[u] = [\lim_{k\to\infty} u_k] \in \operatorname{Im}\Bigl[\operatorname{ind}_I\colon K_1(R[\rho], I) \longrightarrow K_1(R[\pi]^t, \bar{I})\Bigr].$$

It follows that ind_I is surjective.

To prove (i), let $J = \{\sum r_i g_i : \sum r_i \in p\}$ be the Jacobson radical of $R[\rho]$ (see Example 1.12), and consider the following commutative diagram:

$$\begin{array}{ccccccc} K_1(R[\rho],J) & \longrightarrow & K_1(R[\rho]) & \longrightarrow & K_1(R[\rho]/J) & \longrightarrow & 1 \\ \downarrow{\scriptstyle \mathrm{ind}_J} & & \downarrow{\scriptstyle \mathrm{ind}} & & \downarrow{\scriptstyle \mathrm{ind}_{\pi/\rho}} & & \\ K_1(R[\pi]^t,\bar{J}) & \longrightarrow & K_1(R[\pi]^t) & \longrightarrow & K_1(R[\pi]^t/\bar{J}) & \longrightarrow & 1. \end{array} \qquad (2)$$

The rows in (2) are exact, so ind is onto if $\mathrm{ind}_{\pi/\rho}$ is. Also,

$$R[\rho]/J \cong R/\mathfrak{p} \qquad \text{and} \qquad R[\pi]^t/\bar{J} \cong R/\mathfrak{p}[\pi/\rho]^t \cong M_m(S/\mathfrak{q}),$$

where $m = |\pi/\rho| = [R:S]$. The composite

$$K_1(R/\mathfrak{p}) \xrightarrow{\mathrm{ind}_{\pi/\rho}} K_1(R/\mathfrak{p}[\pi/\rho]^t) \cong K_1(S/\mathfrak{q})$$

is the norm map for an inclusion of finite fields, and hence is onto. □

The next lemma will be used to get control over the kernels of the induction maps studied in Lemma 12.1.

Lemma 12.2 *Let* $\rho \subseteq \pi$, $R \subseteq F$, *and* $t\colon \pi \longrightarrow \mathrm{Aut}(F)$ *be as in Lemma 12.1. Then the following hold.*

(i) $K_1'(R[\rho])$, *with the* π/ρ-*action induced by*

$$\pi/\rho \longrightarrow \mathrm{Aut}(R) \times \mathrm{Out}(\rho),$$

is cohomologically trivial.

(ii) *If* $SK_1(R[\rho]) = 1$, *then there is a sequence* $R[\rho] \supseteq J = I_1 \supseteq I_2 \supseteq \ldots$ *of* π-*invariant ideals, where* J *is the Jacobson radical, such that* $\bigcap_{k=1}^{\infty} I_k = 0$, *and such that for all* k:

$$I_{k+1} \supseteq I_kJ + JI_k \qquad \text{and} \qquad \hat{H}^*(\pi/\rho; K_1(R[\rho]/I_k)) = 1. \qquad (1)$$

Proof Set $E = F^{\pi}$. Since F/E is unramified, there exists $r \in R$ with $\mathrm{Tr}_{F/E}(r) = 1$ (see Proposition 1.8(iii)). If M is any

$R[\pi/\rho]^t$-module, then for all $x \in M$

$$\sum_{g\in\pi/\rho} g(r\cdot(g^{-1}x)) = \sum_{g\in\pi/\rho} (t(g)(r))\cdot x = x.$$

In other words, the identity is a norm in $\mathrm{End}(M)$; and so M is cohomologically trivial (see Cartan & Eilenberg [1, Proposition XII.2.4]).

In particular, if ρ is abelian, then this applies to any power of the Jacobson radical $J \subseteq R[\rho]$; and so

$$\hat{H}^*(\pi/\rho;(1+J^k)/(1+J^{k+1})) \cong \hat{H}^*(\pi/\rho;J^k/J^{k+1}) = 0$$

for all $k \geq 1$. Also, π/ρ acts effectively on the finite field $R[\rho]/J$; and $(R[\pi]/J)^*$ is easily seen to be π/ρ-cohomologically trivial. Thus,

$$K_1(R[\rho]/J^k) \cong (R[\rho])^*/(1+J^k)$$

is cohomologically trivial for all k, and in the limit $K_1(R[\rho]) = (R[\rho])^*$ is cohomologically trivial.

Now assume that ρ is nonabelian. Let $\{z_1,\ldots,z_k\} \subseteq \rho$ be the set of central commutators in ρ of order p, set $\sigma = \langle z_1,\ldots,z_k\rangle \triangleleft \rho$, let $\alpha: \rho \longrightarrow \rho/\sigma$ be the projection, and set $I_\alpha = \mathrm{Ker}\Big[R[\rho] \longrightarrow R[\rho/\sigma]\Big]$. Then $\sigma \triangleleft \pi$; and $\sigma \neq 1$ by Lemma 6.5. We may assume inductively that the lemma holds for $R[\rho/\sigma]$.

Define

$$\mathcal{I} = \Big\{I \subseteq R[\rho] \;:\; I\ \pi\text{-invariant};\ I = p^2 I_0 + \sum_{\ell=1}^{k} (1-z_i)I_\ell,\ \text{some } I_\ell \subseteq R[\rho]\Big\}:$$

a family of ideals in $R[\rho]$. For all $I \in \mathcal{I}$, the group

$$K_1'(R[\rho],I) = \mathrm{Im}\Big[K_1(R[\rho],I) \longrightarrow K_1(F[\rho])\Big] \subseteq \mathrm{Ker}\Big[K_1'(R[\rho]) \longrightarrow K_1'(R/p^2[\rho/\sigma])\Big]$$

is torsion free: $\mathrm{tors}(K_1'(R[\rho])) = \mathrm{tors}(R^*) \times \rho^{ab}$ by Theorem 7.3, and this injects into $K_1'(R/p^2[\rho/\sigma])$. Hence, by Theorem 2.8 and Proposition 6.4,

$$\log_I : K_1'(R[\rho],I) \xrightarrow{\cong} \bar{H}_0(\rho;I) = \mathrm{Im}\Big[H_0(\rho,I) \longrightarrow H_0(\rho;R[\rho])\Big]$$

is an isomorphism. In particular, $K_1'(R[\rho],I)$ is π/ρ-cohomologically trivial for $I \in \mathcal{I}$, since $\bar{H}_0(\rho;I)$ is an $R[\pi/\rho]^t$-module.

By Proposition 8.1, the map

$$SK_1(R\alpha) : SK_1(R[\rho]) \twoheadrightarrow SK_1(R[\rho/\sigma]) \tag{3}$$

is surjective. Hence, $K_1'(R[\rho/\sigma]) \cong K_1'(R[\rho])/K_1'(R[\rho],I_\alpha)$. Both $K_1'(R[\rho/\sigma])$ and $K_1'(R[\rho],I_\alpha)$ are cohomologically trivial: the first by the induction hypothesis and the second since $I_\alpha \in \mathcal{I}$. So $K_1'(R[\rho])$ is cohomologically trivial.

If $SK_1(R[\rho]) = 1$, then $SK_1(R[\rho/\sigma]) = 1$ by (3). So we may assume inductively that there are ideals $J(R[\rho/\sigma]) = I_1' \supseteq I_2' \supseteq \dots$ which satisfy (1). Fix m such that $I_m' \subseteq p^2R[\rho/\sigma]$, and set $I_k = (R\alpha)^{-1}(I_k')$ for $1 \le k \le m$. In particular, $I_\alpha \subseteq I_m \subseteq I_\alpha + p^2R[\rho]$. So if we set $I_k = J^{k-m}\cdot I_m$ for $k \ge m$, then $I_k \in \mathcal{I}$ for all such k, and

$$K_1(R[\rho]/I_k) \cong K_1(R[\rho])/K_1'(R[\rho],I_k)$$

is cohomologically trivial. But $K_1(R[\rho]/I_k)$ is cohomologically trivial for $k \le m$ by assumption; and hence the I_k satisfy conditions (1). □

This will now be applied to describe the functors SK_1, K_1, and K_1' on twisted group rings.

<u>Theorem 12.3</u> *Fix a prime* p, *let* F *be any finite extension of* $\hat{\mathbb{Q}}_p$, *and let* $R \subseteq F$ *be the ring of integers. Let* $t\colon \pi \longrightarrow \mathrm{Gal}(F/\hat{\mathbb{Q}}_p)$ *be any homomorphism such that* π *is a p-group, and such that the extension* F/F^π *is unramified. Set* $\rho = \mathrm{Ker}(t)$, *and let* $R[\pi]^t$ *denote the induced twisted group ring. Then the inclusion* $R[\rho] \subseteq R[\pi]^t$ *induces isomorphisms*

(1) $\mathrm{ind}_{SK} : H_0(\pi/\rho; SK_1(R[\rho])) \xrightarrow{\cong} SK_1(R[\pi]^t)$

(2) $\mathrm{ind}_K : H_0(\pi/\rho; K_1(R[\rho])) \xrightarrow{\cong} K_1(R[\pi]^t)$

(3) $\mathrm{ind}_{K'} : H_0(\pi/\rho; K_1'(R[\rho])) \xrightarrow{\cong} K_1'(R[\pi]^t)$

(4) $\mathrm{trf}_{K'} : K_1'(R[\pi]^t) \xrightarrow{\cong} H^0(\pi/\rho; K_1'(R[\rho]))$.

<u>Proof</u> Using Lemma 8.3(ii), choose an extension

$$1 \longrightarrow \sigma \longrightarrow \tilde{\pi} \xrightarrow{\alpha} \pi \longrightarrow 1$$

of p-groups, where $\tilde{\rho} = \alpha^{-1}(\rho)$ and $\alpha_0 = \alpha|\tilde{\rho}$, such that $\sigma \subseteq Z(\rho)$ and $H_2(\alpha_0) = 0$. In particular, by Lemma 8.9, $SK_1(R[\tilde{\rho}]) = 1$.

The composites

$$K_1(R[\tilde{\rho}]) \xrightarrow{\mathrm{ind}} K_1(R[\tilde{\pi}]^t) \xrightarrow{\mathrm{trf}} K_1(R[\tilde{\rho}])$$

$$K_1'(R[\rho]) \xrightarrow{\mathrm{ind}} K_1'(R[\pi]^t) \xrightarrow{\mathrm{trf}} K_1'(R[\rho])$$

are induced by tensoring with $R[\tilde{\pi}]^t$ or $R[\pi]^t$ as bimodules (see Proposition 1.18), and are hence the norm homomorphisms $N_{\pi/\rho}$ for the π/ρ-actions. So $\mathrm{Ker}(\mathrm{ind}) \subseteq \mathrm{Ker}(N_{\pi/\rho})$ in both cases. Since $K_1(R[\tilde{\rho}])$ ($\cong K_1'(R[\tilde{\rho}])$) and $K_1'(R[\rho])$ are cohomologically trivial by Lemma 12.2,

$$\mathrm{Ker}(N_{\pi/\rho}) = \langle g(x)\cdot x^{-1} : g \in \pi/\rho,\ x \in K_1'(R[\rho])\rangle$$

$$\subseteq \mathrm{Ker}\Big[\mathrm{ind}\colon K_1'(R[\rho]) \longrightarrow K_1'(R[\pi]^t)\Big]$$

and similarly for $K_1(R[\tilde{\rho}])$. Also, since $\hat{H}^0(\pi/\rho; K_1'(R[\rho])) = 1$, $H^0(\pi/\rho; K_1'(R[\rho])) = \mathrm{Im}(N_{\pi/\rho})$. The induction maps are onto by Lemma 12.1, and it now follows that

$$\begin{aligned} \mathrm{ind}_{\tilde{K}} &: H_0(\pi/\rho;K_1(R[\tilde{\rho}])) \xrightarrow{\cong} K_1(R[\tilde{\pi}]^t), \\ \mathrm{ind}_{K'} &: H_0(\pi/\rho;K_1'(R[\rho])) \xrightarrow{\cong} K_1'(R[\pi]^t), \quad \text{and} \\ \mathrm{trf}_{K'} &: K_1'(R[\pi]^t) \xrightarrow{\cong} H^0(\pi/\rho;K_1'(R[\rho])) \end{aligned} \tag{5}$$

are isomorphisms.

Now set $I = \mathrm{Ker}\big[R[\tilde{\rho}] \longrightarrow R[\rho]\big]$ and $\bar{I} = \sum_{g\in\pi} I\cdot g$, and consider the following diagrams with exact rows:

$$\begin{array}{ccccccc} K_1(R[\rho],I) & \longrightarrow & H_0(\pi/\rho;K_1(R[\tilde{\rho}])) & \longrightarrow & H_0(\pi/\rho;K_1(R[\rho])) & \longrightarrow & 1 \\ \downarrow{\scriptstyle \mathrm{ind}_I} & & \cong\downarrow{\scriptstyle \mathrm{ind}_{\tilde{K}}} & & \downarrow{\scriptstyle \mathrm{ind}_K} & & \\ K_1(R[\pi]^t,I) & \longrightarrow & K_1(R[\tilde{\pi}]^t) & \longrightarrow & K_1(R[\pi]^t) & \longrightarrow & 1 \end{array}$$

$$\begin{array}{ccccccccc} 1 & \longrightarrow & H_0(\pi/\rho;SK_1(R[\rho])) & \longrightarrow & H_0(\pi/\rho;K_1(R[\rho])) & \longrightarrow & H_0(\pi/\rho;K_1'(R[\rho])) & \longrightarrow & 1 \\ & & \downarrow{\scriptstyle \mathrm{ind}_{SK}} & & \downarrow{\scriptstyle \mathrm{ind}_K} & & \cong\downarrow{\scriptstyle \mathrm{ind}_{K'}} & & \\ 1 & \longrightarrow & SK_1(R[\pi]^t) & \longrightarrow & K_1(R[\pi]^t) & \longrightarrow & K_1'(R[\pi]^t) & \longrightarrow & 1. \end{array}$$

Then $\mathrm{ind}_{\tilde{K}}$ and $\mathrm{ind}_{K'}$ are isomorphisms by (5), ind_I is onto by Lemma 12.1, and hence ind_K and ind_{SK} are also isomorphisms. □

Theorem 12.3 applies in particular to twisted group rings $R\zeta_n[\pi]^t$ of the form occurring in Theorem 11.8: $F\zeta_n/F$ is unramified if $p\nmid n$ by Theorem 1.10(i). So Theorem 11.9 now implies as an immediate corollary:

<u>Theorem 12.4</u> *If p is any prime, and if R is the ring of integers in any finite extension of $\hat{\mathbb{Q}}_p$, then the functors $SK_1(R[G])$, $K_1(R[G])_{(p)}$, and $K_1'(R[G])_{(p)}$ are all computable with respect to induction from p-elementary subgroups, and $K_1'(R[G])_{(p)}$ is computable with respect to restriction to p-elementary subgroups.* □

As another application of Theorem 12.3, the decomposition formulas of

Theorem 11.8, when applied to $SK_1(R[G])$ and tors $K_1'(R[G])_{(p)}$, take the following form:

Theorem 12.5 *Fix a prime* p, *let* F *be any finite extension of* $\hat{\mathbb{Q}}_p$, *and let* $R \subseteq F$ *be the ring of integers. For any finite group* G, *let* $g_1,\ldots,g_k$ *be* F-*conjugacy class representatives for elements in* G *of order prime to* p, *and set*

$$N_i = N_G^F(g_i) = \{x \in G:\ xg_ix^{-1} = g_i^a,\ \text{some } a \in \mathrm{Gal}(K\zeta_{n_i}/K)\} \qquad (n_i = |g_i|)$$

and $Z_i = C_G(g_i)$. *Then*

$$\text{(i)}\quad SK_1(R[G]) \cong \bigoplus_{i=1}^{k} H_0(N_i/Z_i;\ H_2(Z_i)/H_2^{ab}(Z_i))_{(p)};\quad \text{and}$$

$$\text{(ii)}\quad \mathrm{tors}(K_1'(R[G]))_{(p)} \cong [(\mu_F)_p]^k \oplus \bigoplus_{i=1}^{k} H^0(N_i/Z_i;Z_i^{ab})_{(p)}.$$

__Proof__ Set $n_i = |g_i|$, and let $\mathcal{P}(N_i)$ and $\mathcal{P}(Z_i)$ be the sets of p-subgroups. Then by Theorem 11.8,

$$\begin{aligned}
SK_1(R[G]) &\cong \bigoplus_{i=1}^{k} \varinjlim_{\pi\in\mathcal{P}(N_i)} SK_1(R\zeta_{n_i}[\pi]^t)\\
&\cong \bigoplus_{i=1}^{k} \varinjlim_{\pi\in\mathcal{P}(N_i)} H_0\Big(\pi/(\pi\cap Z_i);SK_1(R\zeta_{n_i}[\pi\cap Z_i])\Big)\\
&\cong \bigoplus_{i=1}^{k} H_0\Big(N_i/Z_i;\ \varinjlim_{\rho\in\mathcal{P}(Z_i)} SK_1(R\zeta_{n_i}[\rho])\Big)\\
&\cong \bigoplus_{i=1}^{k} H_0\Big(N_i/Z_i;\ \varinjlim_{\rho\in\mathcal{P}(Z_i)} H_2(\rho)/H_2^{ab}(\rho)\Big)
\end{aligned}$$

$$\cong \bigoplus_{i=1}^{k} H_0\Big(N_i/Z_i;\ H_2(Z_i)/H_2^{ab}(Z_i)\Big)_{(p)}.$$

Here, the last step follows since $H_2(-)_{(p)}$ and $H_2^{ab}(-)_{(p)}$ both are computable for induction from p-subgroups by Theorem 11.1: they are Green modules over the functor $H^0(-;\mathbb{Z}_{(p)})$.

The formula for $\mathrm{tors}_p K_1'(R[G])$ is derived in a similar fashion, but using inverse limits (and Theorem 7.3). □

In contrast to the results for induction, the following example shows that $SK_1(\hat{\mathbb{Z}}_p[G])$ is *not* in general detected by p-elementary restriction.

<u>Example 12.6</u> *Fix a prime* p, *and let* ρ *be any p-group such that* $SK_1(\hat{\mathbb{Z}}_p[\rho]) \neq 1$. *Set* $n = p^p - 1$, *let* $H = C_n \rtimes C_p$ *be the semidirect product induced by the action of* $C_p \cong \mathrm{Gal}(\hat{\mathbb{Q}}_p(\zeta_n)/\hat{\mathbb{Q}}_p)$ *on* $\langle\zeta_n\rangle$, *and set* $G = \rho \times H$. *Then* $SK_1(\hat{\mathbb{Z}}_p[G])$ *is not detected by restriction to p-elementary subgroups of* G.

<u>Proof</u> By Theorem 11.9, it suffices to show that the transfer map

$$\mathrm{trf} : SK_1(\hat{\mathbb{Z}}_p\zeta_n[\rho\times C_p]^t) \longrightarrow SK_1(\hat{\mathbb{Z}}_p\zeta_n[\rho])$$

is not injective. Since the conjugation action of C_p on $SK_1(\hat{\mathbb{Z}}_p\zeta_n[\rho]) \cong H_2(\rho)/H_2^{ab}(\rho)$ is trivial, the inclusion induces an isomorphism

$$SK_1(\hat{\mathbb{Z}}_p\zeta_n[\rho\times C_p]^t) \cong SK_1(\hat{\mathbb{Z}}_p\zeta_n[\rho]) \neq 1$$

by Theorem 12.3. The composite

$$SK_1(\hat{\mathbb{Z}}_p\zeta_n[\rho]) \xrightarrow[\cong]{\text{ind}} SK_1(\hat{\mathbb{Z}}_p\zeta_n[\rho\times C_p]^t) \xrightarrow{\text{trf}} SK_1(\hat{\mathbb{Z}}_p\zeta_n[\rho])$$

is the norm homomorphism for the C_p-action on $SK_1(\hat{\mathbb{Z}}_p\zeta_n[\rho])$ (use Proposition 1.18); is hence multiplication by p, and not injective. □

The next proposition gives some very general conditions for showing that $SK_1(\hat{\mathbb{Z}}_p[G]) = 1$. Note, for example, that it applies to the groups $SL(2,q)$ and $PSL(2,q)$ for any prime power q.

Proposition 12.7 *Let p be any prime, and let G be any finite group. Then $SK_1(\hat{\mathbb{Z}}_p[G]) = 1$ if $SK_1(\hat{\mathbb{Z}}_p[\pi]) = 1$ for all p-subgroups $\pi \subseteq G$. In particular, $SK_1(\hat{\mathbb{Z}}_p[G]) = 1$ if the p-Sylow subgroup $S_p(G)$ has a normal abelian subgroup with cyclic quotient.*

Proof By Theorem 12.5(i), $SK_1(\hat{\mathbb{Z}}_p[G]) = 1$ if $H_2(\pi)/H_2^{ab}(\pi) = 1$ for all p-subgroups $\pi \subseteq G$; and this holds if $SK_1(\hat{\mathbb{Z}}_p[\pi]) = 1$ for all such π. If a p-group π contains a normal abelian subgroup with cyclic quotient, then $SK_1(\hat{\mathbb{Z}}_p[\pi]) = 1$ by Corollary 7.2. □

As a second, more specialized example, we now consider the symmetric and alternating groups. Note that Proposition 12.7 cannot be applied in this case, since any p-group is a subgroup of some S_n.

Example 12.8 *For any $n \geq 1$ and any prime p,*

$$SK_1(\hat{\mathbb{Z}}_p[S_n]) \cong SK_1(\hat{\mathbb{Z}}_p[A_n]) \cong 1.$$

Proof For any $g \in S_n$ of order prime to p, the centralizer $C_{S_n}(g)$ is a product of wreath products:

$$C_{S_n}(g) = C_{m_1} \wr S_{n_1} \times \dots \times C_{m_k} \wr S_{n_k};$$

where for each i, $m_i \mid |g|$ and hence $p \nmid m_i$. So by Theorem 12.5(i), $SK_1(\hat{\mathbb{Z}}_p[S_n]) = SK_1(\hat{\mathbb{Z}}_p[A_n]) = 1$, if $H_2(G)/H_2^{ab}(G) = 0$ whenever G is a product of symmetric groups, or is of index 2 in such a product.

The groups $H_2(S_n)$ and $H_2(A_n)$ have been computed by Schur in [1, Abschnitt 1]. It follows from the description there that for any $n \geq 4$, the maps

$$H_2(C_2\times C_2) \longrightarrow H_2(A_4) \longrightarrow H_2(A_n)_{(2)} \longrightarrow H_2(S_n)$$

(induced by inclusion) are all isomorphisms. Furthermore, $H_2(A_n)$ is a 2-group unless $n = 6$ or 7, $H_2(A_6)$ and $H_2(A_7)$ both have order 6, and A_6 and A_7 have abelian 3-Sylow subgroups. So for all n,

$$H_2(A_n)/H_2^{ab}(A_n) \cong H_2(S_n)/H_2^{ab}(S_n) = 0.$$

By Proposition 8.12, the functor H_2/H_2^{ab} is multiplicative with respect to direct products of groups. Thus, $H_2(G)/H_2^{ab}(G) = 0$ whenever G is a product of symmetric or alternating groups. If G is a semidirect product

$$G = (A_{n_1}\times\ldots\times A_{n_k}) \rtimes (C_2)^{k-1} \subseteq S_{n_1}\times\ldots\times S_{n_k};$$

then since $H_1(A_{n_1}\times\ldots\times A_{n_k})$ has odd order, $H_2(G)$ is generated by

$$H_2(A_{n_1}\times\ldots\times A_{n_k}) \quad\text{and}\quad H_2((C_2)^{k-1}).$$

Thus, $H_2(G)/H_2^{ab}(G) = 0$ for such G, and this finishes the proof. □

Example 12.8 was the last step when showing that $Wh(S_n) = 1$ for all n. We have already seen that $Wh'(S_n)$ is finite (Theorem 2.6) and torsion free (Theorem 7.4); and that $Cl_1(\mathbb{Z}[S_n]) = 1$ (Theorem 5.4). The computation of $SK_1(\mathbb{Z}[A_n]) = Cl_1(\mathbb{Z}[A_n])$ will be carried out in Theorem 14.6.

To end the chapter, we now want to give some alternative, and more direct, descriptions, of $SK_1(\hat{\mathbb{Z}}_p[G])$, $tors_p K_1'(\hat{\mathbb{Z}}_p[G])$, and $K_1'(\hat{\mathbb{Z}}_p[G])$ for arbitrary finite G. For any G, and any fixed prime p, G_r will denote the set of *p-regular* elements in G: i. e., elements of order prime to p. For any $g \in G$, $g_r, g_s \in G$ will denote the unique elements such that $g_r \in G_r$, g_s has p-power order, $g = g_r g_s$, and $[g_r, g_s] = 1$ (note that $g_r, g_s \in \langle g\rangle$). For any R, $H_n(G;R(G_r))$ denotes the homology group induced by the conjugation action of G on $R(G_r)$. It will be convenient to represent elements of $H_1(G;R(G_r))$ via the bar resolution:

$$H_1(G;R(G_r)) \cong H_1\Big(\mathbb{Z}[G]\otimes_{\mathbb{Z}}\mathbb{Z}[G]\otimes_{\mathbb{Z}}R(G_r) \xrightarrow{\partial_2} \mathbb{Z}[G]\otimes_{\mathbb{Z}}R(G_r) \xrightarrow{\partial_1} R(G_r)\Big),$$

where $\partial_1(g\otimes x) = x - gxg^{-1}$ and $\partial_2(g\otimes h\otimes x) = h\otimes x - gh\otimes x + g\otimes hxh^{-1}$.

When R is the ring of integers in a finite unramified extension of $\hat{\mathbb{Q}}_p$, then Φ denotes the automorphism of $H_n(G;R(G_r))$ induced by the map $\Phi(\sum r_i g_i) = \sum \varphi(r_i) g_i^p$ on coefficients. As usual, we write

$$H_n(G;R(G_r))_\Phi = H_n(G;R(G_r))/(1-\Phi); \qquad H_n(G;R(G_r))^\Phi = \mathrm{Ker}(1-\Phi) \subseteq H_n(G;R(G_r)).$$

<u>Theorem 12.9</u> *Fix a prime* p, *an unramified extension* $F \supseteq \hat{\mathbb{Q}}_p$, *and a finite group* G. *Let* $R \subseteq F$ *be the ring of integers. Define*

$$\omega_{RG} : H_0(G;R[G]) \longrightarrow H_1(G;R(G_r)) \quad \text{and} \quad \theta_{RG} : H_0(G;R[G]) \longrightarrow H_0(G;R/2(G_r))$$

by setting, for $r_i \in R$ *and* $g_i \in G$:

$$\omega(\textstyle\sum r_i g_i) = \sum g_i \otimes r_i (g_i)_r \quad \text{and} \quad \theta(\sum r_i g_i) = \sum \bar{r}_i (g_i)_r \quad (\bar{r}_i \in R/2).$$

Then

(i) *There are unique homomorphisms*

$$\nu_{RG} : K_1'(R[G]) \longrightarrow H_1(G;R(G_r)) \quad \text{and} \quad \hat{\theta}_{RG} : K_1'(R[G]) \longrightarrow H_0(G;R/2(G_r)),$$

which are natural with respect to group homomorphisms, and which are characterized as follows. For any $u \in GL(R[G])$, *write* $u = \sum_i r_i g_i$ *and* $u^{-1} = \sum_j s_j h_j$, *where* $r_i, s_j \in M_n(R)$ *and* $g_i, h_j \in G$. *Then*

$$\nu([u]) = \sum_{i,j} g_i \otimes \mathrm{Tr}(s_j r_i)\cdot(h_j g_i)_r \in H_1(G;R(G_r)). \qquad (\mathrm{Tr}: M_n(R) \to R)$$

If $p=2$, *then for any commuting pair of subgroups* $H, \pi \subseteq G$, *where* $|H|$

is odd and π *is a 2-group, and any* $x \in J(R[H\times\pi])$, $\hat{\theta}(1+x)$ *is the image of* x *under the composite*

$$J(R[H\times\pi]) \xrightarrow{pr_H} J(R[H]) = 2R[H] \xrightarrow[\cong]{/2} R[H] \subseteq R(G_r) \longrightarrow H_0(G;R/2(G_r)).$$

(ii) The sequence

$$1 \longrightarrow K_1'(R[G])_{(p)} \xrightarrow{(\Gamma,\upsilon,\Phi\hat{\theta})} H_0(G;R[G]) \oplus H_1(G;R(G_r)) \oplus H_0(G;R/2(G_r))$$

$$\xrightarrow{\begin{pmatrix}\omega & \Phi-1 & 0\\ \theta & 0 & \Phi-1\end{pmatrix}} H_1(G;R(G_r)) \oplus H_0(G;R/2(G_r)) \longrightarrow 0$$

is exact.

(iii) There is an exact sequence

$$0 \longrightarrow H_1(G;R(G_r))^{\Phi} \oplus H_0(G;R/2(G_r))^{\Phi} \longrightarrow K_1'(R[G])_{(p)}$$

$$\xrightarrow{\Gamma} H_0(G;R[G]) \longrightarrow H_1(G;R(G_r))_{\Phi} \oplus H_0(G;R/2(G_r))_{\Phi} \longrightarrow 0.$$

In particular,

$$\operatorname{tors}_p K_1'(R[G]) \cong H_1(G;R(G_r))^{\Phi} \oplus H_0(G;R/2(G_r))^{\Phi}.$$

<u>Proof</u> Using the relation $gh\otimes x = h\otimes x + g\otimes hxh^{-1}$, for $g,h \in G$ and $x \in R(G_r)$, one easily checks that the map $\upsilon_{RG}\colon GL(R[G]) \longrightarrow H_1(G;R(G_r))$ defined in (i) is a homomorphism. Hence, this factors through $K_1(R[G]) = GL(R[G])^{ab}$. If G is p-elementary, then $H_1(G;R(G_r)) \cong H_1(G^{ab};R((G^{ab})_r))$ and $SK_1(R[G^{ab}]) = 1$, so that $SK_1(R[G]) \subseteq \operatorname{Ker}(\upsilon_{RG})$. Since $SK_1(R[G])$ is generated by p-elementary induction, this shows that υ_{RG} factors through $K_1'(R[G]) = K_1(R[G])/SK_1(R[G])$ for arbitrary finite G.

To see that $\hat{\theta}_{RG}$ is well defined when $p=2$, assume first that $G=H\times\pi$ where $|H|$ is odd and π is a 2-group. Then $J(R[H]) = 2R[H]$, and so

$$K_1(R/4[H])_{(2)} \cong K_1(R/4[H],2) \cong H_0(H;2R[H]/4R[H]) \cong H_0(G;R/2(G_r))$$

by Theorem 1.15. This shows that $\hat{\theta}_{RG}$ is well defined in this case; and in particular when G is 2-elementary. Since $K_1'(R[-])_{(2)}$ is 2-elementary computable, $\hat{\theta}$ now automatically extends to a homomorphism defined for arbitrary finite G.

If G is p-elementary — if $G \cong C_n \times \pi$ where $p\nmid n$ and π is a p-group — then $R[G]$ is isomorphic to a product of rings $R_i[\pi]$ for various unramified extensions R_i/R. So in this case, sequence (ii) is exact by Theorem 6.7, and sequence (iii) by Theorems 6.6 and 7.3.

All terms in sequence (ii) are computable with respect to induction from p-elementary subgroups. Hence, since the direct limits used here are right exact, (ii) is exact except possibly at $K_1'(R[G])_{(p)}$. But then (ii) is exact if and only if (iii) is, if and only if

$$|\mathrm{Ker}(\Gamma)| = |H_1(G;R(G_r))^{\Phi}| \cdot |H_0(G;R/2(G_r))^{\Phi}|. \tag{1}$$

Also, $\mathrm{Ker}(\Gamma) = \mathrm{Ker}(\log) = \mathrm{tors}_p K_1'(R[G])$ (Theorem 2.9), and so (1) follows from a straightforward computation based on Theorem 12.5(ii). For details, see Oliver [8, Theorem 1.7 and Corollary 1.8]. □

The above definition of ν was suggested by Dennis' trace map from K-theory to Hochschild homology (see Igusa [1]). We have been unable to find a correspondingly satisfactory definition for $\hat{\theta}$.

We saw in Theorem 6.8 that a restriction map on $H_0(G;R[G])$ can be defined, which makes Γ_{RG} natural with respect to transfer homomorphisms. Unfortunately, there is no way to define restriction maps on the other terms in sequence (ii) above, to make the whole sequence natural with respect to the transfer. If there were, the proof of the injectivity of $(\Gamma,\nu,\Phi\hat{\theta})$ would be simpler, since inverse limits are left exact.

We now end the chapter with a second description of $SK_1(\hat{\mathbb{Z}}_p[G])$. For any finite G and any unramified R, set

$$H_2(G;R(G_r))_{\Phi} = H_2(G;R(G_r))/(\Phi-1);$$

where Φ is induced by the automorphism $\Phi(rg) = \varphi(r)g^p$ of $R(G_r)$. In analogy with the p-group case, we define

$$H_2^{ab}(G;R(G_r))_\Phi = \mathrm{Im}\Big[\bigoplus_{\substack{H\subseteq G\\ H\ abelian}} H_2(H;R(H_r)) \xrightarrow{\ \mathrm{Ind}\ } H_2(G;R(G_r))_\Phi\Big]$$

$$= \Big\langle (g\wedge h)\otimes rk \in H_2(G;R(G_r))_\Phi \ : \ g,h\in G,\ k\in G_r,\ r\in R,\ \ \langle g,h,k\rangle \ abelian\Big\rangle.$$

The following formula for $SK_1(R[G])$ is easily seen to be abstractly the same as that in Theorem 12.5(i), but it allows a more direct procedure for determining whether or not a given element in $SK_1(R[G])$ vanishes. This procedure is analogous to that in the p-group case described in Proposition 8.4. Note, however, that in this case, once $u \in SK_1(R[G])$ has been lifted to $\tilde{u} \in K_1(R[\tilde{G}])$ for some appropriate $\tilde{G}$, it is necessary to evaluate both $\nu_{\tilde{G}}(\tilde{u})$ and $\Gamma_{\tilde{G}}(\tilde{u})$. Knowing $\Gamma_{\tilde{G}}(\tilde{u})$ alone does not in general suffice to determine whether or not u vanishes in $SK_1(R[G])$ — no matter how large $\tilde{G}$ is.

<u>**Theorem 12.10**</u> *Fix a prime* p, *and let* R *be the ring of integers in any finite unramified extension* F *of* $\hat{\mathbb{Q}}_p$. *Then, for any finite group* G, *there is an isomorphism*

$$\Theta_G \ : \ SK_1(R[G]) \xrightarrow{\ \cong\ } H_2(G;R(G_r))_\Phi/H_2^{ab}(G;R(G_r))_\Phi;$$

which is described as follows. Let $1 \longrightarrow K \longrightarrow \tilde{G} \xrightarrow{\alpha} G \longrightarrow 1$ *be any extension of finite groups such that*

$$\mathrm{Im}\Big[H_2(\tilde{G};R(G_r)) \longrightarrow H_2(G;R(G_r))\Big] \subseteq H_2^{ab}(G;R(G_r)). \qquad (1)$$

Consider the homomomorphisms

$$\begin{array}{ccc}
\operatorname{Ker}\Big[H_0(\tilde{G};R[\tilde{G}]) \longrightarrow H_0(G;R[G])\Big] & & \\
\quad\searrow{\scriptstyle \bar{\omega}_\alpha} & & \\
& \dfrac{K^{ab} \otimes_{\mathbb{Z}G} R(G_r)}{\delta^\alpha\Big(H_2^{ab}(G;R(G_r)) + (1-\Phi)H_2(G;R(G_r))\Big)} \xleftarrow{\ \delta^\alpha_{ab}\ } & \dfrac{H_2(G;R(G_r))_\Phi}{H_2^{ab}(G;R(G_r))_\Phi} \\
\quad\nearrow{\scriptstyle (\Phi-1)\circ\iota^{-1}} & & \\
\operatorname{Ker}\Big[H_1(\tilde{G};R(G_r)) \longrightarrow H_1(G;R(G_r))\Big] & &
\end{array} \qquad (2)$$

Here, $\bar{\omega}_\alpha(r(z-1)g) = z \otimes r\cdot\alpha(g_r)$ *for any* $z \in K$, $r \in R$, *and* $g \in \tilde{G}$; *and* δ^α_{ab} *and* ι *are induced by the five term exact sequence*

$$H_2(\tilde{G};R(G_r)) \xrightarrow{H_2(\alpha)} H_2(G;R(G_r)) \xrightarrow{\delta^\alpha} K^{ab} \otimes_{\mathbb{Z}G} R(G_r)$$
$$\xrightarrow{\ \iota\ } H_1(\tilde{G};R(G_r)) \xrightarrow{H_1(\alpha)} H_1(G;R(G_r)) \qquad (3)$$

of Theorem 8.2. Let $\bar{\nu}_{\tilde{G}}: K_1(R[\tilde{G}]) \longrightarrow H_1(\tilde{G};R(G_r))$ *be induced by the homomorphism* $\nu_{R\tilde{G}}$ *of Theorem 12.9(i). Then, for any* $[u] \in SK_1(R[G])$, *and any lifting to* $[\tilde{u}] \in K_1(R[\tilde{G}])$,

$$\Theta_G([u]) = (\delta^\alpha_{ab})^{-1}\Big(\bar{\omega}_\alpha\circ\Gamma_{\tilde{G}}(\tilde{u}) + (\Phi-1)(\iota^{-1}\bar{\nu}_{\tilde{G}}(\tilde{u}))\Big) \in H_2(G;R(G_r))_\Phi/H_2^{ab}(G;R(G_r))_\Phi.$$

<u>Proof</u> By (1) and (3), δ^α_{ab} and $(\Phi-1)\circ\iota^{-1}$ are well defined, and δ^α_{ab} is a monomorphism. To see that $\bar{\omega}_\alpha$ is well defined, first set

$$I_\alpha = \operatorname{Ker}\Big[R[\tilde{G}] \twoheadrightarrow R[G]\Big] \quad\text{and}\quad \bar{H}_0(\tilde{G};I_\alpha) = \operatorname{Ker}\Big[H_0(\tilde{G};R[\tilde{G}]) \longrightarrow H_0(G;R[G])\Big]$$

for convenience. The map $I_\alpha \longrightarrow H_1(K;R(G_r)) \cong K^{ab} \otimes R(G_r)$, defined by sending $(z-1)g$ to $z \otimes \alpha(g_r)$, is easily seen to be well defined; and induces a homomorphism

$$\omega_\alpha : H_0(\tilde{G};I_\alpha) \longrightarrow H_0(\tilde{G};H_1(K;R(G_r))) \cong K^{ab} \otimes_{\mathbb{Z}G} R(G_r).$$

Consider the following commutative diagram:

$$\begin{array}{ccccccc}
H_1(\tilde{G};R[G]) & \xrightarrow{\partial^{\alpha}} & H_0(\tilde{G};I_{\alpha}) & \longrightarrow & H_0(\tilde{G};R[\tilde{G}]) & \longrightarrow & H_0(G;R[G]) \\
\Big\downarrow{\scriptstyle\lambda} & & \Big\downarrow{\scriptstyle\omega_{\alpha}} & & \Big\downarrow{\scriptstyle\omega_{\tilde{G}}} & & \\
H_2(G;R(G_r)) & \xrightarrow{\delta^{\alpha}} & K^{ab}\otimes_{\mathbb{Z}G} R(G_r) & \xrightarrow{\iota} & H_1(\tilde{G};R(G_r)) & &
\end{array} \tag{4}$$

where $\omega_{\tilde{G}}$ is as in Theorem 12.9; and where

$$\lambda(g\otimes rh) = (\alpha(g)\wedge h)\otimes r\cdot h_r \in H_2^{ab}(G;\hat{\mathbb{Z}}_p(G_r))$$

for any $r \in R$, $g \in \tilde{G}$, and $h \in G$ such that $[\alpha(g),h] = 1$. The rows in (4) are exact, and $\overline{H}_0(\tilde{G};I_{\alpha}) \cong \operatorname{Coker}(\partial^{\alpha})$; so ω_{α} factors through a homomorphism $\bar{\omega}_{\alpha}$ as in diagram (2).

For any $\tilde{u} \in K_1(R\alpha)^{-1}(SK_1(R[\tilde{G}]))$, $\bar{\nu}_{\tilde{G}}(\tilde{u}) \in \operatorname{Ker}(H_1(\alpha)) = \operatorname{Im}(\iota)$. By the exact sequence in Theorem 12.9(ii),

$$\bar{\omega}_{\alpha}\circ\Gamma_{\tilde{G}}(\tilde{u}) + (\Phi-1)\circ\iota^{-1}(\bar{\nu}_{\tilde{G}}(\tilde{u}))$$

$$\in \operatorname{Ker}\left[\frac{K^{ab}\otimes_{\mathbb{Z}G} R(G_r)}{\delta^{\alpha}\left(H_2^{ab}(G;R(G_r)) + (1-\Phi)H_2(G;R(G_r))\right)} \xrightarrow{\iota} H_1(\tilde{G};R(G_r))\right] = \operatorname{Im}(\delta^{\alpha}_{ab}).$$

So to see that Θ_G is uniquely defined — with respect to a given α, at least — it remains only to check that

$$\bar{\omega}_{\alpha}\circ\Gamma_{\tilde{G}}(\tilde{u}) + (\Phi-1)(\iota^{-1}\bar{\nu}_{\tilde{G}}(\tilde{u})) = 0 \quad \text{for any } \tilde{u} \in K_1(R[\tilde{G}],I_{\alpha}). \tag{5}$$

To prove this, set $\hat{G} = \{(g,h) \in \tilde{G}: \alpha(g) = \alpha(h)\}$, so that

$$\begin{array}{ccc}
\hat{G} & \xrightarrow{\beta_2} & \tilde{G} \\
\Big\downarrow{\scriptstyle\beta_1} & & \Big\downarrow{\scriptstyle\alpha} \\
\tilde{G} & \xrightarrow{\alpha} & G.
\end{array}$$

is a pullback square. Set $\hat{I} = \operatorname{Ker}(R[\beta_2]) \subseteq R[\hat{G}]$. Since β_2 is split surjective (split by the diagonal map), $\delta^{\beta_2} = 1$, and β_1 induces a homomorphism

$$\beta_{1*} : \mathrm{Ker}\Big[H_1(\hat{G};R(G_r)) \longrightarrow H_1(\tilde{G};R(G_r))\Big] \cong K^{ab} \otimes_{\mathbb{Z}G} R(G_r)$$

$$\longrightarrow \frac{K^{ab} \otimes_{\mathbb{Z}G} R(G_r)}{\delta^{\alpha}\Big(H_2^{ab}(G;R(G_r)) + (1-\Phi)H_2(G;R(G_r))\Big)}.$$

Then any $\tilde{u} \in K_1(R[\tilde{G}],I_\alpha)$ lifts to $\hat{u} \in K_1(R[\hat{G}],\hat{I})$;

$$\bar{\omega}_\alpha \circ \Gamma_{\tilde{G}}(\tilde{u}) + (\Phi-1)(\iota^{-1}\bar{\upsilon}_{\tilde{G}}(\tilde{u})) = \beta_{1*}\Big(\bar{\omega}_{\beta_2} \circ \Gamma_{\hat{G}}(\hat{u}) + (\Phi-1)(\iota^{-1}\bar{\upsilon}_{\hat{G}}(\hat{u}))\Big) = \beta_{1*}(0)$$

by Theorem 12.9(ii); and this proves (5).

We have now shown that there is a well defined epimorphism

$$\Theta_G : SK_1(R[G]) \longrightarrow\!\!\!\!\rightarrow H_2(G;R(G_r))_\Phi/H_2^{ab}(G;R(G_r))_\Phi$$

such that $\Theta_G([u]) = [\delta^{\alpha}_{ab}(\Gamma_{\tilde{G}}(\tilde{u}))]$ for any $[u] \in SK_1(R[G])$ and any lifting to $[\tilde{u}] \in K_1(R[\tilde{G}])$. This is independent of α: given a second surjection α' onto G, the maps Θ_G defined using α and α' can each be compared to the map defined using their pullback. Also, the existence of α satisfying (1) follows from Lemma 8.3.

To show that Θ_G is an isomorphism, it remains to show that the two groups are abstractly isomorphic. But this follows from the formula for $SK_1(R[G])$ in Theorem 12.5; the formula

$$H_2(G;R(G_r)) \cong \bigoplus_{i=1}^{m} H_2(C_G(g_i)) \otimes R(g_i)$$

(when $g_1,\ldots,g_m$ are conjugacy class representatives for G_r); and the description of $N_G^F(g_i)$ in Oliver [8, Lemma 1.5].

Alternatively, since $H_2(G;R(G_r))_\Phi/H_2^{ab}(G;R(G_r))_\Phi$ and $SK_1(R[G])$ are both p-elementary computable, it suffices to show for p-elementary G that Θ_G is an isomorphism. And this is an easy consequence of Theorem 8.6. □

Chapter 13 $Cl_1(\mathbb{Z}[G])$ FOR FINITE GROUPS

The goal now is to reduce as far as possible computations of $Cl_1(\mathbb{Z}[G])_{(p)}$ and $SK_1(\mathbb{Z}[G])_{(p)}$, first to the case where G is p-elementary, and then to the p-group case. The reduction to p-elementary groups is dealt with in Section 13a. The main result in that section, Theorem 13.5, says that $Cl_1(\mathbb{Z}[G])_{(p)}$ and $SK_1(\mathbb{Z}[G])_{(p)}$ are p-elementary computable if p is odd; and that $SK_1(\mathbb{Z}[G])_{(2)}$ can be described in terms of 2-elementary subgroups via a certain pushout square.

Section 13b deals with the reduction from p-elementary groups to p-groups. In particular, explicit formulas for $Cl_1(\mathbb{Z}[G])_{(p)}$, in terms of $Cl_1(\mathbb{Z}[\pi])$ for p-subgroups $\pi \subseteq G$, are given in Theorems 13.9 (p odd) and 13.13 (G abelian). Theorems 13.10 and 13.11 deal with some of the special problems which arise when comparing $Cl_1(R[\pi])$ with $Cl_1(\mathbb{Z}[\pi])$ — when π is a 2-group and R is the ring of integers in an algebraic number field in which 2 is unramified.

In Section 13c, the extension

$$1 \longrightarrow Cl_1(\mathbb{Z}[G]) \longrightarrow SK_1(\mathbb{Z}[G]) \xrightarrow{\ \ell\ } \bigoplus_p SK_1(\hat{\mathbb{Z}}_p[G]) \longrightarrow 1$$

is shown to be naturally split in odd torsion. An example is then constructed (Example 13.16) of a 2-elementary group G for which ℓ has no splitting which is natural with respect to automorphisms of G.

13a. Reduction to p-elementary groups

As seen in Theorem 11.9, reducing calculations to p-elementary groups involves "untwisting" twisted group rings. Before results of this type for $Cl_1(R[\pi]^t)$ and $SK_1(R[\pi]^t)$ can be proven, the other terms in the localization sequence for $SK_1(-)$ must be studied. The main technical results for doing this are in Lemma 13.1 and Proposition 13.3.

<u>Lemma 13.1</u> *Fix a prime* p, *let* F *be any finite unramified extension of* $\hat{\mathbb{Q}}_p$, *and let* $R \subseteq F$ *be the ring of integers. Fix a p-group* π, *and let* $t: \pi \longrightarrow \mathrm{Gal}(F/\hat{\mathbb{Q}}_p)$ *be any homomorphism. Set* $\rho = \mathrm{Ker}(t)$, *and let* $R[\pi]^t$ *denote the induced twisted group ring. Then the following hold.*

(i) For any radical ideal $I \subseteq R[\rho]$ *such that* $gIg^{-1} = I$ *for all* $g \in \pi$, *set* $\bar{I} = \sum_{g\in\pi} I\cdot g \in R[\pi]^t$. *Then the inclusion* $R[\rho] \subseteq R[\pi]^t$ *induces an isomorphism*

$$\mathrm{ind}_I : H_0(\pi/\rho; K_1(R[\rho], I)) \xrightarrow{\cong} K_1(R[\pi]^t, \bar{I}).$$

(ii) If $SK_1(R[\rho]) = 1$, *then the inclusion* $R[\rho] \subseteq R[\pi]^t$ *induces an epimorphism*

$$\mathrm{ind}_{K2} : K_2^c(R[\rho]) \twoheadrightarrow K_2^c(R[\pi]^t).$$

<u>Proof</u> Let $J \subseteq R[\rho]$ denote the Jacobson radical, and set $\bar{\pi} = \pi/\rho$. For any pair $I_0 \subseteq I \subseteq R[\rho]$ of π-invariant ideals of finite index, let

$$\alpha_{I/I_0} : H_0(\pi/\rho; K_1(R[\rho]/I_0, I/I_0)) \longrightarrow K_1(R[\pi]^t/\bar{I}_0, \bar{I}/\bar{I}_0)$$

$$\beta_{I/I_0} : K_2(R[\rho]/I_0, I/I_0) \longrightarrow K_2(R[\pi]^t/\bar{I}_0, \bar{I}/\bar{I}_0)$$

be the homomorphisms induced by the inclusion $R[\rho] \subseteq R[\pi]^t$. The lemma will be proven in three steps. To simplify notation, we write $K_i(I/I_0)$ for $K_i(R[\rho]/I_0, I/I_0)$, etc.

<u>Step 1</u> Assume first that $I_0 \subseteq I \subseteq R[\rho]$ are π-invariant ideals of finite index such that $IJ + JI \subseteq I_0$. We want to show that α_{I/I_0} is an isomorphism, and that β_{I/I_0} is surjective.

Write $\bar{I} = I \oplus \tilde{I}$ and $\bar{I}_0 = I_0 \oplus \tilde{I}_0$, where $\tilde{I} = \sum_{g\in\pi\smallsetminus\rho} I\cdot g$ and

$\tilde{I}_0 = \sum_{g\in\pi\smallsetminus\rho} I_0\cdot g$. By Theorem 1.15,

$$K_1(I/I_0) \cong (I/I_0)/[R[\rho]/I_0, I/I_0] \cong H_0(\rho; I'/I)$$

and (since $R\cdot\pi$ generates $R[\pi]^t$ as an additive group)

$$K_1(\bar{I}/\bar{I}_0) \cong (\bar{I}/\bar{I}_0)/[R[\pi]^t/\bar{I}_0, \bar{I}/\bar{I}_0] \cong H_0(\pi; I/I_0) \oplus H_0(R\cdot\pi; \tilde{I}/\tilde{I}_0).$$

In particular, this shows that α_{I/I_0} is a monomorphism. But α_{I/I_0} is surjective by Lemma 12.1(ii), and is hence an isomorphism.

By Example 1.12, $J = \langle p, g-1 : g\in\rho\rangle$ (as an $R[\rho]$-ideal); and the Jacobson radical $\bar{J} \subseteq R[\pi]^t$ has the same generators as an ideal in $R[\pi]^t$. Hence, by Theorem 3.3, $K_2(\bar{I}/\bar{I}_0)$ is generated by symbols $\{1+p, v\}$ and $\{g, v\}$ for $g \in \rho$ and $v \in 1+\bar{I}/\bar{I}_0$. To show that β_{I/I_0} is onto, it will thus suffice to show that $\{u, 1+\xi g\} = 1$ whenever $u \in (R[\rho])^*$, $\xi \in I/I_0$, and $g \in \pi\smallsetminus\rho$. As in the proof of Lemma 12.1, choose $r \in R$ such that $t(g)(r) \not\equiv r \pmod{pR}$ $(\mathrm{Gal}(F/\hat{\mathbb{Q}}_p) \cong \mathrm{Gal}((R/pR)/\mathbb{F}_p))$; and set

$$s = r^{-1}\cdot t(g)(r) - 1 = r^{-1}\cdot grg^{-1} - 1 \in R^*.$$

Then, since $rur^{-1} = u$,

$$\{u, 1+\xi g\} = \{u\, ,\, 1 + s^{-1}(r^{-1}\cdot grg^{-1} - 1)\xi g\}$$

$$= \{u\, ,\, 1 + r^{-1}(s^{-1}\xi g)r\}\cdot\{u\, ,\, 1 + s^{-1}\xi g\}^{-1} = 1.$$

<u>Step 2</u> In order to prove (i), we first show that α_{I/I_0} is an isomorphism for any pair $I_0 \subseteq I \subseteq R[\rho]$ of π-invariant radical ideals such that $[I:I_0] < \infty$. This will be done by induction on $|I/I_0|$. Fix $I_0 \subseteq I$, set $I_1 = I_0 + IJ + JI$ (so $I_0 \subseteq I_1 \subseteq I$), and consider the following diagram:

$$\begin{array}{ccccccccc}
K_2(I/I_1) & \longrightarrow & H_0(\bar{\pi};K_1(I_1/I_0)) & \longrightarrow & H_0(\bar{\pi};K_1(I/I_0)) & \longrightarrow & H_0(\bar{\pi};K_1(I/I_1)) & \rightarrow & 1 \\
\downarrow{\scriptstyle \beta_{I/I_1}} & & \cong\downarrow{\scriptstyle \alpha_{I_1/I_0}} & & \downarrow{\scriptstyle \alpha_{I/I_0}} & & \cong\downarrow{\scriptstyle \alpha_{I/I_1}} & & \\
K_2(\bar{I}/\bar{I}_1) & \longrightarrow & K_1(\bar{I}_1/\bar{I}_0) & \longrightarrow & K_1(\bar{I}/\bar{I}_0) & \longrightarrow & K_1(\bar{I}/\bar{I}_1) & \rightarrow & 1.
\end{array} \tag{1}$$

The top row is exact, except possibly at $H_0(\bar{\pi};K_1(I_1/I_0))$. By Step 1, α_{I/I_1} is an isomorphism and β_{I/I_1} is onto. Also, α_{I_1/I_0} is an isomorphism by the induction hypothesis, and so α_{I/I_0} is an isomorphism by diagram (1).

Now, for any π-invariant radical ideal $I \subseteq R[\rho]$,

$$K_1(R[\rho],I) \cong \varprojlim K_1(I/I_0) \quad \text{and} \quad K_1(R[\pi]^t,\bar{I}) \cong \varprojlim K_1(\bar{I}/\bar{I}_0)$$

by Theorem 2.10(iii), where the limits are taken over all $I_0 \subseteq I$ of finite index. Also, $H_0(\bar{\pi};-)$ commutes with the inverse limits, since the $K_1(I/I_0)$ are finite. Since the α_{I/I_0} are all isomorphisms, $\alpha_I = \varprojlim \alpha_{I/I_0}$ is also an isomorphism.

<u>Step 3</u> Now assume that $SK_1(R[\rho]) = 1$. By Lemma 12.2(ii), there is a sequence

$$R[\rho] \supseteq J = I_1 \supseteq I_2 \supseteq \cdots$$

of π-invariant ideals, such that $JI_{k-1}+I_{k-1}J \subseteq I_k$ for all k, such that $\bigcap_{k=1}^{\infty} I_k = 0$, and such that $K_1(R[\rho]/I_k)$ is $\bar{\pi}$-cohomologically trivial for all k. We claim that $\beta_{R[\rho]/I_k}$ is surjective for all k; this is clear when $k = 1$ since $K_2(R[\rho]/J) = 1$ (Theorem 1.16).

Fix $k \geq 0$, and assume inductively that $\beta_{R[\rho]/I_{k-1}}$ is onto. Consider the following diagram:

$$\begin{array}{ccccccc}
K_2(I_{k-1}/I_k) & \to & K_2(R[\rho]/I_k) & \to & H_0(\bar{\pi};K_2(R[\rho]/I_{k-1})) & \to & H_0(\bar{\pi};K_1(I_{k-1}/I_k)) \\
\downarrow{\scriptstyle \beta_{I_{k-1}/I_k}} & & \downarrow{\scriptstyle \beta_{R[\rho]/I_k}} & & \downarrow{\scriptstyle \beta_{R[\rho]/I_{k-1}}} & & \cong\downarrow{\scriptstyle \alpha_{I_{k-1}/I_k}} \\
K_2(\bar{I}_{k-1}/\bar{I}_k) & \to & K_2(R[\pi]^t/\bar{I}_k) & \to & K_1(R[\pi]^t/\bar{I}_{k-1}) & \longrightarrow & K_1(\bar{I}_{k-1}/\bar{I}_k)
\end{array} \qquad (2)$$

Since the last two terms in the exact sequence

$$K_2(R[\rho]/I_{k-1}) \longrightarrow K_1(I_{k-1}/I_k) \longrightarrow K_1(R[\rho]/I_k) \longrightarrow K_1(R[\rho]/I_{k-1}) \longrightarrow 1$$

are $\bar{\pi}$-cohomologically trivial, by assumption, the top row in (2) is exact at $H_0(\bar{\pi};K_2(R[\rho]/I_{k-1}))$. Also, by Step 1, α_{I_{k-1}/I_k} is an isomorphism and β_{I_{k-1}/I_k} is onto. We have assumed inductively that $\beta_{R[\rho]/I_{k-1}}$ is onto, and so the same holds for $\beta_{R[\rho]/I_k}$.

In particular, in the limit, $\text{ind}_{K2} = \varprojlim \beta_{R[\rho]/I_k}$ is onto. □

Under certain circumstances, twisted group rings actually become matrix rings. This is the idea behind the next lemma.

__Lemma 13.2__ *Let* $R = \prod_{i=1}^n R_i \subseteq S$ *be rings, and let* $\pi \subseteq S^*$ *be a subgroup such that* $gRg^{-1} = R$ *for all* $g \in \pi$, *and such that this conjugation action of* π *permutes the* R_i *transitively. Assume that* π *generates* S *as a right* R*-module, and that* $gR_1 = R_1$ *for any* $g \in \pi$ *such that* $gR_1g^{-1} = R_1$. *Then there is an isomorphism* $\alpha\colon S \xrightarrow{\cong} M_n(R_1)$ *which sends* R *to the diagonal. More precisely, if* $g_1,\ldots,g_n \in \pi$ *are such that* $g_iR_1g_i^{-1} = R_i$, *then* α *can be defined such that for any* $r = (r_1,\ldots,r_n) \in R$ $(r_i \in R_i)$,

$$\alpha(r) = \alpha(r_1,\ldots,r_n) = \text{diag}(g_1^{-1}r_1g_1,\ldots,g_n^{-1}r_ng_n).$$

__Proof__ Fix elements $g_1,\ldots,g_n \in \pi$, and central idempotents $e_1,\ldots,e_n \in R$, such that $g_iR_1g_i^{-1} = R_i$ and $R_i = Re_i$ for each i. In particular, if $\pi' = \{g \in \pi\colon gR_1g^{-1} = R_1\}$, then the g_i are left coset representatives for π' in π. We first claim that $\{g_1e_1,\ldots,g_ne_1\}$ is a basis for Se_1 as a right R_1-module. The elements generate by assumption ($gR_1 = R_1$ for $g \in \pi'$). To see that the g_ie_i are linearly independent, note for any $r_1,\ldots,r_n \in R_1$ such that $\sum_i g_ir_i = 0$, that

$$e_ig_jr_j = e_i\cdot(g_jr_jg_j^{-1})\cdot g_j = \begin{cases} g_jr_j & \text{if } i = j \\ 0 & \text{if } i \neq j \end{cases} \qquad (g_jr_jg_j^{-1} \in g_jR_1g_j^{-1} = R_j);$$

so that for all i, $g_ir_i = e_ig_ir_i = e_i\cdot\sum_j g_jr_j = 0$.

In particular, if we consider Se_1 as an (S,R_1)-bimodule, this induces a homomorphism

$$\alpha : S \longrightarrow \mathrm{End}_{R_1}(Se_1) \cong M_n(R_1).$$

Furthermore,

$$S = \bigoplus_{i=1}^{n} Se_i = \bigoplus_{i=1}^{n} S\cdot g_ie_1g_i^{-1} = \bigoplus_{i=1}^{n} Se_1\cdot g_i^{-1} = \bigoplus_{i=1}^{n}\bigoplus_{j=1}^{n} g_jR_1g_i^{-1};$$

and for all i, j, and k,

$$\alpha(g_jR_1g_i^{-1})(g_kR_1) = \begin{cases} g_jR_1 & \text{if } k = i \\ 0 & \text{if } k \neq i. \end{cases} \qquad (e_1g_i^{-1}g_ke_1 = g_i^{-1}e_ie_kg_k = 0)$$

This shows that α is an isomorphism. The formula for $\alpha|R$ is clear. □

As is suggested by Theorem 11.9, the goal now is to compare $SK_1(R[\pi]^t)$, where $R[\pi]^t$ is a global twisted group ring and $\rho = \mathrm{Ker}(t)$, with $SK_1(R[\rho])$. The next proposition does this for the other terms in the localization sequence of Theorem 3.15.

Recall the definition of Steinberg symbols in Section 3a:

$$\{u,v\} = \left[\phi^{-1}(\mathrm{diag}(u,u^{-1},1))\ ,\ \phi^{-1}(\mathrm{diag}(v,1,v^{-1}))\right] \in St(R)$$

for any (not necessarily commuting) $u,v \in R^*$. Here, $\phi\colon St(R) \longrightarrow E(R)$

denotes the standard projection. It will be convenient here to extend this by defining, for arbitrary $n \geq 1$ and arbitrary matrices $u,v \in GL_n(R)$, $\{u,v\} \in St(M_n(R)) = St(R)$.

Proposition 13.3 *Fix a prime* p, *a* p-*group* π, *a number field* K *in which* p *is unramified, and a homomorphism* $t: \pi \longrightarrow Gal(K/\mathbb{Q})$. *Let* $R \subseteq K$ *be the ring of integers, set* $\rho = Ker(t)$, *and let* $K[\pi]^t$ *and* $R[\pi]^t$ *be the induced twisted group rings. Let*

$$i_{Cp} : H_0(\pi/\rho;C_p(K[\rho])) \longrightarrow C_p(K[\pi]^t), \quad i_{K2} : K_2^c(\hat{R}_p[\rho]) \longrightarrow K_2^c(\hat{R}_p[\pi]^t),$$

$$i_{SKp} : H_0(\pi/\rho;SK_1(\hat{R}_p[\rho])) \longrightarrow SK_1(\hat{R}_p[\pi]^t)$$

be the homomorphisms induced by the inclusion $R[\rho] \subseteq R[\pi]^t$. *Then*

(i) i_{Cp} *is surjective, and is an isomorphism if* p *is odd, or if* $p = 2$ *and* K^π *has no real embedding;*

(ii) i_{SKp} *is an isomorphism; and*

(iii) *there is an isomorphism*

$$\alpha : H_1(\pi/\rho;SK_1(\hat{R}_p[\rho])) \xrightarrow{\cong} Coker(i_{K2}).$$

In (iii), for any $\xi = \sum_{i=1}^k g_i \otimes a_i \in H_1(\pi/\rho;SK_1(\hat{R}_p[\rho]))$, *where* $g_i \in \pi$, $a_i \in SK_1(\hat{R}_p[\rho])$, *and* $\prod(g_i(a_i)\cdot a_i^{-1}) = 1$, $\alpha(\xi)$ *is defined as follows. Fix matrices* $u_i \in GL(\hat{R}_p[\rho])$ *which represent the* a_i, *and write*

$$[g_1,u_1][g_2,u_2]\cdots[g_k,u_k] = \phi(X) \in E(\hat{R}_p[\rho])$$

for some $X \in St(\hat{R}_p[\rho])$. *Then*

$$\alpha(\xi) = \alpha(\sum_{i=1}^k g_i \otimes a_i) = \{g_1,u_1\}\cdot\{g_2,u_2\}\cdots\{g_k,u_k\}\cdot X^{-1} \in K_2^c(\hat{R}_p[\pi]^t).$$

Proof Point (i) will be proven in Step 1. Point (ii), and point (iii) when $SK_1(\hat{R}_p[\pi]) = 1$, are then shown in Step 2; and the general case of (iii) is shown in Step 3. Recall (Theorem 1.7) that $\hat{K}_p \cong \prod_{p|p}\hat{K}_p$ and $\hat{R}_p \cong \prod_{p|p}\hat{R}_p$, where the products are taken over all prime ideals in R which divide p.

Step 1 Set $n = |\pi/\rho|$ and $K_0 = K^\pi$, for short. Then $K\otimes_{K_0}K$ is a product of n copies of K, $K\otimes_{K_0}K[\rho] \cong (K[\rho])^n$; and the factors are permuted transitively under the conjugation action of π. By Lemma 13.2,

$$K\otimes_{K_0}K[\pi]^t \cong M_n(K[\rho]),$$

and the composite $C_p(K\otimes_{K_0}K[\rho]) \longrightarrow C_p(K\otimes_{K_0}K[\pi]^t) \cong C_p(K[\rho])$ is the transfer map. In the following commutative diagram:

$$\begin{array}{ccc} C_p(K\otimes_{K_0}K[\rho]) & \xrightarrow{\ \mathrm{trf}\ } & C_p(K[\rho]) \cong C_p(K\otimes_{K_0}K[\pi]^t) \\ \Big\downarrow \mathrm{trf} & & \Big\downarrow \mathrm{trf} \\ C_p(K[\rho]) & \xrightarrow{\ \ i_{Cp}\ \ } & C_p(K[\pi]^t), \end{array}$$

the transfer maps are all onto by Lemma 4.17, and so i_{Cp} is onto.

If p is odd, or if $p = 2$ and $K_0 = K^\pi$ has no real embedding, then by Lemma 4.17 again,

$$C_p(K[\pi]^t) \cong H_0(\mathrm{Gal}(K/K_0);C_p(K\otimes_{K_0}K[\pi]^t)) \cong H_0(\mathrm{Gal}(K/K_0);C_p(K[\rho])).$$

To see this when $p = 2$, note that Conjecture 4.14 holds for $K[\pi]^t$ by Theorems 1.10(ii) and 4.13(ii): since $\hat{K}_2[\pi]^t \cong \prod_{p|2}\hat{K}_p[\pi]^t$, and each factor is a summand of some 2-adic group ring of the form $F[C_n \rtimes \pi]$. By the description of the isomorphism $K\otimes_{K_0}K[\pi]^t \cong M_n(K[\rho])$ in Lemma 13.2, the action of $\mathrm{Gal}(K/K_0)$ on $C_p(K[\rho])$ is just the conjugation action of π/ρ. So i_{Cp} is an isomorphism in this case.

<u>Step 2</u> Fix some prime $\mathfrak{p}|p$ in R, and let $\mathfrak{p}=\mathfrak{p}_1,\ldots,\mathfrak{p}_n$ be the orbit of $\mathfrak{p}$ under the conjugation action of π. Let $\pi_i' \subseteq \pi$ be the stabilizer of $\mathfrak{p}_i$ — $\pi_i' = \{g \in \pi:\ g_i\mathfrak{p}_i g_i^{-1} = \mathfrak{p}_i\}$ — and set $\mathfrak{q} = \mathfrak{p}_1\cdots\mathfrak{p}_n$. Then $\hat{R}_{\mathfrak{q}} = \prod_{i=1}^n \hat{R}_{\mathfrak{p}_i}$. Consider the following homomorphisms:

$$H_0(\pi/\rho;SK_1(\hat{R}_{\mathfrak{q}}[\rho])) \xrightarrow{\ f_1\ } H_0(\pi/\rho;\ \bigoplus_{i=1}^{n} SK_1(\hat{R}_{\mathfrak{p}_i}[\pi_i']^t)) \xrightarrow{\ f_2\ } SK_1(\hat{R}_{\mathfrak{q}}[\pi]^t).$$

Here, f_1 is an isomorphism by Theorem 12.3; and f_2 is an isomorphism Lemma 13.2: since $\hat{R}_{\mathfrak{q}}[\pi]^t \cong M_n(\hat{R}_{\mathfrak{p}}[\pi_1']^t)$, and the inclusion $\prod_{i=1}^n \hat{R}_{\mathfrak{p}_i}[\pi_i']^t \subseteq \hat{R}_{\mathfrak{q}}[\pi]^t$ is the inclusion of the diagonal. If $SK_1(\hat{R}_{\mathfrak{p}}[\rho]) = 1$, then by a similar argument,

$$K_2^c(\hat{R}_{\mathfrak{q}}[\rho]) \cong \bigoplus_i K_2^c(\hat{R}_{\mathfrak{p}_i}[\rho]) \twoheadrightarrow \bigoplus_i K_2^c(\hat{R}_{\mathfrak{p}_i}[\pi_i']^t) \twoheadrightarrow K_2^c(\hat{R}_{\mathfrak{q}}[\pi]^t)$$

are surjections by Lemma 13.1. After summing over all π-orbits of primes $\mathfrak{p}|p$ in R, this shows that i_{SKp} is an isomorphism, and that i_{K2} is surjective if $SK_1(\hat{R}_p[\rho]) = 1$.

<u>Step 3</u> Now assume that $SK_1(\hat{R}_p[\rho]) \neq 1$. Using Lemma 8.3(ii), choose an extension

$$1 \longrightarrow \sigma \longrightarrow \tilde{\pi} \xrightarrow{\ \alpha\ } \pi \longrightarrow 1 \qquad \text{where} \qquad \tilde{\rho} = \alpha^{-1}(\rho), \quad \alpha_0 = \alpha|\tilde{\rho},$$

such that $\sigma \subseteq Z(\tilde{\rho})$ and $H_2(\alpha_0) = 0$. Then $SK_1(\hat{R}_p[\tilde{\rho}]) = 1$ by Lemma 8.9. Consider the following commutative diagram:

$$\begin{array}{ccccccc}
K_2^c(\hat{R}_p[\tilde{\rho}]) & \longrightarrow & K_2^c(\hat{R}_p[\rho]) & \xrightarrow{\ \beta\ } & H_0(\pi/\rho;K_1(\hat{R}_p[\tilde{\rho}],I)) & \xrightarrow{\ \gamma\ } & H_0(\pi/\rho;K_1(\hat{R}_p[\tilde{\rho}])) \\
\downarrow{\scriptstyle i_3} & & \downarrow{\scriptstyle i_2} & & {\scriptstyle\cong}\downarrow{\scriptstyle i_4} & & {\scriptstyle\cong}\downarrow{\scriptstyle i_5} \\
K_2^c(\hat{R}_p[\tilde{\pi}]^t) & \longrightarrow & K_2^c(\hat{R}_p[\pi]^t) & \xrightarrow{\ \partial\ } & K_1(\hat{R}_p[\tilde{\pi}]^t,\bar{I}) & \longrightarrow & K_1(\hat{R}_p[\tilde{\pi}]^t).
\end{array}$$

Here, $I = \mathrm{Ker}[\hat{R}_p[\tilde{\rho}] \to \hat{R}_p[\rho]]$ and $\bar{I} = \mathrm{Ker}[\hat{R}_p[\tilde{\pi}]^t \to \hat{R}_p[\pi]^t]$. Then i_3

is onto by Step 2 ($SK_1(\hat{R}_p[\tilde{\rho}]) = 1$), i_4 is an isomorphism by Lemma 13.1(i), and i_6 is an isomorphism by Theorem 12.3. Furthermore, $K_1'(\hat{R}_p[\rho])$ and $K_1(\hat{R}_p[\tilde{\rho}])$ ($\cong K_1'(\hat{R}_p[\tilde{\rho}])$) are π/ρ-cohomologically trivial by Lemma 12.2(i); and so a diagram chase gives isomorphisms

$$\mathrm{Coker}(i_2) \cong \mathrm{Ker}(\gamma)/\mathrm{Im}(\beta) \cong H_1(\pi/\rho;K_1(\hat{R}_p[\rho])) \cong H_1(\pi/\rho;SK_1(\hat{R}_p[\rho])).$$

To check the formula for $\alpha(\xi) = \alpha(\sum g_i \otimes a_i)$, lift g_i, u_i, and X to $\tilde{g}_i \in \tilde{\pi}$, $\tilde{u}_i \in GL(\hat{R}_p[\tilde{\rho}])$, and $\tilde{X} \in St(\hat{R}_p[\tilde{\rho}])$. Then ξ lifts to

$$[\tilde{g}_1,\tilde{u}_1]\cdots[\tilde{g}_k,\tilde{u}_k]\cdot\phi(\tilde{X})^{-1} \in K_1(\hat{R}_p[\tilde{\rho}],I);$$

and as an element of $K_1(\hat{R}_p[\tilde{\pi}]^t,\bar{I})$ this pulls back to

$$\alpha(\xi) = \{g_1,u_1\}\cdots\{g_k,u_k\}\cdot X^{-1} \in K_2^c(\hat{R}_p[\pi]^t). \qquad \square$$

Now recall the functor $SK_1^{[p]}$ of Theorem 11.10. This was defined so that for any $\mathbb{Z}$-order $\mathfrak{A}$, there is a short exact sequence

$$1 \longrightarrow Cl_1(\mathfrak{A})_{(p)} \longrightarrow SK_1^{[p]}(\mathfrak{A}) \longrightarrow SK_1(\hat{\mathfrak{A}}_p)_{(p)} \longrightarrow 1.$$

By Theorem 3.14, $SK_1^{[p]}(R[G]) = SK_1(R[G])_{(p)}$ whenever G is a finite group and R is the ring of integers in a number field.

Theorem 13.4 *Fix a prime* p *and a number field* K *where* p *is unramified, and let* $R \subseteq K$ *be the ring of integers. Let* π *be a* p*-group, fix a homomorphism* $t\colon \pi \to \mathrm{Gal}(K/\mathbb{Q})$, *and set* $\rho = \mathrm{Ker}(t)$. *Let* $R[\pi]^t$ *be the induced twisted group ring. Then*

(i) $i_{Cl} : H_0(\pi/\rho;Cl_1(R[\rho])) \twoheadrightarrow Cl_1(R[\pi]^t)_{(p)}$ *is surjective, and is an isomorphism if* p *is odd; and*

(ii) $i_{SK} : H_0(\pi/\rho;SK_1(R[\rho])) \twoheadrightarrow SK_1^{[p]}(R[\pi]^t)$ *is surjective, and is an isomorphism if* p *is odd or if* K^π *has no real embedding.*

Here, i_{Cl} *and* i_{SK} *are induced by the inclusion* $R[\rho] \subsetneq R[\pi]^t$. *In general, the following square is a pullback square:*

$$\begin{array}{ccc}
H_0(\pi/\rho; C_p(K[\rho])) & \xrightarrow{\partial_1} & H_0(\pi/\rho; SK_1(R[\rho])) \\
\Big\downarrow i_{Cp} & & \Big\downarrow i_{SK} \\
C_p(K[\pi]^t) & \xrightarrow{\partial_2} & SK_1^{[p]}(R[\pi]^t).
\end{array} \qquad (1)$$

<u>**Proof**</u> First consider the following two commutative diagrams, whose rows are exact by Theorems 3.9 and 3.15:

$$\begin{array}{ccccccc}
K_2^c(\hat{R}_p[\rho]) & \longrightarrow & H_0(\pi/\rho; C_p(K[\rho])) & \xrightarrow{\tilde{\partial}_1} & H_0(\pi/\rho; Cl_1(R[\rho])) & \longrightarrow & 1 \\
\Big\downarrow i_{K2} & & \Big\downarrow i_{Cp} & (2b) & \Big\downarrow i_{Cl} & & \\
K_2^c(\hat{R}_p[\pi]^t) & \longrightarrow & C_p(K[\pi]^t) & \xrightarrow{\tilde{\partial}_2} & Cl_1(R[\pi]^t)_{(p)} & \longrightarrow & 1
\end{array} \qquad (2)$$

$$\begin{array}{l}
H_1(\pi/\rho; SK_1(\hat{R}_p[\rho])) \xrightarrow{\partial'} \\
\begin{array}{ccccccccc}
 & & H_0(\pi/\rho; Cl_1(R[\rho])) & \xrightarrow{f} & H_0(\pi/\rho; SK_1(R[\rho])) & \longrightarrow & H_0(\pi/\rho; SK_1(\hat{R}_p[\rho])) & \longrightarrow & 1 \\
 & & \Big\downarrow i_{Cl} & (3a) & \Big\downarrow i_{SK} & & \cong \Big\downarrow i_{SKp} & & \\
1 & \longrightarrow & Cl_1(R[\pi]^t)_{(p)} & \longrightarrow & SK_1^{[p]}(R[\pi]^t) & \longrightarrow & SK_1(\hat{R}_p[\pi]^t)_{(p)} & \longrightarrow & 1.
\end{array}
\end{array} \qquad (3)$$

Here, by Proposition 13.3, i_{Cp} is surjective and i_{SKp} is an isomorphism. It follows that i_{Cl} and i_{SK} are surjective.

From the exactness of the rows in (2) and (3), we see that (3a) is a pushout square, and that (2b) is a pushout if i_{K2} is surjective. Thus, the "obstruction" to (2b) being a pushout square is $\mathrm{Coker}(i_{K2}) \cong H_1(\pi/\rho; SK_1(\hat{R}_p[\rho]))$ (Proposition 13.3(iii)). On the other hand, $H_1(\pi/\rho; SK_1(\hat{R}_p[\rho]))$ also occurs in (3), where it generates $\mathrm{Ker}(f)$. So if we somehow can identify these two occurrences of $H_1(\pi/\rho; SK_1(\hat{R}_p[\rho]))$, then (2b) and (3a) will combine to show that (1) is a pushout square.

To make this precise, consider the following square:

$$\begin{array}{ccc} H_1(\pi/\rho;SK_1(\hat{R}_p[\rho])) & \xrightarrow{\partial'} & H_0(\pi/\rho;Cl_1(R[\rho])) \\ \alpha\downarrow\cong & & \downarrow \text{proj} \\ \text{Coker}(i_{K2}) & \xrightarrow{\beta} & H_0(\pi/\rho;Cl_1(R[\rho]))/\tilde{\partial}_1(\text{Ker}(i_{Cp})). \end{array} \qquad (4)$$

Here α is the isomorphism of Proposition 13.3(iii), and β is induced by diagram (2). Assume for the moment that (4) commutes. Then

$$\text{Im}(\text{proj}\circ\partial') = \text{Im}(\beta) = \text{Ker}(i_{Cl})/\tilde{\partial}_1(\text{Ker}(i_{Cp})).$$

It follows that

$$\text{Ker}(i_{Cl}) = \tilde{\partial}_1(\text{Ker}(i_{Cp})) + \text{Im}(\partial') = \tilde{\partial}_1(\text{Ker}(i_{Cp})) + \text{Ker}(f);$$

and hence from (3) that

$$\text{Ker}(i_{SK}) = f(\text{Ker}(i_{Cl})) = f\circ\tilde{\partial}_1(\text{Ker}(i_{Cp})) = \partial_1(\text{Ker}(i_{Cp})).$$

Since $\text{Coker}(\partial_1) \cong \text{Coker}(\partial_2)$ by (3), this shows that (1) is a pushout square.

If p is odd, or if $p = 2$ and K^π has no real embedding, then i_{Cp} is an isomorphism by Proposition 13.3(i), and hence i_{SK} is also an isomorphism. If p is odd, then $\partial' = 1$ in (3) — the standard involution fixes $Cl_1(R[\rho])$ (Theorem 5.12) and negates $SK_1(\hat{R}_p[\rho])$ (Theorem 8.6) — and so i_{Cl} is an isomorphism.

It remains to prove that (4) commutes. Fix

$$\xi = \sum_{i=1}^{k} g_i \otimes a_i \in H_1(\pi/\rho;SK_1(\hat{R}_p[\rho])) \qquad \left(\text{so } \prod(g_i(a_i)\cdot a_i^{-1}) = 1 \in SK_1(\hat{R}_p[\rho])\right);$$

and represent each a_i by some $u_i \in GL(R[\rho])$ ($SK_1(R[\rho])$ surjects onto $SK_1(\hat{R}_p[\rho])$ by Theorem 3.9). Write

$$[g_1,u_1]\cdots[g_k,u_k] = \phi(X) \in E(\hat{R}_p[\rho]) \qquad (X \in St(\hat{R}_p[\rho])).$$

By Proposition 13.3(iii),

$$\alpha(\xi) = \{g_1,u_1\}\cdots\{g_k,u_k\}\cdot X^{-1} \in K_2^c(\hat{R}_p[\pi]^t).$$

Also, $SK_1(R[\frac{1}{p}][\rho]) = 1$ by Theorems 4.15 and 3.14, so the u_i can be lifted to $x_i \in St(R[\frac{1}{p}][\rho])$. Then $\alpha(\xi)$ lifts to

$$\eta = g_1(x_1)\cdot x_1^{-1}\cdot g_2(x_2)\cdot x_2^{-1}\cdots g_k(x_k)\cdot x_k^{-1}\cdot X^{-1} \in K_2^c(\hat{R}_p[\rho]);$$

and the elements

$$g_1(x_1)\cdot x_1^{-1}\cdot g_2(x_2)\cdot x_2^{-1}\cdots g_k(x_k)\cdot x_k^{-1} \in St(R[\tfrac{1}{p}][\rho]), \qquad X \in St(\hat{R}_p[\rho])$$

are both liftings of $\prod[g_i,u_i] \in GL(R[\rho])$. From the description of $\tilde{\partial}_1$ in Theorem 3.12, it now follows that

$$\beta\circ\alpha(\xi) = \tilde{\partial}_1(\eta) = [g_1,u_1]\cdots[g_k,u_k] = \prod_{i=1}^{k}(g_i(u_i)\cdot u_i^{-1})$$

$$= \partial'(\xi) \in H_0(\pi/\rho;Cl_1(R[\rho])). \qquad \square$$

Diagram (1) above need not be a pushout square if SK_1 is replaced by Cl_1 (when $p = 2$). This is the basis of Example 13.16 in Section 13c.

Recall that a 2-hyperelementary group $C_n \rtimes \pi$ ($2\nmid n$, π a 2-group) is *2-$\mathbb{R}$-elementary* if $\operatorname{Im}[\pi \xrightarrow{\text{conj}} \operatorname{Aut}(C_n) \cong (\mathbb{Z}/n)^*] \subseteq \{\pm 1\}$. Theorems 13.4, 11.9, and 11.10 now combine to show:

<u>Theorem 13.5</u> *For any finite group* G, $Cl_1(\mathbb{Z}[G])$ *and* $SK_1(\mathbb{Z}[G])$ *are generated by induction from elementary subgroups of* G. *For any odd prime* p, $Cl_1(\mathbb{Z}[G])_{(p)}$ *and* $SK_1(\mathbb{Z}[G])_{(p)}$ *are p-elementary computable;*

while for $p = 2$, $SK_1(\mathbb{Z}[G])_{(2)}$ *is 2-elementary generated and 2-$\mathbb{R}$-elementary computable. Also, if* $\mathcal{E}$ *denotes the set of 2-elementary subgroups of* G, *then the following is a pushout square:*

$$\begin{array}{ccc} \varinjlim_{H\in\mathcal{E}} C_2(\mathbb{Q}[H]) & \xrightarrow{\ \partial\ } & \varinjlim_{H\in\mathcal{E}} SK_1(\mathbb{Z}[H])_{(2)} \\ \downarrow & & \downarrow \\ C_2(\mathbb{Q}[G]) & \longrightarrow & SK_1(\mathbb{Z}[G])_{(2)}. \end{array} \qquad (1)$$

__Proof__ For odd p, this is an immediate consequence of Theorems 13.4 and 11.9. As for 2-torsion, square (1) is a pushout square by Theorem 13.4 and the decomposition formula for $SK_1^{[2]}$ of Theorem 11.10. Note that direct limits are right exact, so a direct limit of pushout squares is again a pushout square.

Recall the formula

$$C(\mathbb{Q}[G]) \cong \left[R_{\mathbb{C}/\mathbb{R}}(G) \otimes \mathbb{Z}/n\right]_{(\mathbb{Z}/n)^*}$$

of Lemma 5.9: where $R_{\mathbb{C}/\mathbb{R}}(G) = R_{\mathbb{C}}(G)/R_{\mathbb{R}}(G)$, $2|n$, and $\exp(G)|n$. The functor $R_{\mathbb{C}/\mathbb{R}}(-)_{(2)}$ is 2-$\mathbb{R}$-elementary computable by Theorem 11.2. Tensoring by $\mathbb{Z}/n$ and taking coinvariants are both right exact functors, so they commute with direct limits; and $C_2(\mathbb{Q}[G]) = C(\mathbb{Q}[G])_{(2)}$ is thus 2-$\mathbb{R}$-elementary computable. Square (1) remains a pushout if the limits are taken over 2-$\mathbb{R}$-elementary subgroups; and so $SK_1(\mathbb{Z}[G])_{(2)}$ is also computable with respect to induction from 2-$\mathbb{R}$-elementary subgroups of G. □

Square (1) above need not be a pushout square if $SK_1(\mathbb{Z}[G])_{(2)}$ is replaced by $Cl_1(\mathbb{Z}[G])_{(2)}$; and $Cl_1(\mathbb{Z}[G])_{(2)}$ is not in general 2-$\mathbb{R}$-elementary computable. Counterexamples to both of these are constructed in Example 13.16 below.

What would be more useful, of course, would be a result that $Cl_1(\mathbb{Z}[G])$ and $SK_1(\mathbb{Z}[G])$ were *detected* by restriction to elementary subgroups. Unfortunately, just as was the case for $SK_1(\hat{\mathbb{Z}}_p[G])$ (Example

12.6), $Cl_1(\mathbb{Z}[G])_{(p)}$ is not in general p-elementary detected.

13b. Reduction to p-groups

The goal now is to compare $Cl_1(R[\pi])$ to $Cl_1(\mathbb{Z}[\pi])$, whenever π is a p-group and R is the ring of integers in any algebraic number field K in which p is unramified. The main results in this section are that $Cl_1(R[\pi]) \cong Cl_1(\mathbb{Z}[\pi])$ if p is odd (Theorem 13.8); and that when $p = 2$, $Cl_1(R[\pi])$ is isomorphic to one of the groups $Cl_1(\mathbb{Z}[\pi])$, $Cl_1(\mathbb{Z}\zeta_7[\pi])$, or $Cl_1(\mathbb{Z}\zeta_3[\pi])$ (Theorem 13.10). The differences between these last three groups (when π is a 2-group) are examined in Theorems 13.11 and 13.12. When p is odd or G is abelian, these results then allow a complete reduction of the computation of $Cl_1(\mathbb{Z}[G])_{(p)}$ to the p-group case.

The main problem here is to get control over the relationship between $K_2^c(\hat{R}_p[\pi])$ and $K_2^c(\hat{\mathbb{Z}}_p[\pi])$ in the above situation. In fact, these two groups can be compared using Proposition 13.3(iii) from the last section. But first, some new homomorphisms, which connect $K_2^c(\hat{\mathbb{Z}}_p[\pi])$ with $H_2(\pi)$, must be defined.

For any group π,

$$\lambda_\pi : H_2(\pi) \longrightarrow K_2(\mathbb{Z}[\pi])/\{-1,\pi\}$$

will denote the homomorphism constructed by Loday [1]. One way to define λ_π is to fix any extension $1 \longrightarrow R \longrightarrow F \xrightarrow{\alpha} \pi \longrightarrow 1$, where F is the free group on elements $a_2,\ldots,a_n$; and let

$$\Lambda: F \longrightarrow St(\mathbb{Z}[\pi])$$

be the homomorphism defined by setting $\Lambda(a_i) = h_{1i}(\alpha(a_i))$. In particular, $\phi(\Lambda(a_i)) \in E(\mathbb{Z}[\pi])$ is a diagonal matrix with entries $\alpha(a_i)$ and $\alpha(a_i)^{-1}$ in the first and i-th positions (and 1 elsewhere). Then for any $a \in R$, $\phi(\Lambda(a)) = \mathrm{diag}(1,\alpha(a_2)^{j_2},\ldots,\alpha(a_n)^{j_n})$ for some $j_i \in \mathbb{Z}$; and so $\Lambda([R,F]) \subseteq \langle\{g,g\} = \{-1,g\}: g \in G\rangle = \{-1,\pi\}$ by Theorem 3.1(i,iv). Also, $\Lambda(R\cap[F,F]) \subseteq \mathrm{Ker}(\phi) = K_2(\mathbb{Z}[\pi])$, and so Λ induces a homomorphism

$$\lambda_\pi : H_2(\pi) \cong (R\cap[F,F])/[R,F] \longrightarrow K_2(\mathbb{Z}[\pi])/\{-1,\pi\}.$$

Note that $\lambda_\pi(g\wedge h) = \{g,h\}$ for any commuting pair $g,h \in \pi$.

When π is a p-group for some prime p, we let $\hat{\lambda}_\pi$ denote the composite

$$\hat{\lambda}_\pi : H_2(\pi) \xrightarrow{\lambda_\pi} K_2(\mathbb{Z}[\pi])/\{-1,\pi\} \xrightarrow{(\mathbb{Z}\hookrightarrow\hat{\mathbb{Z}}_p)} K_2^c(\hat{\mathbb{Z}}_p[\pi])/\{-1,\pm\pi\}.$$

A splitting map for $\hat{\lambda}_\pi$ is constructed in the next lemma. This map $\theta_\pi: K_2^c(\hat{\mathbb{Z}}_p[\pi]) \longrightarrow H_2(\pi)$ can be thought of as a K_2 version of the homomorphism

$$\upsilon_\pi : K_1(\hat{\mathbb{Z}}_p[\pi]) \longrightarrow \pi^{ab}$$

of Theorem 6.7: defined by setting $\upsilon_\pi(\sum r_ig_i) = \left(\prod g_i^{r_i}\right)^{1/\sum r_i}$ for any unit $\sum r_ig_i \in (\hat{\mathbb{Z}}_p[\pi])^*$. One can also define θ_π using Dennis' trace map from K-theory to Hochschild homology (see Igusa [1]); but for the purposes here the following (albeit indirect) construction is the easiest to use.

<u>Lemma 13.6</u> *Fix a prime* p *and a p-group* π. *Then there is a unique homomorphism*

$$\theta = \theta_\pi : K_2^c(\hat{\mathbb{Z}}_p[\pi]) \longrightarrow H_2(\pi);$$

such that for any central extension $1 \longrightarrow \sigma \longrightarrow \tilde{\pi} \xrightarrow{\alpha} \pi \longrightarrow 1$ *of p-groups, the following diagram commutes:*

$$\begin{array}{ccccc} K_2^c(\hat{\mathbb{Z}}_p[\pi]) & \xrightarrow{\partial} & K_1(\hat{\mathbb{Z}}_p[\tilde{\pi}], I_\alpha) & \longrightarrow & K_1(\hat{\mathbb{Z}}_p[\tilde{\pi}]) \\ \downarrow{\scriptstyle \theta_\pi} & & \downarrow{\scriptstyle \upsilon_\alpha} & & \downarrow{\scriptstyle \upsilon_{\tilde{\pi}}} \\ H_2(\pi) & \xrightarrow{\delta^\alpha} & \sigma & \longrightarrow & \tilde{\pi}^{ab}. \end{array} \tag{1}$$

Here, $I_\alpha = \mathrm{Ker}\left[\hat{\mathbb{Z}}_p[\tilde{\pi}] \to \hat{\mathbb{Z}}_p[\pi]\right]$, $\upsilon_{\tilde{\pi}}$ *is the map defined above, and for* $r_i \in \hat{\mathbb{Z}}_p$, $g_i \in \tilde{\pi}$, *and* $z_i \in \sigma$,

$$\upsilon_\alpha\left(1 + \sum r_i(z_i-1)g_i\right) = \prod z_i^{r_i}.$$

In addition, the following two relations hold for θ_π:

(i) θ_π *factors through* $K_2^c(\hat{\mathbb{Z}}_p[\pi])/\{-1,\pm\pi\}$, *and the composite*

$$H_2(\pi) \xrightarrow{\hat{\lambda}_\pi} K_2^c(\hat{\mathbb{Z}}_p[\pi])/\{-1,\pm\pi\} \xrightarrow{\theta_\pi} H_2(\pi)$$

is the identity.

(ii) *For any* $g \in \pi$, *any* $\rho \subseteq \pi$ *such that* $[g,\rho] = 1$, *and any* $u \in (\hat{\mathbb{Z}}_p[\rho])^*$, $\theta_\pi(\{g,u\}) = g\wedge\upsilon_\rho(u)$.

<u>Proof</u> For any central extension $1 \longrightarrow \sigma \longrightarrow \tilde{\pi} \xrightarrow{\alpha} \pi \longrightarrow 1$, let I_α be as above, and let $I = \mathrm{Ker}\left[\hat{\mathbb{Z}}_p[\tilde{\pi}] \twoheadrightarrow \hat{\mathbb{Z}}_p\right]$ be the augmentation ideal. Then υ_α factors as a composite

$$K_1(\hat{\mathbb{Z}}_p[\tilde{\pi}], I_\alpha) \xrightarrow{\text{proj}} K_1(\hat{\mathbb{Z}}_p[\tilde{\pi}]/II_\alpha, I_\alpha/II_\alpha) \cong H_0(\tilde{\pi}; I_\alpha/II_\alpha) \xrightarrow{\omega_\alpha} \sigma \tag{2}$$

where the middle isomorphism follows from Theorem 1.15, and where $\omega_\alpha(\sum r_i(z_i-1)g_i) = \prod z_i^{r_i}$ for $r_i \in \hat{\mathbb{Z}}_p$, $z_i \in \sigma$, and $g_i \in G$. In particular, this shows that υ_α is well defined. Since the rows in (1) are exact, θ_π can be defined uniquely to make (1) commute whenever δ^α is injective. There exist central extensions with δ^α injective by Lemma

8.3(i); and two such central extensions are seen to induce the same θ_π by comparing them with their pullback over π.

It remains to prove the last two points.

(i) Let $\tilde{\pi} \xrightarrow{\alpha} \pi$ be such that $\delta^\alpha\colon H_2(\pi) \longrightarrow \sigma = \mathrm{Ker}(\alpha)$ is injective. Fix $x \in H_2(\pi)$, and write

$$\delta^\alpha(x) = [g_1,h_1]\cdots[g_k,h_k] \in \sigma \cap [\tilde{\pi},\tilde{\pi}].$$

By the above definition of $\hat{\lambda}_\pi$ (and Theorem 3.1(iv)):

$$\hat{\lambda}_\pi(x) \equiv \{\alpha(g_1),\alpha(h_1)\}\cdots\{\alpha(g_k),\alpha(h_k)\} \in K_2^c(\hat{\mathbb{Z}}_p[\pi])/\{-1,\pm\pi\}.$$

Then

$$\delta^\alpha(\theta_\pi\circ\hat{\lambda}_\pi(x)) = \nu_\alpha\circ\partial\circ\hat{\lambda}_\pi(x) = \nu_\alpha([g_1,h_1]\cdots[g_k,h_k]) \qquad \Big([g_i,h_i] \in K_1(\hat{\mathbb{Z}}_p[\tilde{\pi}],I_\alpha)\Big)$$

$$= [g_1,h_1]\cdots[g_k,h_k] = \delta^\alpha(x) \in \sigma;$$

and so $\theta_\pi\circ\hat{\lambda}_\pi(x) = x$.

(ii) Now let g and ρ be such that $[g,\rho] = 1$, and fix $u = \sum r_ih_i \in (\hat{\mathbb{Z}}_p[\rho])^*$. Let $1 \longrightarrow \sigma \longrightarrow \tilde{\pi} \xrightarrow{\alpha} \pi \longrightarrow 1$ be as before, choose liftings $\tilde{g},\tilde{h}_i \in \tilde{\pi}$ of g and h_i; and set $\tilde{u} = \sum r_i\tilde{h}_i$. Then $\partial(\{g,u\}) = [\tilde{g},\tilde{u}]$ in diagram (1); and

$$[\tilde{g},\tilde{u}] = 1 + (\tilde{g}\tilde{u}\tilde{g}^{-1} - \tilde{u})\cdot\tilde{u}^{-1} \equiv 1 + (\textstyle\sum r_i)^{-1}(\tilde{g}\tilde{u}\tilde{g}^{-1} - \tilde{u}) \qquad (\mathrm{mod}\ I\cdot I_\alpha),$$

(where I again denotes the augmentation ideal). So

$$\theta_\pi(\{g,u\}) = (\delta^\alpha)^{-1}(\nu_\alpha([\tilde{g},\tilde{u}])) = (\delta^\alpha)^{-1}\Big(\nu_\alpha\Big(1 + (\textstyle\sum r_i)^{-1}(\widetilde{gug^{-1}} - \tilde{u})\Big)\Big) \qquad \text{(by (2))}$$

$$= (\delta^\alpha)^{-1}\Big(\nu_\alpha\Big(1 + (\textstyle\sum r_i)^{-1}\cdot\sum r_i(\tilde{g}\tilde{h}_i\tilde{g}^{-1}-\tilde{h}_i)\Big)\Big)$$

$$= (\delta^\alpha)^{-1}\Big(\textstyle\prod[\tilde{g},\tilde{h}_i^{-1}]^{r_i}\Big)^{1/\Sigma r_i} = \Big(\textstyle\prod(g\wedge h_i)^{r_i}\Big)^{1/\Sigma r_i} = g\wedge\nu_p(u). \qquad \square$$

As was hinted above, θ_π is needed here mainly as a tool for describing the cokernel of certain transfer homomorphisms in K_2^c.

<u>Lemma 13.7</u> *Fix a prime* p, *and let* R *be the ring of integers in any algebraic number field in which* p *is unramified. For each prime* $\mathfrak{p}|p$ *in* R, *set*

$$k_\mathfrak{p} = \mathrm{ord}_p([R/\mathfrak{p}:\mathbb{Z}/p]) = \max\Big\{i : p^i\,|\,[R/\mathfrak{p}:\mathbb{Z}/p] = [\hat{K}_\mathfrak{p}:\hat{\mathbb{Q}}_p]\Big\};$$

and set $k = \min_{\mathfrak{p}|p}(k_\mathfrak{p})$. *Then for any p-group* π, *the sequence*

$$K_2^c(\hat{R}_p[\pi]) \xrightarrow{\ \mathrm{trf}\ } K_2^c(\hat{\mathbb{Z}}_p[\pi]) \xrightarrow{\ \theta''_k\ } \mathbb{Z}/p^k \otimes (H_2(\pi)/H_2^{ab}(\pi)) \longrightarrow 0$$

is exact, where θ''_k *is the reduction of* θ_π.

<u>Proof</u> Since $\hat{R}_p = \prod_{\mathfrak{p}|p}\hat{R}_\mathfrak{p}$ (Theorem 1.7), it suffices to show that

$$K_2^c(\hat{R}_\mathfrak{p}[\pi]) \xrightarrow{\ \mathrm{trf}\ } K_2^c(\hat{\mathbb{Z}}_p[\pi]) \xrightarrow{\ \theta''_{k_\mathfrak{p}}\ } \mathbb{Z}/p^{k_\mathfrak{p}} \otimes (H_2(\pi)/H_2^{ab}(\pi)) \longrightarrow 0 \qquad (1)$$

is exact for each $\mathfrak{p}$. Since $\hat{K}_\mathfrak{p}/\hat{\mathbb{Q}}_p$ is unramified, this involves only cyclic Galois extensions of $\hat{\mathbb{Q}}_p$. If $\hat{\mathbb{Q}}_p \subseteq F \subseteq \hat{K}_\mathfrak{p}$, $p\nmid[\hat{K}_\mathfrak{p}:F]$, and $S \subseteq F$ is the ring of integers, then

$$\mathrm{trf} : K_2^c(\hat{R}_\mathfrak{p}[\pi]) \longrightarrow K_2^c(S[\pi])$$

is surjective, since the composite $\mathrm{trf}\circ\mathrm{incl}$ is multiplication by

$[\hat{R}_p:F]$ on the pro-p-group $K_2^c(S[\pi])$. In particular, it suffices to prove the exactness of (1) when $[\hat{R}_p:\hat{\mathbb{Q}}_p] = p^k$ for some k.

In this case, write $G = Gal(\hat{R}_p/\hat{\mathbb{Q}}_p)$, and consider the twisted group ring $\hat{R}_p[\pi\times G]^t$. Then $\hat{R}_p[G]^t$ is a maximal order (see Reiner [1, Theorem 40.14]), and so $\hat{R}_p[G]^t \cong M_{p^k}(\hat{\mathbb{Z}}_p)$ by Theorem 1.9. The transfer thus factors as a composite

$$trf : K_2^c(\hat{R}_p[\pi]) \xrightarrow{incl} K_2^c(\hat{R}_p[\pi\times G]^t) \cong K_2^c(M_{p^k}(\hat{\mathbb{Z}}_p[\pi])) \cong K_2^c(\hat{\mathbb{Z}}_p[\pi]).$$

Proposition 13.3(iii) now applies to show that

$$\begin{aligned} \mathrm{Coker}\Big[trf: K_2^c(\hat{R}_p[\pi]) \longrightarrow K_2^c(\hat{\mathbb{Z}}_p[\pi])\Big] &\cong H_1(G;SK_1(\hat{R}_p[\pi])) \\ &\cong G\otimes (H_2(\pi)/H_2^{ab}(\pi)) \qquad \text{(Theorem 8.6)} \\ &\cong \mathbb{Z}/p^k\otimes (H_2(\pi)/H_2^{ab}(\pi)). \end{aligned}$$

The exactness of (1) will now follow, once we have shown that the composite

$$K_2^c(\hat{R}_p[\pi]) \xrightarrow{trf} K_2^c(\hat{\mathbb{Z}}_p[\pi]) \xrightarrow{\theta''_\pi} \mathbb{Z}/p^k\otimes (H_2(\pi)/H_2^{ab}(\pi))$$

vanishes. To see this, assume that $H_2(\pi)/H_2^{ab}(\pi) \neq 0$, and fix an extension $1 \longrightarrow \langle z\rangle \longrightarrow \tilde{\pi} \xrightarrow{\alpha} \pi \longrightarrow 1$ such that $\delta^\alpha: H_2(\pi) \twoheadrightarrow \langle z\rangle \cong C_p$ is surjective and $Ker(\delta^\alpha) \supseteq H_2^{ab}(\pi)$ (use Lemma 8.3(i)). Then z is not a commutator in $\tilde{\pi}$. The induced map $SK_1(\alpha): SK_1(\hat{\mathbb{Z}}_p[\tilde{\pi}]) \longrightarrow SK_1(\hat{\mathbb{Z}}_p[\pi])$ is injective by Theorem 7.1, and its image has index p by Proposition 8.1. We can thus assume, by induction on $|SK_1(\hat{\mathbb{Z}}_p[\pi])|$, that the result holds for $\tilde{\pi}$.

Consider the following commutative diagram with exact rows:

$$\begin{array}{ccccccc}
K_2^c(\hat{R}_p[\tilde{\pi}]) & \xrightarrow{i_R} & K_2^c(\hat{R}_p[\pi]) & \xrightarrow{\partial_R} & K_1(\hat{R}_p[\tilde{\pi}],(1-z)) & \longrightarrow & K_1(\hat{R}_p[\tilde{\pi}]) \\
\downarrow t_1 & & \downarrow t_2 & & \downarrow t_3 & & \\
K_2^c(\hat{\mathbb{Z}}_p[\tilde{\pi}]) & \xrightarrow{i_{\mathbb{Z}}} & K_2^c(\hat{\mathbb{Z}}_p[\pi]) & \xrightarrow{\partial_{\mathbb{Z}}} & K_1(\hat{\mathbb{Z}}_p[\tilde{\pi}],(1-z)); & &
\end{array}$$

where the t_i are transfer homomorphisms. By Proposition 6.4, the only torsion in $K_1(\hat{R}_p[\tilde{\pi}],(1-z))$ is $\langle z\rangle$ ($H_0(\tilde{\pi};(1-z)\hat{R}_p[\tilde{\pi}])$ is torsion free since z is not a commutator); and so this subgroup generates $\operatorname{Im}(\partial_R)$. It follows that $K_2^c(\hat{R}_p[\pi])$ is generated by $i_R(K_2^c(\hat{R}_p[\tilde{\pi}]))$ and $\hat{\lambda}_\pi(H_2(\pi)) \subseteq K_2^c(\hat{\mathbb{Z}}_p[\pi])$; and hence that

$$\operatorname{Im}(t_2) = \langle i_{\mathbb{Z}}(\operatorname{Im}(t_1)),\ K_2^c(\hat{\mathbb{Z}}_p[\pi])^{p^k}\rangle.$$

Clearly, $K_2^c(\hat{\mathbb{Z}}_p[\pi])^{p^k} \subseteq \operatorname{Ker}(\theta''_\pi)$; and $i_{\mathbb{Z}}(\operatorname{Im}(t_1)) \subseteq i_{\mathbb{Z}}(\operatorname{Ker}(\theta''_{\tilde{\pi}})) \subseteq \operatorname{Ker}(\theta''_\pi)$ by the induction assumption. □

Lemma 13.7 will now be applied to compare $Cl_1(R[\pi])$ with $Cl_1(\mathbb{Z}[\pi])$, when π is a p-group and p is unramified in R. As usual, this is easiest when p is odd.

<u>Theorem 13.8</u> *Fix an odd prime p, a p-group π, and a number field K in which p is unramified. Let $R \subseteq K$ be the ring of integers. Then the transfer homomorphism*

$$\operatorname{trf} : Cl_1(R[\pi]) \longrightarrow Cl_1(\mathbb{Z}[\pi])$$

is an isomorphism.

<u>Proof</u> Consider the following commutative diagram of localization sequences (see Theorem 3.15):

$$\begin{array}{ccccccc} K_2^c(\hat{R}_p[\pi]) & \longrightarrow & C_p(K[\pi]) & \longrightarrow & Cl_1(R[\pi]) & \longrightarrow & 1 \\ \downarrow{\scriptstyle trf_{K2}} & & \cong\downarrow{\scriptstyle trf_{Cp}} & & \downarrow{\scriptstyle trf_{Cl}} & & \\ K_2^c(\hat{\mathbb{Z}}_p[\pi]) & \longrightarrow & C_p(\mathbb{Q}[\pi]) & \longrightarrow & Cl_1(\mathbb{Z}[\pi]) & \longrightarrow & 1. \end{array} \tag{1}$$

Here, trf_{Cp} is onto by Lemma 4.17. For any simple summand A of $\mathbb{Q}[\pi]$ with center F, $F \cong \mathbb{Q}(\xi_n)$ ($\xi_n = \exp(2\pi i/p^n)$) for some n by Theorem 9.1; and since p is unramified in K, $K \otimes_{\mathbb{Q}} F = Z(K \otimes_{\mathbb{Q}} A)$ is a field with the same p-th power roots of unity as F. So $C_p(K[\pi]) \cong C_p(\mathbb{Q}[\pi])$ by Theorem 4.13; and trf_{Cp} is an isomorphism.

It follows from diagram (1) that trf_{Cl} is onto, and also (using Lemma 13.7) that there is a surjection

$$\mathbb{Z}/p^k \otimes (H_2(\pi)/H_2^{ab}(\pi)) \cong \mathrm{Coker}(trf_{K2}) \xrightarrow{\;f\;} \mathrm{Ker}(trf_{Cl}).$$

The standard involution is the identity on $Cl_1(R[\pi])$ by Theorem 5.12; and is (-1) on $\mathrm{Coker}(trf_{K2})$ by Lemma 13.7 (and the description of θ in Lemma 13.6). Hence $f = 1$, and trf_{Cl} is injective. □

This can now be combined with Theorems 11.10 and 13.4, to give the following explicit description of $Cl_1(\mathbb{Z}[G])_{(p)}$ in terms of p-groups.

<u>Theorem 13.9</u> *Fix a finite group G and an odd prime p, and let $\sigma_1,\ldots,\sigma_k \subseteq G$ be conjugacy class representatives for the cyclic subgroups of order prime to p. For each i, set $N_i = N_G(\sigma_i)$, $Z_i = C_G(\sigma_i)$, and let $\mathcal{P}(Z_i)$ be the set of p-subgroups. Then*

$$Cl_1(\mathbb{Z}[G])_{(p)} \cong \bigoplus_{i=1}^{k} H_0\Big(N_i/Z_i;\ \varinjlim_{\pi\in\mathcal{P}(Z_i)} Cl_1(\mathbb{Z}[\pi])\Big). \qquad \square$$

The formula in Theorem 13.9 gives a quick way of computing $Cl_1(\mathbb{Z}[G])_{(p)}$ as an abstract group, but it is clearly not as useful if one wants to detect a given element. The best thing would be to find a generalization of the formula $Cl_1(\mathbb{Z}[G]) \cong \mathrm{Coker}(\psi_G)$ for p-groups in Theorem 9.5. The main problem in doing this is to find a satisfactory definition of ψ_G in the general case. The closest we have come is to show that for any finite G,

$$Cl_1(\mathbb{Z}[G])[\tfrac{1}{2}] \cong \mathrm{Coker}\Big(H_1(G;\mathbb{Z}[G]) \xrightarrow{\psi_G} \big[R_{\mathbb{C}/\mathbb{R}}(G) \otimes \mathbb{Z}/n\big]_{(\mathbb{Z}/n)^*}\Big)[\tfrac{1}{2}]$$

for any n such that $\exp(G)|n$, where $\psi_G(g\otimes h)$ is defined for any commuting pair $g,h \in G$ as follows. Let $V_1,\ldots,V_m$ be the distinct irreducible $\mathbb{C}[G]$-representations. For each i, let $V_i^h \subseteq V_i$ be the subspace fixing h, and let $V_i^h\langle g\rangle \subseteq V_i^h$ be the sum of the $\exp(2\pi i/d)$-eigenspaces for $g\colon V_i^h \longrightarrow V_i^h$, for all $d|n$. Then

$$\psi_G(g\otimes h) = \sum_{i=1}^m \dim_{\mathbb{C}}(V_i^h\langle g\rangle)\cdot[V_i] \in \big[R_{\mathbb{C}/\mathbb{R}}(G) \otimes \mathbb{Z}/n\big]_{(\mathbb{Z}/n)^*}.$$

Under the isomorphism $\big[R_{\mathbb{C}/\mathbb{R}}(G) \otimes \mathbb{Z}/n\big]_{(\mathbb{Z}/n)^*} \cong C(\mathbb{Q}[G])$ of Lemma 5.9, this is easily seen to be equivalent to the definition of ψ_G in Definition 9.2 when G is a p-group. Also, ψ_G is natural with respect to inclusions of groups; and so the isomorphism $Cl_1(\mathbb{Z}[G])[\tfrac{1}{2}] \cong \mathrm{Coker}(\psi_G)[\tfrac{1}{2}]$ follows from Theorems 9.5, 13.5, and 13.8.

In contrast to Theorem 13.8, when π is a 2-group, it turns out that there can be up to three different values for $Cl_1(R[\pi])$ for varying R (in which 2 is unramified). These are described more precisely in the following theorem.

<u>Theorem 13.10</u> *Fix a 2-group* π *and a number field* K *where* 2 *is unramified; and let* $R \subseteq K$ *be the ring of integers. Consider the maps*

$$\varphi : K_2^c(\hat{\mathbb{Z}}_2[\pi]) \longrightarrow C_2(\mathbb{Q}[\pi]) \quad \text{and} \quad \tilde{\varphi} : K_2^c(\hat{\mathbb{Z}}_2[\pi]) \longrightarrow K_2^c(\hat{\mathbb{Q}}_2[\pi]).$$

Let

$$\theta'' : K_2^c(\hat{\mathbb{Z}}_2[\pi]) \longrightarrow \mathbb{Z}/2 \otimes (H_2(\pi)/H_2^{ab}(\pi))$$

be the homomorphism induced by θ_π. *Then*

(*i*) $Cl_1(R[\pi]) \cong C_2(\mathbb{Q}[\pi])/\mathrm{Im}(\varphi)$ *if* K *has a real embedding;*

(*ii*) $Cl_1(R[\pi]) \cong K_2^c(\hat{\mathbb{Q}}_2[\pi])/\mathrm{Im}(\tilde{\varphi})$ *if* K *is purely imaginary and* $[R/p:\mathbb{F}_2]$ *is odd for some prime* $p|2$ *in* R; *and*

(*iii*) $Cl_1(R[\pi]) \cong K_2^c(\hat{\mathbb{Q}}_2[\pi])/\tilde{\varphi}(\mathrm{Ker}(\theta''))$ *if* K *is purely imaginary and* $[R/p:\mathbb{F}_2]$ *is even for all primes* $p|2$ *in* R.

<u>Proof</u> For any simple summand A of $\mathbb{Q}[\pi]$ with center F, $F \subseteq \mathbb{Q}(\xi_n)$ for some n ($\xi_n = \exp(2\pi i/2^n)$) by Theorem 9.1. In particular, since 2 is unramified in K, $K \otimes_{\mathbb{Q}} F = Z(K \otimes_{\mathbb{Q}} A)$ is a field with the same 2-power roots of unity as F, $\hat{F}_2$, and $\hat{K}_p \otimes_{\mathbb{Q}} F$ for any prime $p|2$ in K. Furthermore, $K \otimes_{\mathbb{Q}} F$ has a real embedding if and only if K and F both do. It follows that

$$C_2(K[\pi]) \cong C_2(\mathbb{Q}[\pi]) \qquad \text{if K has a real embedding; and}$$
$$C_2(K[\pi]) \cong K_2^c(\hat{K}_p[\pi])_{(2)} \cong K_2^c(\hat{\mathbb{Q}}_2[\pi]) \qquad \text{if K is purely imaginary.} \tag{1}$$

In the first case, the isomorphism is induced by the transfer map (which is onto by Lemma 4.17). In the second case, we define an isomorphism α to be the composite

$$\alpha : C_2(K[\pi]) \cong \mathrm{Coker}\Big[K_2(K[\pi]) \longrightarrow \bigoplus_p K_2^c(\hat{K}_p[\pi])\Big]_{(2)} \xleftarrow[\cong]{\text{proj}} K_2^c(\hat{K}_p[\pi])_{(2)} \xrightarrow[\cong]{\text{trf}} K_2^c(\hat{\mathbb{Q}}_2[\pi])$$

for any prime $p|2$ in K. To see that α is independent of the choice of p, note that for any simple summand A of $\mathbb{Q}[\pi]$, either

$$C_2(K\otimes_{\mathbb{Q}} A) \cong \mathbb{Z}/2 \cong K_2^c(\hat{A}_2)$$

(and so there is only one possible isomorphism $C_2(K\otimes_{\mathbb{Q}} A) \cong K_2^c(\hat{A}_2)$); or else $\alpha|C_2(K\otimes_{\mathbb{Q}} A)$ is the composite

$$C_2(K\otimes_{\mathbb{Q}} A) \xrightarrow[\cong]{\text{trf}} C_2(A) \xleftarrow[\cong]{\text{proj}} K_2^c(\hat{A}_2).$$

Now consider the following diagram:

$$\begin{array}{ccccc} & & H_2(\pi) & & \\ & & \downarrow{\scriptstyle \hat{\lambda}_\pi} & \searrow{\scriptstyle \text{proj}} & \\ K_2^c(\hat{R}_2[\pi]) & \xrightarrow{t_1} & K_2^c(\hat{\mathbb{Z}}_2[\pi])/\{-1,\pm\pi\} & \xrightarrow{\theta''_k} & \mathbb{Z}/2^k\otimes(H_2(\pi)/H_2^{ab}(\pi)) \longrightarrow 0, \end{array} \tag{2}$$

where $k = \max\{i: 2^i|[R/p:\mathbb{F}_2],\ \text{all } p|2 \text{ in } R\}$, and t_1 is the transfer map. By Lemma 13.7, the row in (2) is exact, and the triangle commutes. Furthermore, $\hat{\lambda}_\pi$ factors through $K_2(\mathbb{Z}[\pi])$, and the composite

$$K_2(\mathbb{Z}[\pi]) \longrightarrow K_2^c(\hat{\mathbb{Z}}_2[\pi]) \xrightarrow{\varphi} \varprojlim_n SK_1(\mathbb{Z}[\pi],n)_{(2)} = C_2(\mathbb{Q}[\pi])$$

vanishes by construction ($(K_2^c(\hat{\mathbb{Z}}_p[\pi])_{(2)} = 1$ for odd p). It follows that

$$K_2^c(\hat{\mathbb{Z}}_2[\pi]) = \left\langle \text{Im}(t_1),\ \text{Im}(\hat{\lambda}_\pi)\right\rangle = \left\langle \text{Im}(t_1),\ \text{Ker}(\varphi)\right\rangle. \tag{3}$$

Similarly, since $\text{Ker}(\varphi)$ and $\text{Ker}(\tilde{\varphi})$ differ by exponent 2 (Theorem 9.1),

$$\text{Im}(t_1) \subseteq \text{Ker}(\theta'') = \text{Ker}(\theta''_1) \subseteq \left\langle \text{Ker}(\theta''_k),\ 2\cdot\text{Ker}(\varphi)\right\rangle \subseteq \left\langle \text{Im}(t_1),\ \text{Ker}(\tilde{\varphi})\right\rangle. \tag{4}$$

When K has a real embedding, there is a commutative diagram

$$\begin{array}{ccccccc} K_2^c(\hat{R}_2[\pi]) & \longrightarrow & C_2(K[\pi]) & \longrightarrow & Cl_1(R[\pi]) & \longrightarrow & 1 \\ \downarrow{\scriptstyle t_1} & & \cong\downarrow{\scriptstyle t_2} & & \downarrow{\scriptstyle t_3} & & \\ K_2^c(\hat{\mathbb{Z}}_2[\pi]) & \xrightarrow{\varphi} & C_2(\mathbb{Q}[\pi]) & \longrightarrow & Cl_1(\mathbb{Z}[\pi]) & \longrightarrow & 1, \end{array}$$

where the t_i are transfer maps, and t_2 is an isomorphism by (1). Then $\mathrm{Im}(\varphi) = \mathrm{Im}(\varphi\circ t_1)$ by (3), and so t_3 is an isomorphism.

If K has no real embedding, then we use the diagram

$$\begin{array}{ccccccc}
K_2^c(\hat{R}_2[\pi]) & \xrightarrow{\varphi_{R\pi}} & C_2(K[\pi]) & \longrightarrow & Cl_1(R[\pi]) & \longrightarrow & 1 \\
\downarrow{\scriptstyle t_1} & & \cong\downarrow{\scriptstyle \alpha} & & & & \\
K_2^c(\hat{\mathbb{Z}}_2[\pi]) & \xrightarrow{\tilde{\varphi}} & K_2^c(\hat{\mathbb{Q}}_2[\pi]) & & & &
\end{array}$$

where the top row is exact, and the square commutes by definition of α. If $[R/p:\mathbb{F}_2]$ is odd for some $p|2$ in R (i. e., if $k=0$), then t_1 is onto by (2), and so $Cl_1(R[\pi]) \cong \mathrm{Coker}(\tilde{\varphi})$. And if $k>0$, then by (4),

$$Cl_1(R[\pi]) \cong K_2^c(\hat{\mathbb{Q}}_2[\pi])/\mathrm{Im}(\tilde{\varphi}\circ t_1) \cong K_2^c(\hat{\mathbb{Q}}_2[\pi])/\tilde{\varphi}(\mathrm{Ker}(\theta'')). \qquad \square$$

In particular, for any 2-group π and any R (such that 2 is unramified in R), $Cl_1(R[\pi])$ is isomorphic to $Cl_1(\mathbb{Z}[\pi])$ (in case (i)), $Cl_1(\mathbb{Z}\zeta_7[\pi])$ (case (ii)), or $Cl_1(\mathbb{Z}\zeta_3[\pi])$ (case (iii)). Theorem 13.10 gives algorithms for computing these groups, and the "unknown quantity" in all of them is $K_2^c(\hat{\mathbb{Z}}_2[\pi])$. This is why one can hope that any procedure for describing $Cl_1(\mathbb{Z}[\pi])$ will also extend to the other two cases. Note that $Cl_1(\mathbb{Z}\zeta_7[\pi]) \cong Cl_1(\mathbb{Z}\zeta_3[\pi])$ if $H_2(\pi) = H_2^{ab}(\pi)$ — in particular, if π is abelian.

We now want to carry these results farther, and get lower bounds, at least, for the differences between these groups $Cl_1(R[\pi])$. We first consider the case where π is abelian (so $Cl_1(R[\pi]) = SK_1(R[\pi])$).

<u>**Theorem 13.11**</u> *Let* π *be an abelian 2-group, and set* $k = \mathrm{rk}(\pi)$. *Then*

$$SK_1(\mathbb{Z}\zeta_3[\pi]) \cong SK_1(\mathbb{Z}\zeta_7[\pi]) \cong SK_1(\mathbb{Z}[\pi]) \oplus SK_1(\mathbb{Z}\zeta_3[\pi/\pi^2])$$

$$\cong SK_1(\mathbb{Z}[\pi]) \oplus (\mathbb{Z}/2)^{2^k-1-k-\binom{k}{2}}.$$

Here, $\pi^2 = \{g^2 : g \in \pi\}$. *If* $\pi \cong (C_2)^k$ *(so* $\pi^2 = 1$*), then the following triangle commutes:*

$$\begin{array}{ccc} K_2^c(\hat{\mathbb{Z}}_2[\pi]) & \xrightarrow{\theta_\pi} & H_2(\pi) \\ & \searrow^{\tilde{\varphi}} & \downarrow V \\ & & K_2^c(\hat{\mathbb{Q}}_2[\pi])/\{-1, \pm\pi\} \cong (\mathbb{Z}/2)^{2^k-1-k}; \end{array} \tag{1}$$

where $V(g \wedge h) = \{g,h\}$ *for* $g,h \in \pi$ *and is injective.*

<u>Proof</u> For convenience, set $\bar{\pi} = \pi/\pi^2$, and let $\alpha: \pi \longrightarrow \bar{\pi}$ be the projection. For each simple summand A of $\mathbb{Q}[\pi]$, either $A \cong \mathbb{Q}$ and is a simple summand of $\mathbb{Q}[\bar{\pi}]$, in which case $C(A) = 1$; or $A \cong \mathbb{Q}(\xi_i)$ for some $i \geq 2$, and $C(A) \cong \langle \xi_i \rangle \cong K_2^c(\hat{A}_2)$. See the table in Theorem 9.1 for more details. In particular, this shows that

$$f_1 \oplus f_2 : K_2^c(\hat{\mathbb{Q}}_2[\pi]) \xrightarrow{\cong} C(\mathbb{Q}[\pi]) \oplus K_2^c(\hat{\mathbb{Q}}_2[\bar{\pi}])$$

is an isomorphism; where f_1 is the usual projection and f_2 is induced by α.

Let $J = \langle 2, 1-g : g \in \pi \rangle \subseteq \hat{\mathbb{Z}}_2[\pi]$ be the Jacobson radical (Example 1.12). From the relation

$$(g-1) + (h-1) = (gh-1) - (g-1)(h-1) \equiv (gh-1) \pmod{J^2} \qquad (\text{for } g,h \in \pi)$$

we get that $(\hat{\mathbb{Z}}_2[\pi])^* = 1 + J = \langle \pm g,\ u : g \in \pi,\ u \in 1 + J^2 \rangle$. So by Corollary 3.4,

$$K_2^c(\hat{\mathbb{Z}}_2[\pi]) = \langle \{\pm g, \pm h\},\ \{g,u\} : g,h \in \pi,\ u \in 1 + J^2 \rangle. \tag{2}$$

For any $g,h \in \pi$,

$$\{\pm g, \pm h\} \in \mathrm{Im}\Big[K_2(\mathbb{Z}[\pi]) \longrightarrow K_2^c(\hat{\mathbb{Z}}_2[\pi])\Big] \subseteq \mathrm{Ker}(f_1 \circ \tilde{\varphi}_\pi)$$

by definition of $C(\mathbb{Q}[\pi])$. Also, $\{\pi, 1+J^2\} \subseteq \operatorname{Ker}(f_2\circ\tilde{\varphi}_\pi)$, since for any $u \in 1+J^2$, $\{g,u\}$ maps to $\{\pm1, 1+4\hat{\mathbb{Z}}_2\} = 1$ at each simple summand $\hat{\mathbb{Q}}_2$ of $\hat{\mathbb{Q}}_2[\bar{\pi}]$ $(\cong (\hat{\mathbb{Q}}_2)^{2^k})$. So by Theorem 13.10, (f_1, f_2) induces an isomorphism

$$SK_1(\mathbb{Z}\zeta_3[\pi]) \cong SK_1(\mathbb{Z}\zeta_7[\pi]) \cong \operatorname{Coker}\Big[\tilde{\varphi}_\pi\colon K_2^c(\hat{\mathbb{Z}}_2[\pi]) \longrightarrow K_2^c(\hat{\mathbb{Q}}_2[\pi])\Big]$$

$$\cong \operatorname{Coker}\Big[K_2^c(\hat{\mathbb{Z}}_2[\pi]) \xrightarrow{\varphi_\pi} C(\mathbb{Q}[\pi])\Big] \oplus \operatorname{Coker}\Big[K_2^c(\hat{\mathbb{Z}}_2[\bar{\pi}]) \xrightarrow{\tilde{\varphi}_{\bar{\pi}}} K_2^c(\hat{\mathbb{Q}}_2[\bar{\pi}])\Big]$$

$$\cong SK_1(\mathbb{Z}[\pi]) \oplus SK_1(\mathbb{Z}\zeta_3[\bar{\pi}]).$$

Now assume that $\pi = \bar{\pi} \cong (C_2)^k$, and consider triangle (1). This clearly commutes on symbols $\{\pm g, \pm h\}$. For any $u \in 1+J^2$ and any $g \in \pi$, we have seen that $\tilde{\varphi}(\{g,u\}) = 1$; and $\theta_\pi(\{g,u\}) = g\wedge\nu_\pi(u) = 0$ by Lemma 13.6(ii). This shows that (1) commutes; and hence by Theorem 13.10 that

$$SK_1(\mathbb{Z}\zeta_3[\pi]) \cong SK_1(\mathbb{Z}\zeta_7[\pi]) \cong \operatorname{Coker}(\tilde{\varphi}) \cong \operatorname{Coker}(V).$$

Since $\hat{\mathbb{Q}}_2[\pi] \cong (\hat{\mathbb{Q}}_2)^{2^k}$, $K_2^c(\hat{\mathbb{Q}}_2[\pi]) \cong (\mathbb{Z}/2)^{2^k}$ by Theorem 4.4. So the remaining claims — the injectivity of V and the ranks of $\operatorname{Coker}(V)$ and $K_2^c(\hat{\mathbb{Q}}_2[\pi])/\{-1,\pm\pi\}$ — will all follow, once we have shown, for any basis $\{g_1,\ldots,g_k\}$ for π, that the set

$$\mathcal{S} = \Big\{\{-1,-1\}, \{-1,g_i\}, \{g_i,g_j\} \in K_2^c(\hat{\mathbb{Q}}_2[\pi]) : 1 \le i < j \le k\Big\}$$

is linearly independent in $K_2^c(\hat{\mathbb{Q}}_2[\pi])$.

To see this, define for each $s \in \mathcal{S}$ a character $\chi^s\colon \pi \longrightarrow \{\pm1\}$ as follows:

$s = \{-1,-1\}$: $\quad \chi^s(g_\ell) = 1 \quad$ (all ℓ)

$s = \{-1,g_i\}$: $\quad \chi^s(g_i) = -1, \quad \chi^s(g_\ell) = 1 \quad$ (all $\ell \ne i$)

$$s = \{g_i, g_j\}: \qquad \chi^s(g_i) = \chi^s(g_j) = -1, \quad \chi^s(g_\ell) = 1 \quad (\text{all } \ell \neq i,j).$$

Then, if $\chi^s_*: K_2^c(\hat{\mathbb{Q}}_2[\pi]) \longrightarrow K_2^c(\hat{\mathbb{Q}}_2) \cong \{\pm 1\}$ denotes the homomorphism induced by χ^s, we see that $\chi^s_*(s) = -1$ for each s, while (under an obvious ordering) $\chi^t_*(s) = 1$ for all $t < s$ in $\mathcal{S}$. This shows that $\mathcal{S}$ is linearly independent in $K_2^c(\hat{\mathbb{Q}}_2[\pi])$. □

For nonabelian π, the best we can do in general is to give lower bounds for the "differences" between the groups $Cl_1(\mathbb{Z}[\pi])$, $Cl_1(\mathbb{Z}\zeta_7[\pi])$, and $Cl_1(\mathbb{Z}\zeta_3[\pi])$. Recall that for any 2-group π, the *Frattini subgroup* $Fr(\pi) \subseteq \pi$ is the subgroup generated by commutators and squares in π; i. e., the subgroup such that $\pi/Fr(\pi) \cong \mathbb{Z}/2 \otimes \pi^{ab}$.

__Theorem 13.12__ *Let π be any 2-group, and set $k = rk(\pi/Fr(\pi))$. Set*

$$R = rk\Big(Im\Big[H_2(\pi) \to H_2(\pi/Fr(\pi))\Big]\Big); \qquad S = rk\Big(Im\Big[H_2^{ab}(\pi) \to H_2(\pi/Fr(\pi))\Big]\Big),$$

so that $S \leq R \leq \binom{k}{2}$. Then there are surjections

$$Cl_1(\mathbb{Z}\zeta_7[\pi]) \longrightarrow\!\!\!\!\!\rightarrow Cl_1(\mathbb{Z}[\pi]) \oplus (\mathbb{Z}/2)^{2^k-1-k-R}$$

and

$$Cl_1(\mathbb{Z}\zeta_3[\pi]) \cong Cl_1(\mathbb{Z}\zeta_{21}[\pi]) \longrightarrow\!\!\!\!\!\rightarrow Cl_1(\mathbb{Z}\zeta_7[\pi]) \oplus (\mathbb{Z}/2)^{R-S}.$$

In particular, $Cl_1(\mathbb{Z}\zeta_3[\pi]) \not\cong Cl_1(\mathbb{Z}\zeta_7[\pi])$ if $R > S$.

__Proof__ Set $\bar{\pi} = \pi/Fr(\pi) \cong (C_2)^k$, and let $\alpha: \pi \longrightarrow \bar{\pi}$ be the projection. Let

$$\theta_\pi : K_2^c(\hat{\mathbb{Z}}_2[\pi]) \longrightarrow\!\!\!\!\!\rightarrow H_2(\pi), \qquad \theta_{\bar{\pi}} : K_2^c(\hat{\mathbb{Z}}_2[\bar{\pi}]) \longrightarrow\!\!\!\!\!\rightarrow H_2(\bar{\pi})$$

be the homomorphisms of Lemma 13.6. Consider the following commutative diagram:

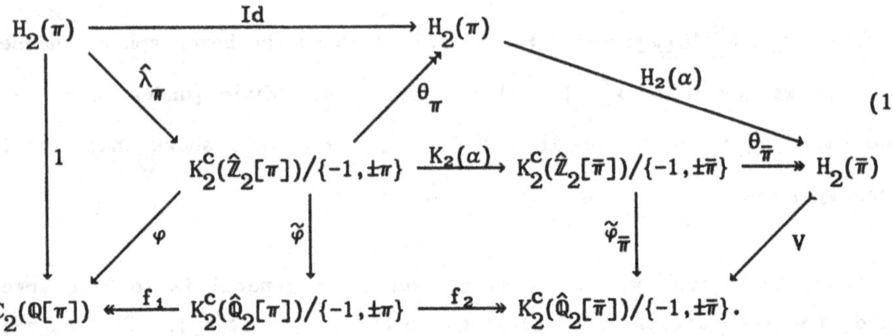

(1)

Here, $V(g\wedge h) = \{g,h\}$, and is injective by Theorem 13.11. Note in particular the following three points:

(a) $K_2^c(\hat{\mathbb{Q}}_2[\bar{\pi}])/\{-1,\pm\bar{\pi}\} \cong (\mathbb{Z}/2)^{2^k-1-k}$, and $V\circ\theta_{\bar{\pi}} = \tilde{\varphi}_{\bar{\pi}}$, by Theorem 13.11.

(b) $\theta_\pi\circ\hat{\lambda}_\pi = \mathrm{Id}$ by Lemma 13.6(i).

(c) $\varphi\circ\hat{\lambda}_\pi = 1$ since $\hat{\lambda}_\pi$ factors through $K_2(\mathbb{Z}[\pi])$, and the composite

$$K_2(\mathbb{Z}[\pi]) \longrightarrow K_2^c(\hat{\mathbb{Z}}_2[\pi]) \xrightarrow{\varphi} \varprojlim_n SK_1(\mathbb{Z}[\pi],n)_{(2)} = C_2(\mathbb{Q}[\pi])$$

vanishes by construction $(K_2^c(\hat{\mathbb{Z}}_p[\pi])_{(2)} = 1$ for odd $p)$.

Now consider the homomorphism

$$(f_1,f_2)\colon K_2^c(\hat{\mathbb{Q}}_2[\pi])/\{-1,\pm\pi\} \longrightarrow C_2(\mathbb{Q}[\pi]) \oplus K_2^c(\hat{\mathbb{Q}}_2[\bar{\pi}])/\{-1,\pm\bar{\pi}\}.$$

Write $\mathbb{Q}[\pi] = A \times \mathbb{Q}[\bar{\pi}]$, where A is the product of all simple summands of $\mathbb{Q}[\pi]$ not isomorphic to $\mathbb{Q}$. Then, since $C_2(\mathbb{Q}[\pi]) = C_2(A)$ (Theorem 4.13), (f_1,f_2) factors through a product of epimorphisms

$$\Big[K_2^c(\hat{A}_2)/\{-1,\pm\pi\} \twoheadrightarrow C_2(A)\Big] \times \Big[K_2^c(\hat{\mathbb{Q}}_2[\bar{\pi}])/\{-1,\pm\bar{\pi}\} \xrightarrow[\cong]{\mathrm{Id}} K_2^c(\hat{\mathbb{Q}}_2[\bar{\pi}])/\{-1,\pm\bar{\pi}\}\Big].$$

This shows that (f_1, f_2) is onto; and hence (using Theorem 13.10(ii)) that there are surjections

$$Cl_1(\mathbb{Z}\zeta_7[\pi]) \cong \operatorname{Coker}(\tilde{\varphi}) \twoheadrightarrow \operatorname{Coker}((f_1,f_2)\circ\tilde{\varphi}) \qquad \text{(by (d))}$$

$$\twoheadrightarrow \operatorname{Coker}(f_1\circ\tilde{\varphi}) \oplus \operatorname{Coker}(f_2\circ\tilde{\varphi})$$

$$= \operatorname{Coker}(\varphi) \oplus \operatorname{Coker}\Big[H_2(\pi) \xrightarrow{H_2(\alpha)} H_2(\bar{\pi}) \xrightarrow{V} K_2^c(\hat{\mathbb{Q}}_2[\pi])/\{-1, \pm\bar{\pi}\}\Big]$$

$$\cong Cl_1(\mathbb{Z}[\pi]) \oplus (\mathbb{Z}/2)^{2^k-1-k-R}. \qquad \text{(by (a))}$$

To compare $Cl_1(\mathbb{Z}\zeta_3[\pi])$ with $Cl_1(\mathbb{Z}\zeta_7[\pi])$, set

$$D = \operatorname{Ker}\Big[\theta'' : H_2(\pi) \twoheadrightarrow \mathbb{Z}/2 \otimes (H_2(\pi)/H_2^{ab}(\pi))\Big],$$

so that $Cl_1(\mathbb{Z}\zeta_3[\pi]) \cong K_2^c(\hat{\mathbb{Q}}_2[\pi])/\varphi(\theta_\pi^{-1}(D))$ by Theorem 13.10(iii). Fix any splitting β: $K_2^c(\hat{\mathbb{Q}}_2[\bar{\pi}])/\{-1, \pm\bar{\pi}\} \twoheadrightarrow \operatorname{Im}(f_2 \circ \tilde{\varphi})$ of the inclusion ($K_2^c(\hat{\mathbb{Q}}_2[\bar{\pi}])$ has exponent 2). Then there is a surjection

$$Cl_1(\mathbb{Z}\zeta_3[\pi]) \cong K_2^c(\hat{\mathbb{Q}}_2[\pi])/\tilde{\varphi}(\theta_\pi^{-1}(D)) \xrightarrow{(\text{proj},\beta\circ f_2)} \operatorname{Coker}(\tilde{\varphi}) \oplus \frac{\operatorname{Im}(f_2\circ\varphi)}{f_2\circ\varphi(\theta_\pi^{-1}(D))}$$

$$\cong \operatorname{Coker}(\tilde{\varphi}) \oplus \frac{\operatorname{Im}(V\circ H_2(\alpha))}{V\circ H_2(\alpha)(D)} \cong \operatorname{Coker}(\tilde{\varphi}) \oplus \frac{\operatorname{Im}[H_2(\pi) \longrightarrow H_2(\bar{\pi})]}{\operatorname{Im}[H_2^{ab}(\pi) \longrightarrow H_2(\bar{\pi})]}$$

$$\cong Cl_1(\mathbb{Z}\zeta_7[\pi]) \oplus (\mathbb{Z}/2)^{R-S}. \qquad \square$$

The groups constructed in Example 8.11 (when $p = 2$) have the property that $R > S$ in the above theorem, and hence that $Cl_1(\mathbb{Z}\zeta_3[\pi]) \not\cong Cl_1(\mathbb{Z}\zeta_7[\pi])$ for such π. This difference is the basis for the construction in Example 13.16 below of a group G for which the inclusion $Cl_1(\mathbb{Z}[G]) \subseteq SK_1(\mathbb{Z}[G])$ has no natural splitting.

When G is a finite abelian group, Theorems 13.8 and 13.11 yield as a corollary the following formula for $SK_1(\mathbb{Z}[G])_{(p)}$ $(= Cl_1(\mathbb{Z}[G])_{(p)})$. Note that this reduces the computation of $SK_1(\mathbb{Z}[G])$ (for abelian G) to the p-group case — which is handled by Theorems 9.5 and 9.6.

Theorem 13.13 *Fix an abelian group* G *and a prime* $p \mid |G|$. *Write* $G = H \times \pi$, *where* π *is a p-group and* $p \nmid |H|$. *Set* $k = rk(\pi)$, *and let* n *denote the number of simple summands of* $\mathbb{Q}[H]$. *Then*

$$SK_1(\mathbb{Z}[G])_{(p)} \cong \begin{cases} \bigoplus^{n} SK_1(\mathbb{Z}[\pi]) & \textit{if } p \textit{ is odd} \\ \bigoplus^{n} SK_1(\mathbb{Z}[\pi]) \oplus (\mathbb{Z}/2)^{(n-1)\cdot(2^k-1-k-\binom{k}{2})} & \textit{if } p = 2. \end{cases}$$

Proof Identify $\mathbb{Q}[H] = \prod_{i=1}^{n} K_i$, where the K_i are fields. Let $R_i \subseteq K_i$ be the ring of integers. Then $\mathfrak{M} = \prod_{i=1}^{n} R_i$ is the maximal order in K[H], and $[\mathfrak{M}[\pi]: \mathbb{Z}[G]]$ is prime to p by Theorem 1.4(v). Hence

$$SK_1(\mathbb{Z}[G])_{(p)} \cong \bigoplus_{i=1}^{n} SK_1(R_i[\pi])$$

by Corollary 3.10, and the result follows from the formula in Theorem 13.11 $(p = 2)$ or Theorem 13.8 (p odd). □

13c. Splitting the inclusion $Cl_1(\mathbb{Z}[G]) \subseteq SK_1(\mathbb{Z}[G])$

So far, all results about $SK_1(\mathbb{Z}[G])$ deal with its components $Cl_1(\mathbb{Z}[G])$ and $SK_1(\hat{\mathbb{Z}}_p[G])$ separately. It is also natural to consider the extension

$$1 \longrightarrow Cl_1(\mathbb{Z}[G]) \longrightarrow SK_1(\mathbb{Z}[G]) \xrightarrow{\ \ell\ } \bigoplus_p SK_1(\hat{\mathbb{Z}}_p[G]) \longrightarrow 1;$$

and in particular to try to determine when it is split. The key to doing this, in odd torsion at least, is the standard involution.

Theorem 13.14 *For any finite group* G, *there is a homomorphism*

$$s_G : SK_1(\mathbb{Z}[G]) \longrightarrow Cl_1(\mathbb{Z}[G]),$$

natural in G, *whose restriction to* $Cl_1(\mathbb{Z}[G])$ *is multiplication by* 2.

Proof This will be shown for $SK_1(\mathbb{Z}[G])_{(p)}$, one prime p at a time. Fix p, and consider the short exact sequence

$$1 \longrightarrow Cl_1(\mathbb{Z}[G])_{(p)} \longrightarrow SK_1(\mathbb{Z}[G])_{(p)} \xrightarrow{\ell_G} SK_1(\hat{\mathbb{Z}}_p[G]) \longrightarrow 1$$

of Theorem 3.15.

Step 1 Assume first that G is p-elementary: $G = C_n \times \pi$, where π is a p-group and $p \nmid n$. Instead of the usual involution, we consider the antiinvolution τ on $\mathbb{Q}[G]$ defined by:

$$\tau(\sum a_i r_i g_i) = \sum a_i r_i g_i^{-1} \qquad (a_i \in \mathbb{Q},\ \ r_i \in C_n,\ \ g_i \in \pi).$$

We claim that τ_* acts via the identity on $Cl_1(\mathbb{Z}[G])_{(p)}$, and via negation on $SK_1(\hat{\mathbb{Z}}_p[G])$.

Step 1A Let $\alpha \in Aut(G)$ be the automorphism: $\alpha(rg) = r^{-1}g$ for $r \in C_n$ and $g \in \pi$. Then τ is the composite of $\mathbb{Q}[\alpha]$ with the usual involution on $\mathbb{Q}[G]$. In particular, by Theorem 5.12, $\tau_* = \alpha_*$ on $Cl_1(\mathbb{Z}[G])_{(p)}$ and $C_p(\mathbb{Q}[G])$.

By construction, $\mathbb{Q}[\alpha]$ fixes all p-th power roots of unity in the center of $\mathbb{Q}[G]$. So by Theorem 4.13, α_* is the identity on $C_p(\mathbb{Q}[G])$, and hence (by the localization sequence of Theorem 3.15) on $Cl_1(\mathbb{Z}[G])_{(p)}$. It follows that $\tau_* = id$ on $Cl_1(\mathbb{Z}[G])_{(p)}$.

Step 1B Write $\hat{\mathbb{Q}}_p[C_n] = \prod_{i=1}^k F_i$, and $\hat{\mathbb{Z}}_p[C_n] = \prod_{i=1}^k R_i$, where the F_i are unramified field extensions of $\hat{\mathbb{Q}}_p$, and $R_i \subseteq F_i$ is the ring of integers. Then this induces decompositions

$$\hat{\mathbb{Z}}_p[G] \cong \prod_{i=1}^{k} R_i[\pi] \quad \text{and} \quad SK_1(\hat{\mathbb{Z}}_p[G]) \cong \prod_{i=1}^{k} SK_1(R_i[\pi]).$$

By construction, τ leaves each of these summands invariant, and acts on each one via the identity on coefficients and by inverting elements of π. By Theorem 8.6, τ_* acts on each $SK_1(R_i[\pi])$, and hence on $SK_1(\hat{\mathbb{Z}}_p[G])$, by negation.

<u>Step 1C</u> Now define $s'_G\colon\ SK_1(\hat{\mathbb{Z}}_p[G]) \longrightarrow SK_1(\mathbb{Z}[G])_{(p)}$ as follows: given $x \in SK_1(\hat{\mathbb{Z}}_p[G])$, lift x to $\tilde{x} \in SK_1(\mathbb{Z}[G])_{(p)}$, and set

$$s'_G(x) = \tilde{x}\cdot\tau_*(\tilde{x})^{-1}.$$

This is independent of the choice of lifting by Step 1A, and its composite with the projection to $SK_1(\hat{\mathbb{Z}}_p[G])$ is multiplication by 2 (i. e., squaring) by Step 1B. By construction, s'_G is natural with respect to homomorphisms between p-elementary groups.

<u>Step 2</u> Now let G be an arbitrary finite group, and let $\mathcal{E}$ be the set of p-elementary subgroups of G. By Theorem 12.4,

$$SK_1(\hat{\mathbb{Z}}_p[G]) \cong \varinjlim_{H\in\mathcal{E}} SK_1(\hat{\mathbb{Z}}_p[H]),$$

where the limit is taken with respect to inclusion and conjugation. Hence, by Step 1, there is a well defined homomorphism

$$s'_G = \varinjlim_{H\in\mathcal{E}} s'_H : SK_1(\hat{\mathbb{Z}}_p[G]) \longrightarrow \varinjlim_{H\in\mathcal{E}} SK_1(\mathbb{Z}[H])_{(p)} \xrightarrow{\text{Ind}} SK_1(\mathbb{Z}[G])_{(p)};$$

where s'_G is natural and $\ell_G \circ s'_G$ is multiplication by 2. So $s_G\colon SK_1(\mathbb{Z}[G])_{(p)} \longrightarrow Cl_1(\mathbb{Z}[G])_{(p)}$ can be defined by setting:

$$s_G(x) = x^2\cdot(s'_G \circ \ell_G(x))^{-1}. \qquad \square$$

An immediate corollary to Theorem 13.14 is:

<u>Theorem 13.15</u> *For any finite group* G *and any odd prime* p, *the p-power torsion in* $SK_1(\mathbb{Z}[G])$ *splits naturally as a direct sum*

$$SK_1(\mathbb{Z}[G])_{(p)} \cong Cl_1(\mathbb{Z}[G])_{(p)} \oplus SK_1(\hat{\mathbb{Z}}_p[G]). \qquad \square$$

The problem remains to describe the extension $Cl_1(\mathbb{Z}[G]) \subseteq SK_1(\mathbb{Z}[G])$ in 2-torsion, in general. It seems likely that examples exist of 2-groups where the inclusion $Cl_1(\mathbb{Z}[G]) \subseteq SK_1(\mathbb{Z}[G])$ has no splitting at all. This problem is closely related to Conjecture 9.7 above, and the discussion following the conjecture. In particular, the splitting of the inclusion $Cl_1(\mathbb{Z}[G]) \subseteq SK_1(\mathbb{Z}[G])$ seems likely to be closely related to the splitting of $H_2^{ab}(G) \subseteq H_2(G)$.

The following example shows, at least, that the inclusion $Cl_1(\mathbb{Z}[G]) \subseteq SK_1(\mathbb{Z}[G])$ need have no *natural* splitting in 2-torsion: more precisely, no splitting which commutes with the action of the automorphism group Aut(G). At the same time, it illustrates how Theorem 13.5 can fail if $SK_1(-)$ is replaced by $Cl_1(-)$.

<u>Example 13.16</u> *Let* π *be any 2-group with the property that*

$$\mathrm{Im}\Big[H_2(\pi) \longrightarrow H_2(\pi/\mathrm{Fr}(\pi))\Big] \supsetneqq \mathrm{Im}\Big[H_2^{ab}(\pi) \longrightarrow H_2(\pi/\mathrm{Fr}(\pi))\Big].$$

Set $G = C_7 \times S_3 \times \pi$, *and* $G_0 = C_7 \times C_3 \times \pi \lhd G$ $(S_3 \cong C_3 \rtimes C_2)$. *Then*

(i) $Cl_1(\mathbb{Z}[G])_{(2)}$ *is not 2-$\mathbb{R}$-elementary computable;*

(ii) *the square*

$$\begin{array}{ccc} \varinjlim_{H\in\mathcal{E}} C_2(\mathbb{Q}[H]) & \longrightarrow\!\!\!\!\rightarrow & \varinjlim_{H\in\mathcal{E}} Cl_1(\mathbb{Z}[H])_{(2)} \\ \big\downarrow & & \big\downarrow \\ C_2(\mathbb{Q}[G]) & \longrightarrow\!\!\!\!\rightarrow & Cl_1(\mathbb{Z}[G])_{(2)} \end{array} \qquad (1)$$

is not a pushout square, where $\mathcal{E}$ *denotes the set of 2-elementary subgroups of* G; *and*

(iii) the extension

$$1 \longrightarrow Cl_1(\mathbb{Z}[G_0]) \longrightarrow SK_1(\mathbb{Z}[G_0]) \longrightarrow SK_1(\hat{\mathbb{Z}}_2[G_0]) \longrightarrow 1 \qquad (2)$$

has no splitting which is natural with respect to automorphisms of G_0.

__Proof__ Set $\sigma = Gal(\mathbb{Q}\zeta_{21}/\mathbb{Q}\zeta_7) \cong C_2$, and let $\mathbb{Z}\zeta_{21}[\pi\times\sigma]^t$ be the induced twisted group ring. Then σ acts trivially on $Cl_1(\mathbb{Z}\zeta_{21}[\pi])$ (it acts trivially on $C_2(\mathbb{Q}\zeta_{21}[\pi])$ by Theorem 4.13). Furthermore, there is an inclusion

$$\mathbb{Z}\zeta_{21}[\pi\times\sigma]^t \subseteq M_2(\mathbb{Z}\zeta_7[\pi])$$

of odd index (see Reiner [1, Theorem 40.14]); and so by Corollary 3.10 and Theorem 13.12:

$$Cl_1(\mathbb{Z}\zeta_{21}[\pi\times\sigma]^t) \cong Cl_1(\mathbb{Z}\zeta_7[\pi]) \ncong Cl_1(\mathbb{Z}\zeta_{21}[\pi]) \cong H_0(\sigma;Cl_1(\mathbb{Z}\zeta_{21}[\pi])). \quad (3)$$

Note that $G = C_{21}\rtimes(\pi\times\sigma)$. Just as in the proof of Theorem 11.9, this shows that $Cl_1(\mathbb{Z}[G])_{(2)}$ is not computable with respect to induction from 2-$\mathbb{R}$-elementary subgroups. Also, since $C_2(\mathbb{Z}[G])$ is 2-$\mathbb{R}$-elementary computable, this shows that square (1) above is not a pushout square.

Now, by Theorem 13.4,

$$SK_1^{[2]}(\mathbb{Z}\zeta_{21}[\pi\times\sigma]^t) \cong H_0(\sigma;SK_1(\mathbb{Z}\zeta_{21}[\pi]));$$

and similarly (by Proposition 13.3(ii)) for $SK_1(\hat{\mathbb{Z}}_2\zeta_{21}[\pi\times\sigma]^t)$. Together with (3) above, this shows that the sequence

$$1 \to H_0(\sigma;Cl_1(\mathbb{Z}\zeta_{21}[\pi])) \to H_0(\sigma;SK_1(\mathbb{Z}\zeta_{21}[\pi])) \to H_0(\sigma;SK_1(\hat{\mathbb{Z}}_2\zeta_{21}[\pi])) \to 1$$

is not exact. This implies in turn that the exact sequence

$$1 \longrightarrow Cl_1(\mathbb{Z}\zeta_{21}[\pi]) \longrightarrow SK_1(\mathbb{Z}\zeta_{21}[\pi]) \longrightarrow SK_1(\hat{\mathbb{Z}}_2\zeta_{21}[\pi]) \longrightarrow 1 \qquad (4)$$

has no splitting which commutes with the action of σ. But (4) is a direct summand of sequence (2) above by Corollary 3.10 ($\hat{\mathbb{Z}}_2\zeta_{21}[\pi]$ is a direct summand of $\hat{\mathbb{Z}}_2[G_0]$); and so (2) has no natural splitting. □

More concretely, consider the group

$$\pi = \langle a,b,c,d \mid a^2 = b^2 = c^2 = d^2 = 1 = [\pi,[\pi,\pi]],\ [a,b][c,d] = 1\rangle.$$

Then $\pi^{ab} \cong (C_2)^4$, and $[\pi,\pi] = Z(\pi) \cong (C_2)^5$ (see Example 8.11). If $\alpha: \pi \longrightarrow \pi^{ab}$ is the projection, then $H_2(\alpha)$ has image of rank one (generated by $a\wedge b+c\wedge d$); while its restriction to $H_2^{ab}(\pi)$ is zero. So π satisfies the hypotheses of Example 13.16.

Chapter 14 EXAMPLES

We now list some examples of calculations of $SK_1(\mathbb{Z}[G])$. These illustrate a variety of techniques, and apply many of the results from earlier chapters.

We have already seen, in Theorem 5.4, that $Cl_1(\mathbb{Z}[G]) = 1$ if $\mathbb{R}[G]$ is a product of matrix algebras over $\mathbb{R}$. The first theorem extends this to some conditions which imply that $SK_1(\mathbb{Z}[G]) = 1$ or $Wh(G) = 1$. It shows, for example, not only that the Whitehead group of any symmetric group vanishes, but also that $Wh(G)$ vanishes whenever G is a product of symmetric groups, or a product of wreath products $S_m \wr S_n$, etc.

Theorem 14.1 *Define classes* $\mathfrak{R}$, $\mathcal{Q}$, $\mathfrak{D}$ *of finite groups by setting:*

$$\mathfrak{R} = \left\{G : \mathbb{R}[G] \text{ is a product of matrix algebras over } \mathbb{R}\right\};$$

$$\mathcal{Q} = \left\{G : \mathbb{Q}[G] \text{ is a product of matrix algebras over } \mathbb{Q}\right\} \subseteq \mathfrak{R}; \text{ and}$$

$$\mathfrak{D} = \left\{G : H_2(C_G(g))/H_2^{ab}(C_G(g)) = 0, \text{ all } g \in G\right\}.$$

Then

(i) $Wh(G) = 1$ *for any* $G \in \mathcal{Q} \cap \mathfrak{D}$, $SK_1(\mathbb{Z}[G]) = 1$ *for any* $G \in \mathfrak{R} \cap \mathfrak{D}$, *and* $SK_1(\mathbb{Z}[G]) = Cl_1(\mathbb{Z}[G])$ *for any* $G \in \mathfrak{D}$;

(ii) all symmetric groups lie in $\mathcal{Q} \cap \mathfrak{D}$, *and all dihedral and symmetric groups lie in* $\mathfrak{R} \cap \mathfrak{D}$; *and*

(iii) all three of the classes $\mathfrak{R}$, $\mathcal{Q}$, *and* $\mathfrak{D}$ *are closed under products, and under wreath products with any symmetric group* S_n.

<u>Proof</u> By Theorem 12.5(i), $SK_1(\mathbb{Z}[G])/Cl_1(\mathbb{Z}[G]) = \oplus_p SK_1(\hat{\mathbb{Z}}_p[G]) = 1$ for any $G \in \mathfrak{D}$. If $G \in \mathfrak{R}$, then $Cl_1(\mathbb{Z}[G]) = 1$ by Theorem 5.4. If $G \in \mathcal{Q}$, then $Wh'(G) = Wh(G)/SK_1(\mathbb{Z}[G])$ is torsion free (Theorem 7.4) and has rank zero (Theorem 2.5); and so $Wh(G) = SK_1(\mathbb{Z}[G])$ in this case.

For convenience, write $\mathcal{H}(G) = H_2(G)/H_2^{ab}(G)$ for any G. By Proposition 8.12, $\mathcal{H}$ is multiplicative; and so $\mathfrak{D}$ is closed under taking products (note that $C_{G\times H}(g,h) = C_G(g) \times C_H(h)$). When checking that $\mathcal{H}(C_{G\wr S_n}(g)) = 0$ for any $G \in \mathfrak{D}$ and any $g \in G\wr S_n$, we are quickly reduced to the following two cases:

(a) $g = (g_1,\ldots,g_n) \in G^n \subseteq G\wr S_n$: then $C_{G\wr S_n}(g)$ is a product of wreath products (by symmetric groups) over the centralizers $C_G(g_i)$.

(b) $g = (g_1,\ldots,g_n)\cdot\sigma \in G\wr S_n$, where $\sigma = (1\,2\,\ldots\,n) \in S_n$: then $C_{G\wr S_n}(g) = \langle g, C_{G^n}(g)\rangle$, and $\mathcal{H}(C_{G\wr S_n}(g)) \cong \mathcal{H}(C_{G^n}(g)) \cong \mathcal{H}(C_G(g_1\cdots g_n)) = 0$.

Using Proposition 8.12 again, we see that $\mathcal{H}(G\wr S_n) = 0$ if $\mathcal{H}(G) = 0$ (any p-Sylow subgroup of $G\wr S_n$ is contained in a product of wreath products $G\wr C_p\wr\ldots\wr C_p$). Together, these relations show that $G\wr S_n \in \mathfrak{D}$ if $G \in \mathfrak{D}$.

Clearly, $\mathfrak{R}$ and $\mathcal{Q}$ are closed under products. Also, $\mathbb{Q}[S_n]$ is a product of matrix rings over $\mathbb{Q}$ (see James & Kerber [1, Theorem 2.1.12]); and so $S_n \in \mathcal{Q} \subseteq \mathfrak{R}$. Using this, it is an easy exercise in manipulating twisted group rings to check that $\mathbb{Q}$ (or $\mathbb{R}$) is a splitting field for $\mathbb{Q}[G\wr S_n]$ for all n, if it is a splitting field for $\mathbb{Q}[G]$.

Finally, for each n, $D(2n) \in \mathfrak{D}$, since $\mathcal{H}(G) = 0$ whenever G contains an abelian subgroup of prime index (see Proposition 12.7). And $\mathbb{R}[D(2n)]$ is easily seen to be a product of matrix algebras over $\mathbb{R}$. □

The condition that $\mathbb{Q}[G]$ be a product of matrix algebras over $\mathbb{Q}$ does not by itself guarantee that $Wh(G) = 1$. The simplest counterexample to this is the central extension

$$1 \longrightarrow (C_2)^4 \longrightarrow G \longrightarrow (C_2)^4 \longrightarrow 1;$$

defined by the relations:

$$G = \Big\langle a,b,c,d \;:\; a^2 = b^2 = c^2 = d^2 = 1 = [G,[G,G]] = [b,ac][c,d] = [a,cd][b,d] \Big\rangle.$$

A straightforward check shows that $\mathbb{Q}[G]$ is a product of copies of $M_r(\mathbb{Q})$ for $r = 1,2,4$; and so $Wh(G) \cong SK_1(\hat{\mathbb{Z}}_2[G])$ by the same arguments as in the above proof. But using Lemma 8.9, applied with $\hat{G} = G/[G,G] \cong (C_2)^4$, one can show that $SK_1(\hat{\mathbb{Z}}_2[G]) \cong (\mathbb{Z}/2) \times (\mathbb{Z}/2)$.

The next theorem gives necessary and sufficient conditions for when $SK_1(\mathbb{Z}[G]) = 1$ in the case of an abelian group G. Note that while it also gives some conditions for when $SK_1(\mathbb{Z}[G])$ or $Cl_1(\mathbb{Z}[G])$ does or does not vanish for nonabelian G, a comparison of Theorems 14.1 and 14.2 indicates that a complete answer to this question is quite unlikely.

<u>Theorem 14.2</u> *Fix a finite group* G.

(i) If each Sylow subgroup of G *has the form* C_{p^n} *or* $C_p \times C_{p^n}$ *(any* $n \geq 0$*), then* $SK_1(\mathbb{Z}[G])_{(p)} = 1$.

(ii) If G *is a p-group for some prime* p*, and if* $Cl_1(\mathbb{Z}[G]) = 1$*, then either* $G \cong C_{p^n}$ *or* $C_p \times C_{p^n}$ *for some* n*, or* $p = 2$ *and* $G^{ab} \cong (C_2)^k$ *for some* k.

(iii) If G *is abelian, then* $SK_1(\mathbb{Z}[G]) = 1$ *if and only if either (a) each Sylow subgroup of* G *has the form* C_{p^n} *or* $C_p \times C_{p^n}$ *for some* n*; or (b)* $G \cong (C_2)^k$ *for some* k.

<u>Proof</u> (i) By Theorem 5.3, $SK_1(\mathbb{Z}[G])_{(p)}$ is generated by induction from p-elementary subgroups. Hence, it suffices to show for all $n \geq 1$

that $SK_1(\mathbb{Z}[C_n]) \cong SK_1(\mathbb{Z}[C_p \times C_n]) = 1$. This follows from Example 9.8 when n is a power of p; and from Theorem 13.13 in general.

(ii) For nonabelian G, this was shown in Example 9.9. In the abelian case, recall first that a surjection $G \twoheadrightarrow G'$ of finite groups induces a surjection $Cl_1(\mathbb{Z}[G]) \twoheadrightarrow Cl_1(\mathbb{Z}[G'])$ (Corollary 3.10). So $Cl_1(\mathbb{Z}[G])$ is nonvanishing if G surjects onto $C_{p^2} \times C_{p^2}$ (Example 9.8(ii)), onto $C_4 \times C_2 \times C_2$ (Example 5.1), or onto $(C_p)^3$ if p is odd (Alperin et al [3, Theorem 2.4]). The only abelian p-groups which do not surject onto one of these groups are C_{p^n}, $C_p \times C_{p^n}$, and $(C_2)^k$.

(iii) By Theorem 13.13, for any finite abelian group G and any prime $p \mid |G|$, $SK_1(\mathbb{Z}[G])_{(p)} = 1$ if and only if $SK_1(\mathbb{Z}[S_p(G)]) = 1$ and (if $p = 2$ and G is not a 2-group) $rk(S_2(G)) \leq 2$. By (i) and (ii), this holds if and only if $S_p(G) \cong C_{p^n}$ or $C_p \times C_{p^n}$, or $G \cong (C_2)^n$. □

Note that the exact exponent of $SK_1(\mathbb{Z}[G])$, for arbitrary abelian G, is computed in Alperin et al [3, Theorem 4.8] (see Example 5 in the introduction).

We next give a direct application of the results about twisted group rings in Chapter 13. We want to describe the 2-power torsion in $SK_1(\mathbb{Z}[G])$ when $S_2(G)$ is dihedral, quaternionic, or semidihedral. Note that this includes all groups with periodic cohomology, in particular, all groups which can act freely on spheres — and that was the original motivation for studying this class. The following lemma deals with the twisted group rings which arise.

<u>Lemma 14.3</u> *Let R be the ring of integers in an algebraic number field K in which 2 is unramified. Let π be any dihedral, quaternionic, or semidihedral 2-group. Let $t\colon \pi \longrightarrow Gal(K/\mathbb{Q})$ be any homomorphism, set $\rho = Ker(t)$, and let $K[\pi]^t$ and $R[\pi]^t$ be the induced twisted group rings. Then*

$$Cl_1(R[\pi]^t)_{(2)} \cong \begin{cases} \mathbb{Z}/2 & \text{if } \rho \text{ is nonabelian and } K^{\pi} \not\subseteq \mathbb{R} \\ 1 & \text{otherwise.} \end{cases}$$

<u>Proof</u> Assume first that $t = 1$. If K has a real embedding, then $Cl_1(R[\pi]) = 1$ by Example 5.8. If K has no real embedding, then $|Cl_1(R[\pi])| \le 2$ by Example 5.8 again; while $|Cl_1(R[\pi])| \ge 2$ by Theorem 13.12 ($H_2(\pi)$ maps trivially to $H_2(\pi^{ab}) \cong \mathbb{Z}/2$).

Now assume that $t: \pi \longrightarrow Gal(K/\mathbb{Q})$ is nontrivial, and set $\rho = Ker(t)$. By Theorem 13.4, $Cl_1(R[\rho])$ surjects onto $Cl_1(R[\pi]^t)$, and

$$Cl_1(R[\pi]^t) \cong H_0(\pi/\rho; Cl_1(R[\rho])) \cong \mathbb{Z}/2$$

if K^π has no real embedding. So it remains only to consider the case where ρ is nonabelian, where $K^\pi \subseteq \mathbb{R}$, but where $K \not\subseteq \mathbb{R}$.

Assume this, and consider the pushout square of Theorem 13.4:

$$\begin{array}{ccc} H_0(\pi/\rho; C_2(K[\rho])) & \xrightarrow{\partial_\rho} & H_0(\pi/\rho; Cl_1(R[\rho])) \cong \mathbb{Z}/2 \\ \downarrow{\scriptstyle i_{C2}} & & \downarrow \\ C_2(K[\pi]^t) & \longrightarrow & Cl_1(R[\pi]^t)_{(2)}. \end{array}$$

We saw, when computing $Cl_1(R[\rho])$ in Example 5.8, that ∂_ρ can be identified with the composite

$$C_2(K[\rho]) \longrightarrow C_2(K[\rho^{ab}]) \cong \bigoplus^4 C_2(K) \xrightarrow{\text{sum}} C_2(K) \cong \mathbb{Z}/2;$$

(note that $\rho^{ab} \cong C_2 \times C_2$). Also, $\pi/[\rho,\rho] \cong D(8)$, the dihedral group of order 8: since $[\pi:\rho] = 2$, $|\pi^{ab}| = 4$, and $D(8)$ is the only nonabelian group of order 8 which contains $C_2 \times C_2$. The pair $K[\rho^{ab}] \subseteq K[\pi/[\rho,\rho]]^t$ now splits as a product of inclusions

$$(K\times K) \times (K\times K) \subseteq (M_2(K)) \times (M_2(K^\pi) \times M_2(K^\pi)).$$

Since $C_2(K^\pi) = 1$ ($K^\pi \subseteq \mathbb{R}$), this shows that $Ker(i_{C2}) \not\subseteq Ker(\partial_\rho)$ in (2); and hence that $Cl_1(R[\pi]^t)_{(2)} = 1$. □

Applying this to integral group rings is now straightforward.

Example 14.4 *Let G be a finite group such that the 2-Sylow subgroups of G are dihedral, quaternionic, or semidihedral. Then*

$$SK_1(\mathbb{Z}[G])_{(2)} = Cl_1(\mathbb{Z}[G])_{(2)} \cong (\mathbb{Z}/2)^k,$$

where k *is the number of conjugacy classes of cyclic subgroups* $\sigma \subseteq G$ *such that (a)* $|\sigma|$ *is odd, (b)* $C_G(\sigma)$ *has nonabelian 2-Sylow subgroup, and (c) there is no* $g \in N_G(\sigma)$ *with* $gxg^{-1} = x^{-1}$ *for all* $x \in \sigma$.

Proof Note first that $SK_1(\hat{\mathbb{Z}}_2[G]) = 1$ by Proposition 12.7, so that $SK_1(\mathbb{Z}[G])_{(2)} \cong Cl_1(\mathbb{Z}[G])_{(2)}$. By Theorem 11.10, if $\sigma_1,\ldots,\sigma_k$ are conjugacy class representatives of cyclic subgroups of G of odd order, and if $n_i = |\sigma_i|$ and $N_i = N_G(\sigma_i)$, then

$$Cl_1(\mathbb{Z}[G])_{(2)} \cong \bigoplus_{i=1}^{k} \varinjlim_{\pi\in\mathcal{P}(N_i)} Cl_1(\mathbb{Z}\zeta_{n_i}[\pi]^t)_{(2)},$$

where $\mathcal{P}(N_i)$ denotes the set of 2-subgroups. The result is now an immediate consequence of Lemma 14.3. □

We now finish by giving examples of two more specialized families of groups for which $SK_1(\mathbb{Z}[G])$ is nonvanishing in general, but still can be computed.

Theorem 14.5 *For any prime power* $q = p^k$,

(i) $SK_1(\mathbb{Z}[PSL(2,q)]) \cong \mathbb{Z}/3$ *and* $SK_1(\mathbb{Z}[SL(2,q)]) \cong \mathbb{Z}/3 \times \mathbb{Z}/3$, *if* $p = 3$ *and* k *is odd,* $k \geq 5$; *and*

(ii) $SK_1(\mathbb{Z}[PSL(2,q)]) \cong SK_1(\mathbb{Z}[SL(2,q)]) = 1$ *otherwise.*

Proof Write $G = PSL(2,q)$ and $\tilde{G} = SL(2,q)$, for short. By Huppert [1, Theorem II.8.27], the only noncyclic elementary subgroups of G and

$\tilde{G}$ are dihedral and quaternionic 2-groups, elementary abelian p-groups, and (in $\tilde{G}$) products of C_2 with elementary abelian p-groups. In particular, $SK_1(\mathbb{Z}[G]) = Cl_1(\mathbb{Z}[G])$, and similarly for $\tilde{G}$, by Proposition 12.7. Furthermore, by Theorem 14.2 and Example 14.4, this list shows that $Cl_1(\mathbb{Z}[H]) = 1$ for all elementary subgroups H in G or $\tilde{G}$, except possibly for p-elementary subgroups when p is odd. Since $Cl_1(\mathbb{Z}[G])$ is generated by elementary induction (Theorem 5.3), $Cl_1(\mathbb{Z}[G])$ and $Cl_1(\mathbb{Z}[\tilde{G}])$ are p-groups, and vanish if $p = 2$ or $k = 1$.

Assume now that p is odd and $k > 1$. Most of the terms vanish in the decomposition formulas for $Cl_1(\mathbb{Z}[G])_{(p)}$ and $Cl_1(\mathbb{Z}[\tilde{G}])_{(p)}$ of Theorem 13.9; leaving only

$$Cl_1(\mathbb{Z}[G]) \cong \varinjlim_{\rho\in\mathcal{P}(G)} Cl_1(\mathbb{Z}[\rho])$$

(where $\mathcal{P}(G)$ is the set of p-subgroups), and

$$Cl_1(\mathbb{Z}[\tilde{G}]) \cong \varinjlim_{\rho\in\mathcal{P}(\tilde{G})} Cl_1(\mathbb{Z}[\rho]) \times \varinjlim_{\rho\in\mathcal{P}(\tilde{G})} Cl_1(\mathbb{Z}[\rho]).$$

Since these limits are all isomorphic, it remains only to show that

$$\varinjlim_{\rho\in\mathcal{P}(G)} Cl_1(\mathbb{Z}[\rho]) \cong \begin{cases} \mathbb{Z}/3 & \text{if } p = 3,\ k \geq 5,\ k \text{ odd} \\ 1 & \text{otherwise.} \end{cases}$$

Furthermore, the p-Sylow subgroups of G are isomorphic to $\mathbb{F}_q$, and any two p-Sylow subgroups of G intersect trivially (any nontrivial element of $SL(2,q)$ of p-power order fixes some unique 1-dimensional subspace of $(\mathbb{F}_q)^2$). Hence, for any p-Sylow subgroup $P \subseteq G$,

$$\varinjlim_{\rho\in\mathcal{P}(G)} Cl_1(\mathbb{Z}[\rho]) \cong H_0\big(N(P)/P;\ Cl_1(\mathbb{Z}[P])\big) \cong H_0\big(\mathbb{F}_q^{*2};\ Cl_1(\mathbb{Z}[\mathbb{F}_q])\big).$$

Here, $\mathbb{F}_q^{*2}$ denotes the group of squares in $\mathbb{F}_q^{*}$.

Now, since $\mathbb{F}_q^{*2}$ has order prime to p,

$$H_0\left(\mathbb{F}_q^{*2};\ Cl_1(\mathbb{Z}[\mathbb{F}_q])\right) \cong Cl_1(\mathbb{Z}[\mathbb{F}_q])^{\mathbb{F}_q^{*2}} \qquad \text{(elements fixed by } \mathbb{F}_q^{*2})$$

$$\cong \operatorname{Coker}\left[\left(\mathbb{F}_q \otimes \mathbb{Z}[\mathbb{F}_q]\right)^{\mathbb{F}_q^{*2}} \xrightarrow{\ \psi\ } C_p(\mathbb{Q}[\mathbb{F}_q])^{\mathbb{F}_q^{*2}}\right].$$

Also, any element $a \in \mathbb{F}_p^*$ acts on $C_p(\mathbb{Q}[\mathbb{F}_q])$ via $(x \mapsto x^a)$, and this leaves no fixed elements if $a \neq 1$. So $H_0(\mathbb{F}_q^{*2};Cl_1(\mathbb{Z}[\mathbb{F}_q])) = 1$ if $\mathbb{F}_p^* \cap \mathbb{F}_q^{*2} \neq 1$; and this is the case if $p \geq 5$, or if $p = 3$ and k is even.

Now assume that $p = 3$, $k \geq 3$, and k is odd. Then $\mathbb{F}_q^{*2}$ permutes the nontrivial summands of $\mathbb{Q}[\mathbb{F}_q]$ simply and transitively; so that

$$C_p(\mathbb{Q}[\mathbb{F}_q])^{\mathbb{F}_q^{*2}} \cong C_p(\mathbb{Q}\zeta_3) \cong \mathbb{Z}/3.$$

Furthermore, by Alperin et al [3, Proposition 2.5], there is an isomorphism

$$\operatorname{Im}\left[\mathbb{F}_q \otimes \mathbb{Z}[\mathbb{F}_q] \xrightarrow{\ \psi\ } C_p(\mathbb{Q}[\mathbb{F}_q])\right] \cong S^p(\mathbb{F}_q)$$

(the p-th symmetric power) which is natural with respect to automorphisms of $\mathbb{F}_q$. This now shows that $H_0(\mathbb{F}_q^{*2};\ Cl_1(\mathbb{Z}[\mathbb{F}_q])) \cong (\mathbb{Z}/3)^r$, where

$$r = 1 - rk_{\mathbb{F}_p}\left(S^p(\mathbb{F}_q)^{\mathbb{F}_q^{*2}}\right).$$

If we regard $V = \mathbb{F}_q$ as an $\mathbb{F}_p[\mathbb{F}_q^{*2}]$-module, tensor up by the splitting field $\mathbb{F}_q$, and then look at eigenvalues in the symmetric product, we see that $S^p(\mathbb{F}_q)$ has a component fixed by $\mathbb{F}_q^{*2}$ if and only if $k = 3$. □

The last example is given by the alternating groups. These show the same phenomenon: the only torsion in their Whitehead groups is at the prime 3.

Theorem 14.6 *Fix* $n > 1$, *and let* A_n *be the alternating group on* n *letters. Then*

$$SK_1(\mathbb{Z}[A_n]) \cong \begin{cases} \mathbb{Z}/3 & \text{if } n = \sum_{i=1}^{r} 3^{m_i} \geq 27,\ m_1 > \ldots > m_r \geq 0,\ \sum(m_i) \text{ odd} \\ 1 & \text{otherwise.} \end{cases}$$

Proof We sketch here the main points in the proof. For more details, see Oliver [3, Theorem 5.6].

(1) $SK_1(\mathbb{Z}[A_n]) = Cl_1(\mathbb{Z}[A_n])$: $SK_1(\hat{\mathbb{Z}}_p[A_n]) = 1$ for all p by Example 12.8.

(2) Since $[S_n:A_n] = 2$, and $\mathbb{Q}[S_n]$ is a product of matrix algebras over $\mathbb{Q}$ (see James & Kerber [1, Theorem 2.1.12]), $\mathbb{Q}[A_n]$ is a product of matrix algebras over fields of degree at most 2 over $\mathbb{Q}$. Hence, if $p \geq 5$, then $C_p(\mathbb{Q}[A_n]) = Cl_1(\mathbb{Z}[A_n])_{(p)} = 1$.

(3) If $\sigma_1, \ldots, \sigma_k$ are conjugacy class representatives for cyclic subgroups of A_n, and if $m_i = |\sigma_i|$, then

$$Z(\mathbb{Q}[A_n]) \cong \prod_{i=1}^{k} (\mathbb{Q}\zeta_{m_i})^{N(\sigma_i)}.$$

To see this, note that both sides are products of fields of degree at most 2 over $\mathbb{Q}$. Hence, it suffices to show that both sides have the same number of simple summands after tensoring by any quadratic extension K of $\mathbb{Q}$. This follows from the Witt-Berman theorem (Theorem 1.6): for each K, the number of irreducible $K[A_n]$-modules equals the number of K-conjugacy classes in A_n.

(4) By Theorem 4.13, $C_2(\mathbb{Q}[A_n])$ has rank equal to the number of purely imaginary field summands of $\mathbb{Q}[A_n]$; and by (3) this is equal to

the number of conjugacy classes of cyclic subgroups $\sigma = \langle g\rangle \subseteq A_n$ such that g is not conjugate to g^{-1}. Each such g is a product of disjoint cycles of lengths $k_1 > \dots > k_s$, such that $\sum k_i = n$, k_i is odd for all i, and $\sum_i (k_i-1)/2$ is odd. In particular, the centralizer $C_{A_n}(\sigma)$ has odd order for each such σ, and so by Theorem 13.4:

$$\varinjlim_{\pi\in\mathcal{P}(N\sigma)} Cl_1(\mathbb{Z}\zeta_k[\pi]^t)_{(2)} = 1 \qquad (k = |\sigma|)$$

($\mathcal{P}(N\sigma)$ denotes here the set of 2-subgroups). On the other hand,

$$\varinjlim_{\pi\in\mathcal{P}(N\sigma)} C_2(\mathbb{Q}\zeta_k[\pi]^t) \cong \varinjlim_{\pi\in\mathcal{P}(N\sigma)} C_2((\mathbb{Q}\zeta_k)^\pi) \cong C_2((\mathbb{Q}\zeta_k)^{N\sigma}) \cong \mathbb{Z}/2$$

since $(\mathbb{Q}\zeta_n)^{N\sigma} \not\subseteq \mathbb{R}$. These terms thus account for all of $C_2(\mathbb{Q}[A_n])$ under the decomposition of Theorem 11.8; and so $Cl_1(\mathbb{Z}[A_n])_{(2)} = 1$.

(5) By (3) again, $C_3(\mathbb{Q}[A_n]) \cong (\mathbb{Z}/3)^s$, where s is the number of conjugacy classes of cyclic $\sigma \subseteq A_n$ such that $3\,|\,|\sigma|$, and such that $C_3 \subseteq \sigma$ is centralized by $N(\sigma)$. An easy check then shows that

$$C_3(\mathbb{Q}[A_n]) \cong \begin{cases} \mathbb{Z}/3 & \text{if } n = \sum_{i=1}^{s} 3^{m_i},\ m_1 > m_2 > \dots > m_s \ge 0,\ \sum m_i \text{ odd} \\ 1 & \text{otherwise.} \end{cases}$$

Assume that $C_3(\mathbb{Q}[A_n]) \cong \mathbb{Z}/3$: write $n = \sum_{i=1}^{s} 3^{m_i}$, where the m_i are as above. Let $P \subseteq A_n$ be the "standard" 3-Sylow subgroup. Then $P = P_1 \times \dots \times P_s$, where P_i is a 3-Sylow subgroup of $A(3^{m_i})$. Also,

$$P^{ab} \cong (C_3)^m \quad \text{and} \quad N_{S_n}(P)/P \cong (C_2)^m.$$

In fact, there are bases $g_1,\dots,g_m$ of P^{ab} and $x_1,\dots,x_m$ of $N_{S_n}(P)/P$ such that in P^{ab}, $[x_i,g_j] = 1$ if $i \ne j$, and $x_i g_i x_i^{-1} = g_i^{-1}$ for all i. For example, if $n = 12$, then $P^{ab} \cong (C_3)^3$ is generated by

$$g_1 = (1\ 2\ 3), \qquad g_2 = (1\ 4\ 7)(2\ 5\ 8)(3\ 6\ 9), \qquad g_3 = (10\ 11\ 12);$$

while $N_{S_n}(P)/P \cong (C_2)^3$ is generated by

$$x_1 = (1\ 2)(4\ 5)(7\ 8), \qquad x_2 = (1\ 4)(2\ 5)(3\ 6), \qquad x_3 = (10\ 11).$$

For any $\epsilon_1,\ldots,\epsilon_m \in \mathbb{Z}/3$, let $V(\epsilon_1,\ldots,\epsilon_m)$ denote the irreducible $\mathbb{Q}[P^{ab}]$-module with character $\chi(g_i) = (\zeta_3)^{\epsilon_i}$. If any $\epsilon_i = 0$, then there is an element of $N_{A_n}(P)/P$ which negates the character, and hence negates the corresponding $\mathbb{Z}/3$ summand in $C_3(\mathbb{Q}[P^{ab}]) \cong (\mathbb{Z}/3)^{(3^m-1)/2}$. The remaining irreducible representations of P^{ab} ($\epsilon_i = \pm 1$ for all i) are permuted simply and transitively by $N_{A_n}(P)/P$; and hence

$$H_0(N(P)/P;\ C_3(\mathbb{Q}[P^{ab}])) \cong \mathbb{Z}/3.$$

If $P' \neq P$ is any other 3-Sylow subgroup in A_n, then $P' \cap P$ is contained in the subgroup generated by some proper subset of the g_i. It follows that the induced map

$$C_3(\mathbb{Q}[P' \cap P]) \longrightarrow H_0(N(P)/P;\ C_3(\mathbb{Q}[P^{ab}]))$$

is trivial. Hence, there is a natural epimorphism

$$\varprojlim_{\rho\in\mathfrak{F}(A_n)} C_3(\mathbb{Q}[\rho]) \twoheadrightarrow H_0(N_{A_n}(P)/P;\ C_3(\mathbb{Q}[P^{ab}])) \cong \mathbb{Z}/3.$$

But by Theorem 11.8, this limit is a direct summand of $C_3(\mathbb{Q}[A_n]) \cong \mathbb{Z}/3$. So with the help of Theorem 9.5 we now get

$$Cl_1(\mathbb{Z}[A_n]) \cong \varprojlim_{\rho\in\mathfrak{F}(A_n)} Cl_1(\mathbb{Q}[\rho]) \cong \text{Coker}\Big[K_2^c(\hat{\mathbb{Z}}_3[P]) \longrightarrow \varprojlim_{\rho\in\mathfrak{F}(A_n)} C_3(\mathbb{Q}[\rho])\Big]$$

$$\cong \text{Coker}\Big[H_1(P;\mathbb{Z}[P]) \xrightarrow{\ \Psi\ } H_0(N(P)/P;\ C_3(\mathbb{Q}[P^{ab}])) \cong \mathbb{Z}/3\Big].$$

The calculation now splits into the following cases:

<u>n = 3,4:</u> $P \cong C_3$, so $Cl_1(\mathbb{Z}[A_n])_{(3)} = 1$ by Theorem 14.2(i).

<u>n = 12,13:</u> $\Psi(g_3 \otimes g_1g_2g_3)$ generates $H_0(N(P)/P;\ C_3(\mathbb{Q}[P^{ab}]))$ (where g_i are the elements defined above); so $Cl_1(\mathbb{Z}[A_n]) = 1$.

<u>n = 27,28:</u> The image of any abelian subgroup of P is cyclic in P^{ab}. Hence, $\mathrm{Im}(\Psi) = \Psi(P \otimes 1) = 0$, and $Cl_1(\mathbb{Z}[A_n]) \cong \mathbb{Z}/3$.

<u>n > 28:</u> In this case, $m = \sum m_i \geq 5$. By Alperin et al [3, Proposition 2.5],

$$\mathrm{Im}\Big[K_2^c(\hat{\mathbb{Z}}_3[P^{ab}]) \longrightarrow C_3(\mathbb{Q}[P^{ab}])\Big] \cong S^3(P^{ab}),$$

where $S^3(P^{ab}) \cong S^3(\mathbb{F}_3{}^m)$ denotes the symmetric product. Furthermore, since $m \geq 5$, $S^3(P^{ab})^{N(P)/P} \cong S^3(\mathbb{F}_3{}^m)^{N(P)/P} = 0$ by the above description of $N(P)/P$. There are thus surjections

$$\mathbb{Z}/3 \cong C_3(\mathbb{Q}[A_n]) \twoheadrightarrow Cl_1(\mathbb{Z}[A_n]) \twoheadrightarrow H_0(N(P)/P;\ Cl_1(\mathbb{Z}[P^{ab}]))$$

$$\cong \mathrm{Coker}\Big[S^3(P^{ab})^{N(P)/P} \longrightarrow C_3(\mathbb{Q}[P^{ab}])^{N(P)/P}\Big] \cong \mathbb{Z}/3;$$

and so $Cl_1(\mathbb{Z}[A_n]) \cong \mathbb{Z}/3$ in this case. □

REFERENCES

Alperin, R. et al

1. R. Alperin, R. K. Dennis, and M. Stein, The nontriviality of $SK_1(\mathbb{Z}[\pi])$, Proceedings of the conference on orders, group rings, and related topics (Ohio State Univ., 1972), Lecture notes in math. 353, Springer-Verlag, Berlin (1973), 1-7

2. R. Alperin, R. K. Dennis, and M. Stein, SK_1 of finite abelian groups, I, Invent. Math. 82 (1985), 1-18

3. R. Alperin, R. K. Dennis, R. Oliver and M. Stein, SK_1 of finite abelian groups, II, Invent. Math. 87 (1987), 253-302

Artin, E. and Hasse, H.

1. Die beiden Ergänzungssätze zum Reziprozitätsgesetz der ℓ^n-ten Potenzreste im Körper der ℓ^n-ten Einheitswurzeln, Abh. Math. Sem. Hamburg 6 (1928), 146-162

Bak, A.

1. The involution on Whitehead torsion, General Topology and Appl. 7 (1977), 201-206

2. K-theory of forms, Princeton Univ. Press, Princeton, N. J. (1981)

3. A norm theorem for K_2 of global fields, Proceedings of the conference on algebraic topology (Århus, 1982), Lecture notes in math. 1051, Springer-Verlag, Berlin (1984), 1-7

Bak, A. and Rehmann, U.

1. The congruence subgroup and metaplectic problems for $SL_{n\geq 2}$ of division algebras, J. Algebra 78 (1982), 475-547

Bass, H.

1. K-theory and stable algebra, I. H. E. S. Publications math. 22 (1964), 5-60

2. Algebraic K-theory, Benjamin, New York (1968)

Bass, H. et al

1. H. Bass, J. Milnor, and J.-P. Serre, Solution of the congruence subgroup problem for SL_n $(n \geq 3)$ and Sp_{2n} $(n \geq 2)$, I. H. E. S. Publications math. 33 (1967), 59-137

Bass, H. and Tate, J.

1. The Milnor ring of a global field, Proceedings of the conference on algebraic K-theory (Battelle, 1972), vol. 2, Lecture notes in math. 342, Springer-Verlag, Berlin (1973), 349-446

Brown, K.

1. Cohomology of groups, Springer-Verlag, Berlin (1982)

Cartan, H. and Eilenberg, S.

1. Homological Algebra, Princeton Univ. Press, Princeton, N. J. (1956)

Cassels, J. W. S. and Fröhlich, A.

1. Algebraic number theory, Academic Press, London (1967)

Chase, S. and Waterhouse, W.

1. Moore's theorem on the uniqueness of reciprocity laws, Invent. math. 16 (1972), 267-270

Curtis, C. and Reiner, I.

1. Methods of representation theory, with applications to finite groups and orders, vol. 1, Wiley-Interscience, New York (1981)

Dennis, R. K.

1. Stability for K_2, Proceedings of the conference on orders, group rings, and related topics (Ohio State Univ., 1972), Lecture notes in math. 353, Springer-Verlag, Berlin (1973), 85-94

Dennis, R. K. and Stein, M.

1. K_2 of discrete valuation rings, Adv. in Math. 18 (1975), 182-238

Draxl, P.

1. Skew fields, London Math. Soc. lecture notes 81, Cambridge Univ. Press, Cambridge (1983)

Dress, A.

1. A characterization of solvable groups, Math. Z. 110 (1969), 213-217

2. Induction and structure theorems for orthogonal representations of finite groups, Annals of Math. 102 (1975), 291-325

Galovich, S.

1. The class group of a cyclic p-group, J. Algebra 30 (1974), 368-387

Grayson, D.

1. Higher algebraic K-theory II, Proceedings of the conference on algebraic K-theory (Evanston, 1976), Lecture notes in math. 551,

Springer-Verlag, Berlin (1976), 217-240

Hambleton, I. and Milgram, J.

1. The surgery obstruction groups for finite 2-groups, Invent. Math. 61 (1980), 33-52

Hasse, H.

1. Über p-adische Schiefkörper und ihre Bedeutung für die Arithmetik hyperkomplexer Zahlsysteme, Math. Ann. 104 (1931), 495-534

Higman, G.

1. The units of group rings, Proc. London Math. Soc. (2) 46 (1940), 231-248

Hilbert, D.

1. Die Theorie der algebraischen Zahlkörper, Gesammelte Abhandlungen, Vol. I, Chelsea, New York (1965), 63-363

Hilton, P. and Stammbach, U.

1. A course in homological algebra, Springer-Verlag, Berlin (1971)

Huppert, B.

1. Endliche Gruppen, Springer-Verlag, Berlin (1967)

Igusa, K.

1. What happens to Hatcher and Wagoner's formula for $\pi_0\mathscr{C}(M)$ when the first Postnikov invariant of M is nontrivial, Proceedings of the conference on algebraic K-theory, number theory, geometry, and analysis (Bielefeld, 1982), Lecture notes in math. 1046 (1984), 104-172

James, G. and Kerber, A.

1. The representation theory of the symmetric group, Addison-Wesley, Reading, Mass. (1981)

Janusz, G.

1. Algebraic number fields, Academic Press, London (1973)

Keating, M.

1. Whitehead groups of some metacyclic groups and orders, J. Algebra 22 (1972), 332-349

2. On the K-theory of the quaternion group, Mathematika 20 (1973), 59-62

3. Values of tame symbols on division algebras, J. London Math. Soc. (2) 14 (1976), 25-30

Kervaire, M. and Murthy, M. P.

1. On the projective class group of cyclic groups of prime power order, Comment. Math. Helv. 52 (1977), 415-452

Keune, F.

1. The relativization of K_2, J. Algebra 54 (1978), 159-177

Kolster, M.

1. K_2 of non-commutative local rings, J. Algebra 95 (1985), 173-200

Kuku, A. O.

1. Whitehead groups of orders in p-adic semi-simple algebras, J. Algebra 25 (1973), 415-418

Lam, T.-Y.

1. Induction theorems for Grothendieck groups and Whitehead groups of finite groups, Ann. Sci. École Norm. Sup. (4) 1 (1968), 91-148

Loday, J.-L.

1. K-théorie algébrique et représentations de groupes, Ann. Sci. École Norm. Sup. (4) 9 (1976), 309-377

Maazen, H. and Stienstra, J.

1. A presentation for K_2 of split radical pairs, J. Pure Appl. Algebra 10 (1977), 271-294

Magurn, B.

1. SK_1 of dihedral groups, J. Algebra 51 (1978), 399-415

2. Whitehead groups of some hyperelementary groups, J. London Math. Soc. (2) 21 (1980), 176-188

Magurn, B. et al

1. B. Magurn, R. Oliver, and L. Vaserstein, Units in Whitehead groups of finite groups, J. Algebra 84 (1983), 324-360

Mazur, B.

1. Relative neighborhoods and the theorems of Smale, Annals of Math. 77 (1963), 232-249

Milgram, J. and Oliver, R.

1. P-adic logarithms and the algebraic K,L groups for certain transcendence degree 1 extensions of $\mathbb{Z}\pi$ (preprint)

Milnor, J.

1. Whitehead torsion, Bull. A. M. S. 72 (1966), 358-426

2. Introduction to Algebraic K-theory, Princeton Univ. Press, Princeton, N. J. (1971)

Moore, C.

1. Group extensions of p-adic and adelic linear groups, I. H. E. S. Publications math. 35 (1968), 5-70

Nakayama, T. and Matsushima, Y.

1. Über die multiplikative Gruppe einer p-adischen Divisionsalgebra, Proc, Imp. Acad. Japan 19 (1943), 622-628

Obayashi, T.

1. On the Whitehead group of the dihedral group of order 2p, Osaka J. Math. 8 (1971), 291-297

2. The Whitehead groups of dihedral 2-groups, J. Pure Appl. Algebra 3 (1973), 59-71

Oliver, R.

1. SK_1 for finite group rings: I, Invent. Math. 57 (1980), 183-204

2. SK_1 for finite group rings: II, Math. Scand. 47 (1980), 195-231

3. SK_1 for finite group rings: IV, Proc. London Math. Soc. 46 (1983), 1-37

4. Lower bounds for $K_2^{top}(\hat{\mathbb{Z}}_p\pi)$ and $K_2(\mathbb{Z}\pi)$, J. Algebra 94 (1985), 425-487

5. The Whitehead transfer homomorphism for oriented S^1-bundles, Math. Scand. 57 (1985), 51-104

6. K_2 of p-adic group rings of abelian p-groups, Math. Z. 195 (1987), 505-558

7. SK_1 for finite group rings: V, Comment. Math. Helv. (to appear)

8. Logarithmic descriptions of $K_1'(\hat{\mathbb{Z}}_pG)$ and classgroups of symmetric groups, Math. Annalen 273 (1985), 45-64

9. Central units in p-adic group rings, preprint

Oliver, R. and Taylor, L.

1. Logarithmic descriptions of Whitehead groups and class groups for p-groups (to appear)

Prasad, G. and Raghunathan, M. S.

1. On the congruence subgroup problem: determination of the "metaplectic kernel", Invent. Math. 71 (1983), 21-42

Quillen, D.

1. Higher algebraic K-theory: I, Proceedings of the conference on algebraic K-theory (Battelle, 1972), vol. 1, Lecture notes in math. 341, Springer-Verlag, Berlin (1973), 85-147

Rehmann, U.

1. Zentrale Erweiterungen der speziellen linearen Gruppe eines Schiefkörpers, J. reine ang. Math. 301 (1978), 77-104

2. Central extensions of SL_2 over division rings and some metaplectic theorems, Contemp. Math. 55, Amer. Math. Soc. (1986), 561-607

Rehmann, U. and Stuhler, U.

1. On K_2 of finite dimensional division algebras over arithmetic fields, Invent. Math. 50 (1978), 75-90

Reiner, I.

1. Maximal orders, Academic Press, London (1975)

Riehm, C.

1. The norm 1 group of p-adic division algebra, Amer. J. Math. 92 (1970), 499-523

Roquette, P.

1. Realisierung von Darstellungen endlicher nilpotenter Gruppen, Arch. Math. 9 (1958), 241-250

2. On the Galois cohomology of the projective linear group and its applications to the construction of generic splitting fields of algebras, Math. Annalen 150 (1963), 411-439

Schur, J.

1. Über die Darstellung der symmetrischen und der alternierenden Gruppe durch gebrochene lineare Substitutionen, Journal reine ang. Math. 139 (1911), 155-250

Serre, J.-P.

1. Corps locaux, Hermann, Paris (1968)

2. Linear representations of finite groups, Springer-Verlag, Berlin (1977)

Siegel, C.-L.

1. Discontinuous groups, Annals of Math. 44 (1943), 674-689

Silvester, J. R.

1. Introduction to algebraic K-theory, Chapman and Hall, London (1981)

Stammbach, U.

1. Homology in group theory, Lecture notes in math. 359, Springer-Verlag, Berlin (1973)

Stein, M. and Dennis, R. K.

1. K_2 of radical ideals and semi-local rings revisited, Proceedings of the conference on algebraic K-theory (Battelle, 1972), vol. 2, Lecture notes in math. 342, Springer-Verlag, Berlin (1973), 281-303

Suslin, A. A.

1. Torsion in K_2 of fields, K-theory 1 (1987), 5-29

Swan, R.

1. Induced representations and projective modules, Ann. of Math. 71 (1960), 552-578

2. Excision in algebraic K-theory, J. Pure Appl. Alg. 1 (1971), 221-252

3. K-theory of finite groups and orders, Lecture notes in math. 149, Springer-Verlag, Berlin (1970)

Taylor, M.

1. A logarithmic approach to classgroups of integral group rings, J. Algebra 66 (1980), 321-353

2. On Fröhlich's conjecture for rings of integers of tame extensions, Invent. math. 63 (1981), 41-79

Ullom, S.

1. Class groups of cyclotomic fields and group rings, J. London Math. Soc. (2) 17 (1978), 231-239

Vaserstein, L.

1. On the stabilization of the general linear group over a ring, Math. USSR Sbornik 8 (1969), 383-400

Wagoner, J.

1. Homotopy theory for the p-adic special linear group, Comment. Math. Helv. 50 (1975), 535-559

Wall, C. T. C.

1. Norms of units in group rings, Proc. London Math. Soc. (3) 29 (1974), 593-632

Wang, S.

1. On the commutator group of a simple algebra, Amer. J. Math. 72 (1950), 323-334

Washington, L.

1. Introduction to cyclotomic fields, Springer-Verlag, Berlin (1982)

Whitehead, J. H. C.

1. Simple homotopy types, Amer. J. Math. 72 (1950), 1-57

Witt, E.

1. Die algebraische Struktur des Gruppenringes einer endlichen Gruppe über einem Zahlkörper, J. Reine Angew. Math. 190 (1952), 231-245

Yamada, T.

1. The Schur subgroup of the Brauer group, Lecture notes in math. 397, Springer-Verlag, Berlin (1974)

INDEX